LOSS CONTROL MANAGEMENT

by
Frank E. Bird, Jr.
and
Robert G. Loftus

© Copyright Institute Press
Division of International Loss Control
Institute, 1976.

Printed in U.S.A.

Reprinted in 1978, 1979, 1981 & 1982

All rights reserved. This book may not be reproduced in whole or part by any method without the express permission of the publisher.

Library of Congress Catalogue
Card Number 76-7279
ISBN 0-88061-000-X

Published by
Institute Press
(A Division of International Loss Control Institute)
Highway 78, P.O. Box 345
Loganville, Georgia 30249

PREFACE

Any safety or loss control manager, or in fact any executive or professional regardless of occupation, who takes the survival of his industry, business or family seriously, will be deeply interested in this book.

Loss Control Management provides ideas, tools and inspiration to help keep personal injuries, with the resulting human suffering and severe economic losses, to a minimum. Its pages are filled with information that will allow those concerned to carry out their responsibilities in a better and more comprehensive way.

With rapidly changing environments at work and in leisure hours, and with the additional, frequently new, problems these changes create, it can be very hard to find effective solutions. This book provides many of them, and gives the reader guidance in developing many more.

Its nineteen chapters, provided by experts in their fields, cover the full spectrum of loss control management and fill a void in the approach to such management techniques.

Through my experience and our research in the development of safety-oriented first aid, I have learned the tremendous importance of loss control management in the direct reduction in the number and severity of personal injuries and in minimizing economic waste.

I can strongly recommend Loss Control Management to all who are dedicated to achieving that result.

C. J. Laurin,
Vice Chancellor, Priory of Canada,
St. John Ambulance.

CONTENTS

Preface iii
Cyril J. Laurin

Introduction 1
Robert G. Loftus & Frank E. Bird, Jr.

Chapter One: HISTORY AND PHILOSOPHY 3

Introduction. Perspective Viewpoint. The Worker's Plight. Public Sympathy Rises. Early Attitudes Reflect Practices in Safety. A Demand for Change: Legislative Progress. Organized Help Provided in Canada. Comprehensive Federal Law in U.S. Positive Attitudes—Practices—Results. A New Activism. (Increased Union Activity. Rise in Consumerism. Attitudes of the Courts. Medical Research. Technological Developments. Government Legislation. Escalating Equipment Costs. Customer Service Demands. New Accident Facts. Trend to Professionalism. The Energy Limitation Problem.) What Lessons Have We Learned? (Progress Has Been Made. Past Efforts Were Narrow. System Approach Best.) The Economy and Efficiency of Loss Control. (Performance and Results.) Where Do We Go From Here? Bibliography.

Frank E. Bird & Robert G. Loftus

Chapter Two: THE CAUSES AND EFFECTS OF
LOSS PRODUCING EVENTS 28

Introduction. Incident-Accident Definitions. Further Analyses of Incidents. Accident Ratio Study. What About Near-Loss Incidents? The Sources of Loss Producing (Downgrading) Events. (People, Equipment, Material, Environment) An Updated Domino Sequence (Lack of Control-Management, Basic Cause(s)-Origin(s), Immediate Cause(s)-Symptoms, Incident-Contact, People-Property-Loss). Accident Results. Real Costs of Accidents. Summary. Bibliography.

Frank E. Bird & Robert G. Loftus

Chapter Three: MANAGEMENT CONTROL OF LOSS 51

Introduction. The Profession of Loss Control. (Loss Control Definitions) The Principle of Resistance to Change. The Principle of Definition. The Principle of Reciprocated Interest. The Principle of the Critical Few. The Principle of Recognition. The Principle of Future Characteristics. The Principle of Multiple Causes. The Principle of Application. The Principle of Point of Control. The Principle of Reporting Relationships. The Principle of Management Results. Management Commitment-A Key to Management Control System. Loss Control Performance Evaluation. Managing An Imperfect System. The Pre-Contact Stage. The Contact Stage. The Post-Contact Stage. Summary. Minimum Standards of Loss Control. Bibliography.

Frank E. Bird & Robert G. Loftus

Chapter Four: PROPERTY DAMAGE AND WASTE
CONTROL 93

Introduction. The Size and Scope of the Problem. Workable Definitions — A Starting Point. Roadblocks to Program Implementation. (Injury-Oriented Habits are Difficult to Change. Management's Failure to Recognize the Problem. Acceptance of Damage Reported as Representative of Damage Occurrence. Weak Reporting Relationship of Staff Advisor. Inadequate Overall Capability of Staff Advisors.) Three Significant Studies. A Dual Approach to Property Damage and Waste Control. (Control Through Loss Improvement Project System. The Loss Control Project Team. Maintaining Long-Range Damage Cost Control: 1. Management Training. 2. Supervisory Investigation 3. Rules and Practices. 4. Engineering Controls. 5. Skill Training. 6. General Promotion. 7. Regular Repair Center Audits. Waste and Property Damage Control Inter-Related.) Final Advice on Program Implementation. Chapter Summary and Program Benefits. Bibliography.

Frank E. Bird & Robert G. Loftus

Chapter Five: ECONOMICS IN LOSS CONTROL 139

Introduction. How to Sell Top Management. Motivational Scale Exercise. The Principle of Economic Association. The Principle of the Critical Few. The Principle of Reciprocated Interest. The Principle of Future Characteristics. The Principle of Application. The Principle of Economic Priorities. The Principle of Vested Interest. The Principle of Substantial Evidence. The Principle of Adequate Evidence. The Principle of Dimensional Value. Summary. Bibliography.

Frank E. Bird & Robert G. Loftus

Chapter Six: MEASUREMENT TOOLS FOR MANAGEMENT 167

Introduction. Characteristics of a Good Measurement Tool. Classification of Measurements. Measurements of Consequences. (Actual Loss Measurements. Injury Rates. Definition of Serious Injury. Definition of Reportable Injury. The All-Injury/Accident Frequency Rates. Property Damage Rates. Potential Loss Measurements. The Critical Incident Technique. Incident Recall.) Measurements of Cause. (Measurements of Actual Cause. Standard Cause Measurement. Measurement of Organizational Error. Measurements of Potential Cause. Behavior Sampling. Environmental Sampling. Mathematical Evaluation for Controlling Hazards.) Measurements of Control. (Advantages of Use. A Dynamic System at Work. Practical Utilization of Control Measurements. Common Methods of "Control" Measurement: 1. Random Sampling, 2. Actual Count. 3. Professional Judgment. The Use of Value Factors. Systems Safety Techniques in "Control" Measurements.) General Summary. Bibliography.

Frank E. Bird & Robert G. Loftus

Chapter Seven: EVALUATING INDIVIDUAL PERFORMANCE 198

Introduction. An Opportunity. Current and Past Practices. The Hypothesis and Its Goals. The Basis for Belief. Implementing a Performance Review Program. (Orientation. The Self-Appraisal. Self-Appraisal Follow-up. Records Review). The Performance Review. (Evaluation of Past Performance. Scoring Alternatives. Evaluation of Personal Development Accomplished. Analysis of Needs. Developmental Objectives. Summary. (Review of the Choices.) Bibliography.

Frank E. Bird & Robert G. Loftus

Chapter Eight: INCIDENT RECALL TECHNIQUES 215

Introduction. Incident and Accident Definitions. Why Incident Recall Should Be Utilized. Differences in Techniques. Two Types of Incident Recall. The Planned Incident Recall Interview. (Front-line Supervisor as Interviewer. Privacy Preferred. Selection of People to be Interviewed. Adjusting Interview Time. No-Fault Assurance a Must. Preparation for Interviews. Conducting the Interview. Making Out the Report. Practical Application of the Planned IR Interview.) Informal Incident Recall Techniques. (Shortcomings of Incident/Accident Statistics. Critical Stages in the Program.) Summary. Bibliography. Additional References.

Frank E. Bird & Robert G. Loftus

Chapter Nine: BEHAVIOR MOTIVATION 247

Motivational Analysis. (A Human Factors Formula. Do You Have a Motivational Problem?) Know-How and Beliefs. (Job Attitudes. Views from Opposite Ends of the Telescope. Employee Safety Attitudes.) Can-Do. (Skill Deficiencies in the Man and the Job. Leftover Skills. Does the Job Grab You? The Prisoner's Dilemma. Degree of Involvement. What Keeps Him On the Job? Who Takes the Job Seriously? The Drama of Work. The Drama of Goals. The Space of Free Movement. The Overlapping Situation. Incidents and Outbursts. Handling Tensions. Job Enlargement. Fun, Games and Jobs. Mysterious Accidents.) Will-Do. Is There a Lack of Commitment to Safety? Why Workers Won't Perform Safely. Operant Conditioning. The Technology of Behavior. The Outcomes of Behavior. Does Punishment Work? Is Unsafe Performance Rewarding? The Immediate Effects of Unsafe Acts. Unsafe Acts and Job Specialization. The Principle of Behavior Reinforcement. What to Do About Unsafe Acts. Is Proper Performance Punishing? Eliminating Obstacles to Working Safely. Is Safe Performance Rewarding? Beginnings of the Technique. From Animals to People. Safe Behavior Reinforcement. How Can Management Use Safe Behavior Reinforcement? From Theory to Practice. Bibliography.

Lawrence E. Schlesinger, Ph.D.

Chapter Ten: PRODUCTS LOSS CONTROL 279

Management Leadership. Coordination of the Program. Coordinating Group or Department. Design or New Product Development Department. (Codes and Standards. Human Factors Engineering. Critical Parts. Packaging and Handling. Safety Audit. Warning Labels and Instructions. Design and Changing Technology.) A Partial Checklist of Possible Hazards. Manufacturing. Quality Control. (In-Process Testing. Finished Product Testing. Storage and Shipping.) Service Department. Insurance Department. Sales, Advertising and Instruction Book Writers. Personnel Department. Purchasing Department. Recordkeeping for Products Loss Control. Field Monitoring of Product Complaints, Incidents, Accidents. Summary. Organizations for Developing Product Safety Standards. Bibliography.

Robert E. Shankula

Chapter Eleven: ENVIRONMENTAL HEALTH 317

Introduction. Classification of Occupational Health Hazards. Mode of Entry. (Inhalation. Skin Contact Absorption. Ingestion.) Mode of Action. Severity of Action of Toxic Agents. Exposure Limits for Dangerous Substances. (Threshold Limit Values. Acceptable Concentrations. Emergency Exposure Limits. Criteria Used for Establishing Standards.) Physical Hazards. (Ionizing Radiation.Ultra-Violet Radiation.Visible Light. Infra-Red Radiation. Microwaves and Lasers. Noise. Noise Measurement. Procedure for Noise Survey. Interpretation of Data. Control of Noise. Abnormal Temperature and Humidity. Shock and Vibration.) Control of Hazards. (Identification of Exposures. Control of Exposures. Air Cleaning. Medical Control Program.) Maintenance of Control Procedures. The Need for an Occupational Health Program. Organizing and Managing an Environmental Health Program. (Responsibilities of the Engineering Organization. Responsibilities of the Medical Organization. Responsibilities of the Industrial Hygiene Organizations. Responsibilities of the Safety Organization. Supervisor's Responsibilities. Employee's Responsibilities.) Measuring the Performance of an Environmental Health Program. References. Bibliography.

Joseph F. Stelluto

Chapter Twelve: FIRE LOSS CONTROL 351

Total Loss Control. Professional Assistance. C.O.P.E.: Construction. (How Fires Spread. Vertical Fire Protection. Horizontal Fire Protection.) Occupancy. Protection. (Public Fire Protection. Private Fire Protection.) Types of Extinguishing Agents. (Carbon Dioxide. Dry Chemical Interrupts Combustion Reaction. Foams Provide Blanketing and Cooling Effectiveness. Halons are Used for Explosion Suppression, Sprinklers Do the Humanly Impossible by Being Constantly on Guard Against

Fire. Water Spray Systems Control Burning, Extinguish Fires and Protect Exposures.) Classification of Fires and Recommended Protective Equipment. Exposure. Definitions. Summary. Bibliography.

Peter R. Vallet

Chapter Thirteen: AIR POLLUTION 382

Introduction. General Environment. Working Environment. Pollution Appraisal. Types of Pollutants and Their Sources. (Aerosols. Gaseous Materials.) Air Pollution Effects. (Effects on Man. Economic Effects. Detection, Measurement and Anaylsis.) Ambient Air Quality Sampling. Stack Sampling. Pollution Control. (Hoods. Ductwork, Pollution Control Equipment. Mechanical Separators. Filter Bag Collectors. Scrubbers. Absorption Equipment. Electric Precipitators. Absorption Equipment. Afterburners. Fans. Contaminant Disposal. Dust Suppression. Maintenance of Control Systems.) Summary. Bibliography.

Kenneth H. Suter

Chapter Fourteen: ENGINEERING CONTROLS 410

Identify the Problem. A Workable Plan. Design Guidelines. Plant Layout. System Safety Analysis. (Fault Tree Method. Failure Mode and Effect Method. THERP, Technique for Human Error Prediction. Cost Effectiveness Method.) Human Factors. Purchasing. Summary. Bibliography.

Richard A. Watson

Chapter Fifteen: MOTOR FLEET SAFETY 429

Fleet Efforts and Damage Reduction. Driver Selection. Research Findings in Driving Task Analysis. (Information-Receiving. Decision-Making. Driver-Action. Dilemma: Thinking Versus Habit. Research Findings Identifying the Traits of Safe Drivers.) More Sophisticated Application Blanks. Psychological, Psychophysical, Knowledge and Skill Tests. Standard History Checking Methods. What to Look for in Physical Examinations. More Refined Interviewing Methods. Training. (Fleet Training is Rare.) Driver Rehabilitation. (Assessment and Diagnosis. Remedial Programs. Evaluation and Research.) Driver Improvement Systems. How to Establish Training Programs. Vehicle Inspection and Maintenance. Supervision. (The Supervisor is the Key Figure. All Men Have Needs. Needs are

Frustrated in Daily Living. Clues to Bad Management. Need for Middle-Management Training.) Summary. Bibliography.

Harold L. Henderson, Ph.D.

Chapter Sixteen: SYSTEM SAFETY 464

The System Safety Concept. Key Elements. (Pre-planned. Organized Effort, Conservation of Resource. System, Product or Operation. Pre-Accident Hazard Identification. Timely Incorporation of Safety Inputs to System Requirements. Early Evaluation of Compliance with Safety Requirements and Criteria. Continued Safety Surveillance Throughout System's Life-Span, Including Disposal.) Three Distinct Aspects. (System Safety Engineering. Product Safety Engineering. Product/System Safety Engineering.) System Safety Engineering. System Safety Management. System Safety Analysis. The Basic Elements. (Identification. Evaluation. Communication.) A General Definition. The Encompassing Scope. Analytical Techniques. Preliminary Hazard Analyses. Design Safety Analyses. Deductive Methods. Fault Tree Logic Diagram Construction. Fault Tree Evaluation. Simulation. Resources. (Government Publications. System Safety Conference Proceedings. Associations.) System Safety Education Resources.

Rex B. Gordon

Chapter Seventeen: A LOSS CONTROL PROGRAM
                     FOR ALCOHOLISM 504

The Alcoholic Employee. The Disease Concept of Alcoholism. Industry's Advantage. Cooperation Between Labor and Management. An Effective System of Employee Alcoholism Control. Separation of Management and Treatment Functions. Assignment of Administrative Responsibility. Formulation of Policy. (Company Statement of Policy.) Definition of Management and Labor Functions. (Management Functions. Union Functions. Development of Procedures to Implement Policy.) Training. Referral for Treatment. Insurance Coverage. Recordkeeping and Program Evaluation. Role of Program Administrator. Summary. Bibliography.

Ross A. Von Wiegand

Chapter Eighteen: REHABILITATION 530

What is Rehabilitation and What are its Deterrents? The Rehabilitation Process. The Identification Process. (Congenital Handicaps. Accidents or Illnesses Incurred During One's Life-Time.) Rehabilitation and Medical Management. Bibliography.

George T. Welch

Chapter Nineteen: SECURITY 540

Introduction. Case Histories. Types of Security Problems. Problem Locations. Measurement of Losses. Insurance. Security Plans. Some Methods for Controlling Security. Security Check List. Specimen Industrial Security Manual. (Introduction. Courtesy and Self Control. Telephone Techniques. Fire and Fire Equipment. Security Supervisor's Duties. Security Officer's Duties. Visitor's Passes.) Emergency Condition Security Planning. (Types of Emergencies. Preventive Measures. Emergency Procedures. Duties and Responsibilities) A Formal Emergency Organization. Bombs and Bomb Threats. (Introduction. Planning for Action.) Bibliography.

Staff, IAPA of Ontario

INTRODUCTION

The co-authors of the first eight chapters of this book have been privileged through the years to share in-depth experiences on the subject of loss control with a large number of professionals around the world. In addition, our 55 combined years of personal experiences in safety, loss control and industrial management have provided us with practical insight and understanding on this subject that we hope will prove of value to readers.

We want to express our appreciation to the staff of the Industrial Accident Prevention Association of Ontario for their splendid cooperation on all aspects of this project.

Likewise, we extend heartfelt thanks to the contributing authors who so willingly gave the extensive amount of time required to share their great expertise with others. Their names are listed in the foreward "contents" section.

We have no doubt whatsoever that management's ever increasing need to effectively control loss will provide opportunities galore for those ready to meet the great challenges of this exciting period of time. Our hope is that the thoughts expressed herein will provide additional stimulus for others to contribute knowledge on this subject so vital to the perpetuation of "management control" within the free enterprise system.

                                            Robert G. Loftus
                                            Frank E. Bird, Jr.

History And Philosophy

I.

INTRODUCTION

The practice of loss control by business men is certainly not new. There is recorded evidence to indicate that efforts to minimize damage to property by fire and cargo loss from the ravages of the sea were practiced as early as 4,000 B.C. in Babylonia. It is a logical assumption that even the primitive entrepreneur, thousands of years ago, who first recognized he could trade or sell the meat or skins of animals for useful gain, must also have recognized the need to protect his assets until their value to him was fully realized.

Only in recent years have businessmen begun to recognize the inter-relationship of management effort to control losses involving fire, energy and resources devoted to these problem areas.

This coordinated effort, referred to as "Loss Control Management", will prove of ever-increasing value in the years ahead to the businessman in general industry — and particularly to the smaller businessman.

As the probability of major or catastrophic loss becomes increasingly evident to him, he will more quickly recognize the value of attending to critical areas of potential loss control in a manner that affords the greatest protection at the least possible cost.

This chapter will discuss the historical development and philosophy of the fast-emerging discipline of loss control management, and will show how it is expanding within the organized safety movement in general industry as a result of unusual social, legal, political, and economic factors now bearing upon the businessman.

While each of these important areas of special interest requires certain specific knowledge and expertise, there is a great management efficiency to be realized when they are closely coordinated.

PERSPECTIVE VIEWPOINT

Professional management consultants frequently examine the past to gain insight into problems that exist today, and to forecast what future experience might be expected to be. This perspective of historical developments is thereby presented so that the reader may gain possible lessons to help in the development of loss control solutions for the future.

The extensive use of power machinery initially imported from England during a time referred to as the Industrial Revolution ushered in a period of work deaths and disability never seen before or since. (see Figure 1). While it began around 1837, the greatest upsurge in industrial growth occurred around 1880 with the rapid increase of steel production. During these years, new weaving machines and equipment, first powered by steam and later by electricity, replaced the slow hand-production methods of the home or small handicraft systems. Machines were designed with little or no consideration for the safety or convenience of the operators. Since available labor was untrained and unskilled in the use of this new unguarded machinery, injuries increased greatly. As the source of power changed from manpower or horsepower to steam and electricity, the number of crippling injuries also climbed rapidly. (1)

THE WORKER'S PLIGHT

A number of circumstances that existed around 1900 provided strong motivation for groups that would soon highlight the need for improvement in conditions for the industrial worker. Workday operations averaging 11 to 13 hours increased exposure to accident potential. Facilities for emergency care were horribly inadequate, and professional medical help was seldom available. Injured or disabled workers seldom received compensation sufficient for even a bare existence, and the general practice was for the company to pay only certain doctor bills and medical expenses. The company generally paid the funeral expenses and presented a donation to the widow and dependents. Fellow workers usually made a cash gift, obtained by collection within their group.

While the recourse to obtain additional benefits through legal action was an open possibility to a worker during this period, he seldom sought compensation through this route for several reasons. The biggest reason was probably the jeopardy of his very job if he brought suit against his company. He also knew he bore the burden of proof that the employer was at fault. Accepted court practices around the world were quite similar at this time,

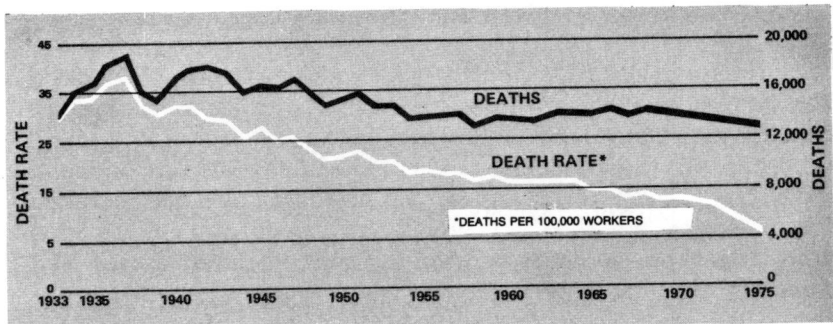

FIGURE 1 – THE NATIONAL DEATH RATE TREND. The white line shows the decline in industrial fatality rates since 1933. The solid black line shows the actual number of industrial fatalities. What conclusion can you draw about accident prevention progress? (National Safety Council Data.)

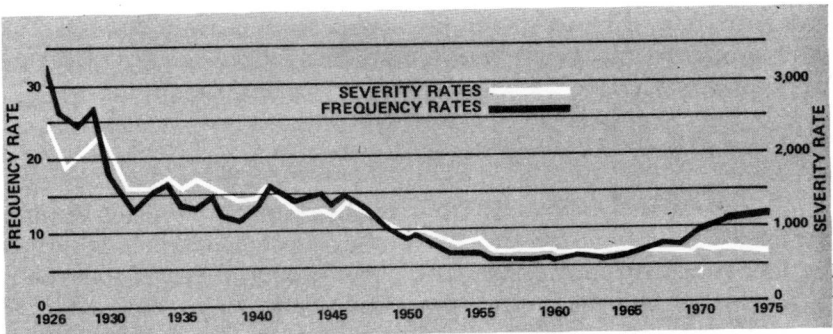

FIGURE 2 – THE NSC FREQUENCY AND SEVERITY RATE TREND. The black line shows how the national frequency rate has declined since 1926. The white line shows the decline of the national severity rate. (National Safety Council Graph.)

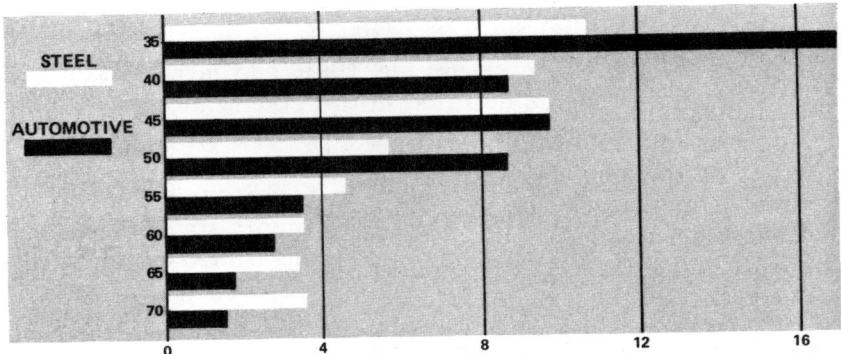

FIGURE 3 – FREQUENCY RATE TRENDS FOR THE AUTOMOBILE AND STEEL INDUSTRY. Each bar represents the average disabling frequency rate for a five year period. Notice how the war years increased frequency rates. What conclusions can you draw from these industry rate trends? (Based on National Safety Council data.)

and usually gave protection to the employer if proof existed that the injured worker's own negligence contributed to the accident's occurrence. Courts operated under the theory that a worker accepted all the customary risks associated with an occupation when he took the job. Coupled with all these one-sided factors was the fear of job loss that strongly influenced any fellow worker who might be called as a witness by the injured person or his family.

Then, too, legal fees took a large share of any award if a worker was fortunate enough to collect. One 1908 study in the state of New York revealed that the average award was $1,600, with $913 going to the lawyer. The remaining amount could not go far, considering that each widow in the study had 3 dependent children. It is easy to understand why the relatively few workers who brought suit against their companies would usually settle out of court for whatever small amount they could obtain.

PUBLIC SYMPATHY RISES

With this picture of the pathetic plight of the injured industrial worker at the turn of the twentieth century, it is easy to understand why powerful public sympathy began to develop. Efforts that brought improved conditions and benefits to workers and their families usually started within the labor organization itself and quite naturally spread through the social agencies that had contact with the distressed families of the dead and injured. As facts were made known, the clergy, educators and writers joined ranks with labor leaders and social workers to campaign vigorously for reforms.

One historic study that had an enormous influence at the time is known as the Pittsburgh Survey. It was completed in Allegheny County, Pennsylvania, in 1909, and revealed that 526 fatal industrial accidents occurred in this county alone during the 12 months of 1906—1907. The study showed further that over 50 percent of the surviving widows and children were left with no source of income. It was calculated that only 30 percent of the settlements exceeded $500. This same report stated that there were 30,000 fatal industrial injuries in the United States in the same year (1909).[3]

Existing records of a Pennsylvania steel company indicate that 1600 of its 2200 employees lost time from work because of injury during a four year period in the early 1900's. Statistically speaking, 75 percent of this plant's entire work force lost time from work because of accidents within a four year period.

EARLY ATTITUDES REFLECT PRACTICES IN SAFETY

Let's examine the thinking and attitudes of employers and employees at the turn of the century (1900) in order to better understand why such unbelievable conditions existed. Leading psychologists have frequently stated that a man's thinking is reflected in the general principles he practices. (The Bible, in Proverbs 23:7 ... "for as he thinketh in his heart, so is he ...", is early evidence of the timelessness of this theory.) There are very few actual writings on thoughts and attitudes regarding loss control subjects during this turn-of-the-century period. However, a number of leaders in the safety and fire protection fields have documented excerpts from numerous periodicals, reports and surveys written then, which give considerable insight into the subject.

Comments made during this period by employers have been reported to reflect their attitudes:

"I don't have money for frills like safety."

"Some people are just accident prone and no matter what you do, they'll hurt themselves some way."

"... 90% of all accidents are caused by just plain carelessness."

"We're not in business for safety."

"... I just can't see wasting money on safety."

"There's no place for sissies in dangerous work."

"These stubs on my hand are just part of doing business."

By and large, attitudes toward safety and other loss control areas such as fire protection, environmental health or rehabilitation were very negative. Employees were usually warned of the risks they took at the time they were hired. During this period, even the courts used a legal expression, "assumption of risk", meaning that the worker accepted all the customary risks associated with the occupation he accepted. Workers were also instructed to protect themselves from special hazards such as heat extremes or molten and sharp metal. Devices such as heat sandals for the feet, split wooden eye shields, smoked glasses, felt nose and mouth covers and hand pads were designed by employees who passed the know-how along to their fellow workers. It's interesting that, even today, employees will make certain pieces of personal protective equipment themselves if market demand has not motivated protective equipment companies to manufacture similar protective items.

Printed safety rules, written job procedures, plant inspections, supervisory investigations and other loss control tools common today were generally unheard of in the early 1900's. Of course, we must keep in mind that very little information on subjects related to safety or loss control was available to employers. Precepts that are almost second nature to management people today (such as "The Value of Good Communications" and "The Basic Principles of Motivation") would have been completely foreign to an employer at that time. In effect, the principles applied in practice reflected the attitudes and thoughts of the typical employer.[1]

Let's not leave this period of time or this discussion of attitudes without some mention of employee outlook on safety and loss control. Researchers also agree that, even allowing for the gross injustices dealt the employee during this period of time, his related thoughts and attitudes strangely paralleled those of his employers. Scars and stumps on fingers and hands were often proudly referred to as a badge of honor. The thought that safety was a matter of "luck" was frequently reflected in such statements as, "I never thought He'd get it; he was always one of the lucky ones," or "When your number's up, there's not much you can do." It has been documented that employees sometimes made chance-taking a game and actually taunted fellow employees into taking serious risks to "prove themselves." Certain loss control historians have suggested that the employee attitudes and outlooks of the time were outward gestures to disguise inner fears and concerns.

Whether any justification is possible for the complete failure of employers to provide even a semblance of an accident prevention program during these early periods is a good topic for class discussion. The recorded fact remains that it was not uncommon for manufacturing plants to have 150 to 200 workers losing time for each million man-hours.

A DEMAND FOR CHANGE: LEGISLATIVE PROGRESS

Public clamor peaked in the United States with the publication of the Pittsburgh study, and additional similar studies were made in a number of states. The pressure on politicians to act began to produce results with the promulgation of improved legislation and the enforcement of existing laws. Parallel efforts were taking place throughout Canada at this time, largely because of the influence of similar happenings in England. It is significant to indicate that almost identical conditions to those described earlier that existed in America had already taken place in England and had forced the first corrective measures resulting in Employer Liability Acts to be passed in the 1870's. These laws were quite significant, since they broadened the employer liability concept and increased the chance of benefits to the injured when employer negligence could be established.

The first true compensation act was passed by the British Parliament in 1897. While this law could be abused by proving that employees were injured because of their own misconduct, it brought about automatic compensation to many workers who would have received little or nothing a few years earlier. Employer liability aspects of the law were broadened again in 1907 by Parliament. While answering a great need, it is ironic that most of these improvements and reforms in the law were directed at benefits for the injured worker or his survivors and not at changing conditions causing the deplorable problems. The British Act of 1907 also served as a model for the first compensation law in the United States (in 1908) that covered employees of the federal government.

The year 1910 was quite significant. A commission to study compensation laws throughout the world was appointed by the Ontario government; in the same year, the state of New York passed its first compensation act, providing certain benefits to employees injured in twelve occupations referred to as dangerous. The Compensation Law of Wisconsin (passed in 1911), became generally acknowledged as the first significant legislation enacted in the United States. Several other states passed compensation laws the same year; in 1912, the first organized "Safety Congress" met in Milwaukee, Wisconsin, under the auspices of the Association of Iron and Steel Electrical Engineers. A year later, at a second meeting of the Congress in New York, the National Council for Industrial Safety was organized; shortly thereafter, this name was changed officially to The National Safety Council.

By 1913, the National Association of Manufacturers in the U.S. reported that 276 member companies had a special man in charge of safety shop organization. The U.S. Department of Labor was created in the same year (1913), "to foster, promote, and develop the welfare of the wage earner of the United States, to foster their working conditions and to advance their opportunities for profitable employment."[1]

Parallel efforts continued in Canada and the United States. In 1913 the Ontario Compensation Study Commission completed its work. Within the next two years, the Ontario Workmen's Compensation Act, which was to serve as a model for similar acts in other Canadian provinces, became law, and thirty states in the U.S. had promulgated legislation. Also in 1915, the American Society of Safety Engineers was organized, with the purpose of advancing activities that would assist safety personnel to develop their knowledge and capability to advise management.

ORGANIZED HELP PROVIDED IN CANADA

The Workmen's Compensation Act of Ontario, which came into effect on January 1, 1915, made provision for employers to form associations for the purpose of education in accident prevention.

Employers took advantage of this provision and, between January of 1915 and July of 1917, twenty-two accident prevention associations were chartered and approved by the Workmen's Compensation Board. Nineteen of these associations federated on July 17, 1917, when the Industrial Accident Prevention Association received its charter. It has continued in operation since that date.

This approach appears to have been unique in North America and it has not yet been paralleled by "Acts" in the United States (including the Occupational Safety and Health Act of 1970).

In Ontario there are eight other accident prevention associations encompassing Forest Products, Pulp and Paper, Mines, Transportation, Electrical, Construction, Hospitals, and Farms. The ninth association, the IAPA (Industrial Accident Prevention Association) is the largest of these, and its classifications are: Woodworkers; Ceramics and Stone; Metal Trades; Chemical Industries; Millers, Feed and Grain; Food Products; Leather, Rubber and Tanners; Textiles; Printing Trades; and Retailers. The nine associations operate independently of each other in the pursuit of their educational endeavors.

Although all provinces in Canada have the same basic Workmen's Compensation legislation, this is not true with respect to their means of providing effective accident prevention programming. For example, Quebec uses a method somewhat similar to that employed in Ontario, but in seven provinces (Alberta, British Columbia, Manitoba, Newfoundland, Nova Scotia, Prince Edward Island and Saskatchewan), a department of the Workmen's Compensation Board is responsible for accident prevention. In New Brunswick, the tenth Canadian province, an Industrial Safety Council promotes accident prevention education by a variety of methods. This council is composed of six members: 2 from employers, 2 from labor, the Deputy Minister of Labour and the Chairman of the Workmen's Compensation Board.

Another major influence on accident prevention in Canada is provincial and federal legislation governing plant hazards. Accident experience has provided the guideposts for developing needed legislation in each province. Similar to standards provided in the U.S. under the Occupational Safety and Health Act, the "Safety Acts" in Canada prescribe requirements that employers must follow to protect employees against hazards in the workplace. These laws also provide for inspections of plants, mines and other locations to assure employer compliance.

A good example of the type of protection afforded employees by such laws could be found in the Ontario Industrial Safety Act of June, 1972. This Act prescribes safety measures that must be taken to protect employees in a wide variety of situations, including such potentially-hazardous operations as entering tanks, vats or pits. It regulates protection standards for employees to assure their safety and health when working with explosives and hazardous gases, fumes and flammables. It sets ventilation and dust control measures that must be followed, and defines precautions and applications of mechanical guarding requirements. It also specifies when employees must be provided with necessary personal protective equipment.

There are also special acts that cover specific industrial hazards such as mining, logging, electrical, pressure vessels, construction and excavation. As analyses of accident causes indicate that new problems have been created, revisions in legislation are made to cover new hazards and situations. In truth, it could be said that nearly every law represents a prevention measure that was learned by accident . . . accidents that will be prevented or minimized in the future by the protection and requirements of such laws.[3]

A rather special approach to keeping a finger on the pulse of industrial accidents is that of The Labour Safety Council in Ontario. This is an advisory body to the Minister of Labour at the policy-making level. Its membership consists of representatives from management (represented by officials elected from industry to the Accident Prevention Associations), organized labor and the Ministry of Labour, Ontario. The Council facilitates two-way communication between management and labor in matters of accident prevention. It conducts or commissions scientific research related to current and pressing problems of the accident prevention movement in Ontario.

In Canada, a National Work Injury Statistics Program is being co-ordinated through Statistics Canada. This program maintains uniform classifications which are compatible with the American National Standards Institute.

One of the great difficulties encountered in statistical comparisons in North America — be it in Canada or the United States — is due to the fact that the statistical base keeps changing. Benefits are continually being improved, waiting periods reduced, grants increased, medical aspects changed and new technological hazards introduced. However, there is an increased awareness of the inter-relationship of downgrading incidents and the potential for their reduction by loss control management techniques.

The National Work Injury Statistics Program in Canada is a step toward making valid comparisons.

This section would not be complete without some recognition of the part that other agencies play in the "Help Provided": St. John Ambulance Association, The Canada Safety Council, The Canadian Manufacturers and other trade associations, other Governmental Departments — such as Health, Transport, etc. Unfortunately the list is too long to include here all the resources focused on loss control, but recognition must certainly be accorded to some of them.

A fine example is provided by the St. John Ambulance Association, which trains industrial employees and others in first aid. It has been proved in Canada that people with first aid training are more safety conscious and less likely to become injured either on or off-the-job.

COMPREHENSIVE FEDERAL LAW IN U.S.

By 1948, all states in the U.S. had workmen's compensation laws. It was not until 1970 that the U.S. Congress passed a comprehensive national safety law, known as the Occupational Safety and Health Act of 1970 (OSHA) . . . "to assure so far as possible every working man and woman in the nation safe and healthful working conditions and to preserve our human resources . . ." The Act (OSHA) covers every employer in a business affecting commerce who has one or more employees. It does not affect workplaces covered under federal laws, such as the Coal Mine Health and Safety Act and the Federal Metal and Nonmetallic Safety Act. Federal, state and local government employees are covered under separate provisions in the Act. The law is enforced by OSHA compliance officers (inspectors) who visit workplaces to determine whether or not standards are being met.

Compliance inspections come under the supervision of an OSHA area director, who assigns compliance officers and industrial hygienists on the basis of an established system of priorities. When violations are discovered, citations may be issued and civil penalties may be proposed. In order of significance, types of violations could vary from a condition that has no direct or immediate relationship to job safety and health (referred to as "de minimus", e.g. — lack of toilet partitions) to a condition where there is reasonable certainty that a hazard exists that could be expected to cause death or serious physical harm, either immediately or before the hazard can be eliminated through regular procedures. This latter condition is classified as an "imminent danger". If the employer fails to abate such conditions immediately, the compliance officer, through his area director, can go directly to the nearest Federal District Court for legal action if necessary.

Other types of violations may result in proposed penalties up to $1,000 per individual violation, depending on the employer's good faith, history of previous violations and size of business.

Employers who disagree with the citation and/or proposed penalty can request an informal meeting with the area director to discuss the case or legally contest the citation through a specific appeal procedure contained in the Act that guarantees full review of the case by an agency separate from the Labor Department.

The Act also contains a section amending the Small Business Act to make it possible for small employers to obtain long-term loans, through the Small Business Administration, to assist them in coming into compliance with the standards.

POSITIVE ATTITUDES — PRACTICES — RESULTS

Having considered attitudes and legislative progress through the years, let's now examine the principles that were applied to reduce accidents . . . and the results achieved. Programs began to develop steadily following the formation of the National Safety Council in the United States and the Industrial Accident Prevention Association in Canada. There are records of departmental safety contests and group safety meetings as far back as 1913, but this type of safety activity was not a common practice in industry. By the early 1920's, however, hundreds of companies in Canada and the U.S. had accident prevention programs that included personal protective equipment, safety rules, group safety meetings, investigation of injuries, recordkeeping and regular plant inspections. Most of the larger companies, which had such great attention focused on them earlier by various studies of death and disability rates, now had personnel with titles such as Safety Director or Safety Supervisor. Certain of these same companies in later years have won national safety recognition for their outstanding performance.[1]

The organized industrial safety movement grew fast enough that by 1928, the National Safety Council's records revealed that those companies reporting to them were, in general, showing positive downward trends in the frequency and severity of disabling injury. Between 1926 and 1967, the frequency rate of these reporting firms had dropped from 31.87 to 7.22, representing a 77.35% improvement for this period of time. (Refer to Figure 2).

The reduction was steady through the years and reflected the effort industry was making to improve safety and health programs for workers. Certain industries with special hazards (like the steel industry and the automotive industry) proved beyond doubt that accident causes can be controlled effectively with well-organized programs. In order to conserve manpower in World War II, the governments in both Canada and the U.S. encouraged safety activities with all contractors. Attempts were made throughout industry to add safety personnel and give greater emphasis to accident prevention programs, as the war effort required ever-expanding needs. (Whether or not safety and health programs kept pace with the fast wartime rate of production is a good point of discussion. Look at Figure 2 and 3 to observe the frequency rate levels during the war years.)

One lesson concerning safety that seemed to be strengthened

greatly because of the war effort was the relationship of accident prevention to the quality of products and to the production of quality products on time.

Management leaders of most large companies became enthusiastic safety motivators and welcomed opportunities to tell their associates about their feelings on this important subject. The direct quotations below reflect the attitudes of operating managers in the leading companies throughout the U.S. and Canada in guiding the program that brought about the more recent safety achievements from the 1940's to the present day:

"When periodically appraising subordinates, each department head should evaluate the efficiency with which his supervisors carry out this important job function."
— R. G. Uhler

"A good safety performance results from the same effective management that produced good quality or delivery on time."
— L. I. Mandich

"The main reason for success in safety is the day-by-day and hour-by-hour consistency with which you apply the principles of our safety program."
— W. D. Taylor

"Knowing and understanding the safety requirements of our area of responsibility does not discharge our obligation until we act in accordance with that knowledge to produce the desired results of decreased accidents, improved quality and general increased operating efficiency."
— R. C. McMichael

"Safety doesn't require extra time. On the contrary, unsafe practices, particularly those resulting in injuries and property damage, are time-consuming and costly. Just as quality comes from doing things the right way the first time, so does safety result from doing things the safe way the first time."
— W. E. Mullestine

With thinking and attitudes like this, it is easy to understand why the principles of accident prevention practiced by the leading industries from the 1940's to the 1950's emerged in scope and added such seemingly-advanced program facets as Job Safety Analysis and Job Safety Observation.

Surely there would seem to be no end to the progress that could be expected, and yet the progress stopped. For reasons that seemed logical to many, a leveling off of disabling injury frequency and severity rates began to occur in the late 1950's and carried into the early 60's. (See Figure 2 for trends). Because rates in certain companies and industries had reached low levels not believed attainable in the past, there was speculation that it was logical that further improvement would just be that much harder. Certain large companies whose disability rates exceeded 100 injured employees per million hours in the early 1900's were now consistently reporting rates less than 2. Some management people speculated that the bottom had been reached in realistically-attainable safety goals; they felt that, from here on, it was a matter of holding ground. In effect, almost everything that could be done for safety and health was being done.

A NEW ACTIVISM

Simultaneously with an upward trend in disabling injury rates that began in the late 60's and continued into the 70's, significant developments took place in a number of areas that brought great changes in philosophy and practice to the movement to control losses in industrial establishments. Let's briefly examine some of these factors having a major impact on industry during the years ahead.

Increased Union Activity

While unions have played a major role through the years in bringing about safety and health improvements in the workplace, a greater-than-ever upsurge of interest has been demonstrated in recent years. The employment of professional safety and health specialists at the national leadership level has increased at a faster rate than ever before. Several unions have bargained for professional union safety specialists to be added to company payrolls at management expense. Pressure on politicians to improve and enforce safety and health legislation has been a major factor in the promulgation of several new comprehensive safety and health related laws. Major independent research and studies are being conducted with union funds to prove the inadequacy of existing legislation and programs to combat safety and health problems. Liability suits against employers are more than infrequently supported by union funds. It has been said by a number of labor relations people that safety and health related items will be a key target for most unions in the years ahead. It would appear that the level of union efforts to improve safety and health conditions in industry today has only been equalled in the years that preceded the early compensation laws.

Rise in Consumerism

Public awareness of the responsibility of a manufacturer to produce a product that will not cause injury or illness to the user has never been greater. Spokesmen for the consumer (like Ralph Nader) have assisted in creating organized demands from the public for improvements in products. This widespread demand for safe products has ushered in a frequency of liability cases in the U.S. courts that has never been greater in history. Over 500,000 cases reached American courts in 1972. This figure was 10 times greater than the number 10 years before, and provoked one products expert to state that liability cases had reached epidemic proportions.

Public demand in the U.S. for improved control of product design and manufacture brought about the passing of the first consumer products safety act in 1972. Newspapers and magazines around the world have increased their coverage of products safety tremendously, indicating that this problem is not restricted to America. Employers have been brought to the sharp realization that the area of safety is much broader than the narrow concept of preventing injuries to employees on the job.

Attitudes of the Courts

Court attitudes have changed greatly since the 1900's, when protection was given to employers who could give proof of negligence by the injured. In 1965, the American Law Institute recommended in its "2nd Restatement of Torts" that manufacturers (as well as wholesalers and retailers) be held strictly liable for accidents caused by their products. Presently, courts in over 40 states have adopted this doctrine referred to as "strict liability". In effect, the legal responsibility of manufacturers in the U.S. has been greatly increased, and traditional defenses have been stripped away.

Important to the purpose of this training guide is the fact that "accident losses upheld by the courts" involve property damage as well as personal injury. In addition to the major change in court attitude toward the fixing of liability, the size of the awards made in recent years has reached record limits in all areas of safety and health liability. Awards of several hundreds of thousands of dollars to individuals are common, and individual awards from 1 to 4 million are made each year. The attitudes of the courts toward liability and the size of awards are certainly not limited to product safety.

Employers are more concerned than ever about the number of former employees (with a wide range of health problems) who can use sympathetic courts to an advantage never before possible. The number and size of awards granted to contractors or to visitors who have been injured on the premises has also grown to be a major concern to employers, particularly in the light of recent changes in the attitudes of the courts.

Medical Research

Perhaps no area is causing more concern among employers than the potential for loss that exists in the area of environmental health. Recent occupational medical research has substantiated the cancer-producing capability of a number of chemical substances that are essential to the manufacturing processes of many large companies, and ongoing research is expected to prove that proper awareness of health hazards on the job by management people has only begun to develop. Unions have charged that the high blue-collar death rate from "natural causes" has been the result of hazardous substances used by their members during their working careers. Dramatic changes are taking place (i.e., in the use of asbestos and common cotton in recent years) due to the serious related health problems.

Problems involved with noise, dusts, vapors, fumes and mists are becoming more common subjects of discussion among operating management people because of the new emphasis and awareness of environmental health hazards.

Technological Developments

Within the past two and a half decades, scientists and engineers have developed or designed more new products and equipment than were created by man during the entire previous history of the human race. It is doubtful that anyone would deny that Neil Armstrong's first step on the lunar surface on July 20, 1969, climaxed the greatest of these new scientific achievements.

It has been stated that what would normally have taken scientists and engineers several decades to accomplish was compressed into one decade to achieve this great success. One has only to recognize that the Apollo program involved some 20,000 companies employing more than 350,000 people in the construction and assembly of 15,000,000 component parts to realize the size of the task involved with this program and the possibilities for accident or failure that human error or mistakes could make. Yet, a flight system was produced that proved to have a reliability exceeding 99.9%.[5] (Perhaps one of the greatest benefits to be achieved from the space program will be the safety related measures that were utilized to insure the success of this great venture.)

Among contributions to this success, the application of a system approach was probably the overriding key. At all stages of conception, design, manufacture and operation, the man-machine-environment-equipment subsystems were considered as interrelated, interdependent components of the overall system. The enormous potential for catastrophic loss brought about the development of a new safety discipline that came to be known as Systems Safety Engineering. The application of new predictive techniques (discussed in this chapter) to determine the probability of component, sub-system or system failure at all stages of the system's life cycle enabled management decisions to be made that could correct difficulties or control them in advance of the systems operation.

The meaning of safety in aerospace no longer represented the simple "freedom from hazard for man" as defined by Webster in the *Intercollegiate Dictionary*. Safety had come to mean "freedom from the man-machine-media interactions that result in damage to the system, degradation of mission success, substantial time loss, or injury to personnel." In effect, the desire to insure the gross safety of the system and the ultimate success of the mission brought about a level of total safety confidence never before realized in the annals of industrial management.

Safety in the space program had a much broader meaning than it is generally given in industry. Largely because of our space activities, industrial management personnel will come to realize that safety relates to the prevention and control of all losses that could involve people, equipment, material and/or environment.

Government Legislation

In addition to the Occupational Safety and Health Act of 1970 and the 1972 Consumer Products Safety Act in the United States, a new comprehensive study of safety and health at work was completed in England by a committee under the chairmanship of Lord Robens. While the recommendations of the "Robens Report" have proved very controversial because of its sweeping reform suggestions, there seems to be little doubt that this report will motivate many changes in British industrial safety and health practices in the near future. A new compensation act in New Zealand provides, among other new benefits, 24-hour compensation coverage to all workers. Just as the new Products Safety Act in the United States points out a relatively-new area of safety concern for management groups in other countries, so the New Zealand Act emphasizes to all nations the importance of giving more serious consideration to off-the-job safety provisions in on-the-job safety programs. No matter how one looks at it, it seems quite clear that governments of the world will continue to pass stronger and more expansive laws to give the worker greater assurance of safe and healthful working conditions.

Escalating Equipment Costs

Just as the system approach to safety (people-equipment-material-environment) in aerospace was a major factor in preventing failures that could wipe the mission out, so the lack of a system approach in industry accounts for billions of dollars of accidental equipment loss each year. Few companies have followed the successes of aerospace by expanding their safety programs to include the prevention or control of property damage accidents and loss. Yet, replacement costs for a damaged piece of machinery or equipment are frequently three to five times greater than its purchase price several years ago. Material shortages brought about by an energy limitation awareness escalated the costs of certain pieces of equipment within a few short years to four to six times their purchase value. It has been stated by fire protection engineers that over 75% of smaller companies that have a major fire today will not be able to afford to rebuild and get back into business, because costs of material and equipment have outstripped insurance and other protective shields of the past.

Customer Service Demands

As never before, customer service is a key to sales. Economists have said that it will be several years before there are adequate supplies of certain materials and that business managers should take every possible measure to conserve their raw materials. This truth could logically be carried to the conservation of all property and products in order that customer needs not be interrupted. One safety director revealed that the president of his corporation had issued a policy statement to all plants to expand their injury-oriented programs to include the conservation of property. The incident that triggered this action by his chief executive was a property damage accident involving a lift truck. The operator was lifting a pallet holding four barrels of chemical, to place it on top of four other barrels at a two-barrel height. In the process of setting the barrels into position, a part of the pallet holding them (or a fork in the truck) snagged one of the barrels on the second tier and caused it to fall and break open. The chemical was contaminated and could not be replaced in time to meet the customer's delivery schedule. Not only was a $90,000 order lost, but a three million dollar a year customer as well. Service is a key to sales around the world, and accident prevention programs that expand their coverage to include safety measures for material and equipment will take a giant step in protecting good customer relations.

New Accident Facts

A good number of large companies with progressive safety

programs experiencing relatively-low disabling injury rates have been maintaining what is referred to as a serious injury index or a serious injury frequency rate. The types of injuries included in the rate may vary somewhat from company to company, since the technique has not been accepted in any standard for measuring industrial injuries, and its use has been a voluntary one for self evaluation only. Those companies using the serious injury rate have usually included in their figures or injury score (1) eye injuries requiring treatment by a doctor, (2) non-disabling fractures, (3) work injuries requiring hospitalization for observation, (4) loss of consciousness, and (5) other work injuries requiring doctor's care, work restriction or assignment to another job. (A guide for using a nonstandard measure of this type can be found in ANSI Standard Z16.1—1967). After 5 to 10 years of experience, many large companies acknowledge serious injury rates that range from 13 to 20 injuries per million man-hours, compared to their disabling injury rates of 1 to 4 per million man-hours. Consulting statisticians have suggested that this method of measuring injury occurrence rates is a far more valid barometer of performance than the disabling injury rates of the past.

The relatively-new Occupational Safety and Health Act in the U.S. requires the reporting by all employers to the Department of Labor all injuries that involve loss of consciousness, restriction of work or motion, transfer to another job, or any medical treatment except for treatment narrowly defined as first aid. This rate is somewhat similar to the serious injury rate described earlier. The government calculates the rate on a basis of reportable injuries per 100 employee man-years, or 200,000 man-hours. For the first year following adoption of the Act in 1970, the national reportable injury rate in the private non-farm sector was above 10 injuries per 200,000 man-hours. By converting this, we could logically assume there were over 50 reportable injuries occurring every million man-hours, according to the U.S. Department of Labor statistics.

It is no secret that the union leaders and public spokesmen like Ralph Nader have for years leveled severe criticisms against management for engaging in practices such as utilizing employees on crutches or walking casts, and providing sitting-down jobs just to avoid lost time, so that the injuries involved would not be counted as disabling. The need to establish a rate that would be a more sensitive barometer of actual injury losses as well as be less susceptible to finagling has been evident to safety leaders for some time. There is no question about it. Greater attention will probably be paid in the future to the reportable or serious injury rates than to the disabling injury rates.

While they are a relative minority compared to the many thousands of industrial firms in the world, companies in Australia, New Zealand, Canada, the United Kingdom, and Finland have accident prevention programs that include the control of property damage as well as personal injury type losses. Several hundred companies have developed their injury-oriented programs into total accident control programs with most gratifying results. The data gathered from over 200 industrial organizations indicate that property damage accident costs far exceed the costs of personal injury type accidents, and property damage frequency is at least 5 times greater than the total number of injuries reported.

Trend to Professionalism

The factors or forces in the 50's and 60's bearing upon management to more scientifically solve its many problems were certainly not limited to the few discussed. Even the smaller businessman found values in getting back to school to learn how to manage his operation more efficiently, in order to compete with big business and his small business associates next door. Management emerged as a professional field, and professional management principles were being taught widely to thousands of front-line supervisors as well as to members of upper management. For the first time, management people were learning that to control safety, quality, production or costs, it is necessary to identify the management work required in each of these areas, to establish standards for required work and to measure compliance with standards, to evaluate how efficiently work has been accomplished. It was this type of professional management thinking that caused numerous safety leaders in the world to ask whether or not the measurement of disabling injuries or any other type of loss results was the only way to measure management performance in safety and loss control.

The Energy Limitation Problem

Economists are saying that most countries of the world will experience shortages of many products for at least several years because of the need to conserve petroleum usage until other sources of energy can be developed. Even those few countries that may not experience petroleum shortages will find limited availability of many products which are dependent upon this source of energy for manufacturing processes. Supervisory personnel will be called upon as never before by upper management to carry out programs that will protect resources from losses such as waste, theft, and property damage.

WHAT LESSONS HAVE WE LEARNED?

Past history does provide lessons from successes as well as from failures; examples of both exist to provide a wealth of information of value to our loss control efforts in the future. Let's consider several lessons among those that will seem obvious to you from the information shared in this chapter.

Progress Has Been Made

There is little doubt among experts that a great reduction in disabling work injuries and deaths has taken place through the years. One can look at the trends in Figures 1 and 2 of this chapter and, with certain known work population figures, estimate the millions of disabilities and the thousands of deaths that would probably have occurred if such progress had not been made. Regardless of any criticisms leveled at management, there is overwhelming historical evidence that a great number of crippling injuries and deaths have been eliminated as a result of the organized safety movement.

Past Efforts Were Narrow

In tune with the old axiom that "necessity is the mother of invention", management's loss control efforts of the past in general industry were largely focused on the most apparent need to improve the high rate of work injuries and deaths.

While certainly not new to industry, improved programs involved with fire prevention, property damage control, environmental health, air and stream pollution, security, products liability, and off-the-job as well as on-the-job safety have all emerged as areas of major concern in the 70's. Without doubt, the pressures and the need for improved controls by management in all these inter-related areas of potential loss have not been greater since the 1900's. Although circumstances involved with these areas of concern vary from country to country, most nations of the world recognize the need for improvement in all of them as a serious (if not urgent) problem.

A view of conditions around the turn of the century indicates the degree of progress that has been made.

System Approach Best

The compensation-oriented manager, focusing his attention largely on the appalling rate of death and disability, concentrated major attention on his people on the job, establishing work injury prevention as his primary target. A look at the successful aerospace program left little doubt that one of the greatest contributions to the attainment of its unbelievably high degree of total safety was the system approach and outlook; all life cycle/stages, the people-equipment-material-environment subsystems, were considered as inter-related interdependent components of the overall system. Perhaps the most significant thing about a system approach and outlook is that, in the end, the safety of people is more assured than when we use the narrower outlook of the past.

THE ECONOMY AND EFFICIENCY OF LOSS CONTROL

As management contemplates the absolute necessity to improve its performance in the major areas of concern mentioned throughout this chapter, it will focus increased thought on specific ways and means to accomplish its goals at minimum costs.

Management personnel in general industry will see much more clearly what their peers in aerospace, petro-chemical, atomic energy, and other high-risk industries have recognized for many years.

Supervisors can inspect for sub-standard conditions related to fire, injury, security, environmental health problems, etc. during their regular inspections. Investigation techniques can be taught that will enable them to determine the causes of loss.

These inter-related problem areas can be considered in each job analysis, job observation or rule development assignment. The identical communication and training techniques used in the safety program can and should be utilized to advance the implementation of these other important areas of concern.

The logic of coordinated effort to minimize the unnecessary costs and inefficiencies of program duplication by specialists in each of these disciplines is self-evident.

Performance and Results

The modern trend for supervisors at all levels to utilize professional management principles involved in the planning-organizing-leading and controlling activities of all their important areas of work suggests that companies examine their methods of evaluating how well they are doing before losses occur. Programs that measure the work performance of supervisors, to determine their level of compliance with established standards of the safety, fire, environmental health and other programs, are in tune with the technological advance of our space age. This application of professional management enables an organization to be predictive rather than reactive in its efforts to control losses.

Professional safety and loss control specialists can anticipate that, more and more, upper management will want to utilize techniques of the types discussed in this book to measure and evaluate work performance efforts, rather than waiting for the results to speak for themselves.

WHERE DO WE GO FROM HERE?

The remaining chapters in this book will discuss subjects of vital importance to the control of the losses mentioned earlier, which are creating major concerns to management groups throughout the world today.

This book has been titled Loss Control Management because it deals with the professional management of a variety of inter-related problems such as injury and illnesses, fire, property damage and products liability. The term "loss control" itself suggests the big picture that management must consider in order to approach problems in a more efficient manner, with the same enthusiasm and organization that proved so successful in the past... yet eliminate the old narrow vision, which was its greatest handicap.

BIBLIOGRAPHY

1. *Accident Prevention Manual for Industrial Operation*, 6th Edition. Chicago: National Safety Council, 1969.

2. "Introduction to Industrial Safety", Bulletin 267, U.S. Government Printing Office, Washington, D.C.

3. Eninger, M. E., *Accident Prevention Fundamentals for Supervisors and Managers*. Toronto, Ontario: Industrial Accident Prevention Association, 1968.

4. "All About OSHA", Booklet No. 2056, U.S. Government Printing Office, Washington, D.C.

5. Lederer, Jerome, "Lessons Learned from NASA's System Safety Program," College of Insurance, New York, N.Y., May 26, 1969.

6. ANSI Z16.1 — 1967 (R-1969). New York: American National Standards Institute.

II.
The Causes and Effects of Loss Producing Events

INTRODUCTION

The importance of this subject in our discussions of loss control management is represented by the forward position it occupies in this book. A sound knowledge of causes and effects of loss producing events is a fundamental prerequisite to a proper understanding of most of the subjects that follow.

This discussion basically expresses the concepts regarding this important subject that are held by the majority of leading safety and loss control practitioners today. It is based on the accepted axiom that "success begets success" and largely deals with those beliefs on this vital subject to be of greatest practical value in every day application.

An attempt has been made to present this information in a form that encourages the reader to consider and utilize the vast amount of valuable related information available in research documents as well as such fields as quality control and other allied areas of industrial activity. It is our hope to provide in this and other chapters the best information available on the subject at hand that can be readily adapted to every day practical usage by our readers.

INCIDENT-ACCIDENT DEFINITIONS

What is an incident?

The term "incident", as it is intended in this book, relates to any undesired or unwanted event that could (or does) result in loss. These "incidents" could be accidents, quality or production problems or even security breaches (such as thefts). One of the major goals of this book is to convey the message that the loss producing (downgrading) incidents that affect production and quality also affect such areas as safety, health and security.

By preventing or controlling incidents through loss control, we protect the overall safety of people—equipment—material and the environment. While the elimination of all loss producing (downgrading) events should be the eventual goal of every supervisor, special emphasis will be given in this book to those incidents historically referred to as "accidents". We must keep in mind that all accidents are incidents, but not all incidents are accidents. If we grasp the meaning and intent of this modern concept, we can take a great step toward the control of all loss producing events, as we learn to control those most frequently referred to as accidents. Let's compare the definitions of "incident" and "accident", and discuss them further.

1. *Definition of incident.*

    "An incident is an undesired event that could (or does) result in loss." This definition could also be expressed as "an undesired event that could (or does) downgrade the efficiency of the business operation.

    Example: The wrong spare part for a machine was received by a firm and therefore, repair time on a maintenance job was delayed four days. Upon investigation, it was found that an improper number had been typed on the order form. The incident in this case was the typing of the wrong number on the order form. This was certainly an undesired event, particularly since it delayed the repair four days, which made it a loss producing downgrading event, or an incident.

2. *Definition of accident.*

    "An accident is an undesired event that results in physical harm to a person or damage to property. It is usually the result of a contact with a source of energy (i.e. kinetic, electrical, chemical, thermal, ionizing radiation, non-ionizing radiation, etc.) above the threshold limit of the body or structure."

    Note: The term "physical harm" in this definition includes both injury and disease as well as adverse mental, neurological or systemic effects resulting from an exposure or circumstance encountered in the course of employment (ANSI 216.2–1962, Rev. 1969). For simplification of purpose, the words "injury" or "illness" will be used hereafter in this book to best define the words, physical harm.

    Example A: A worker was lifting a five gallon can from a storeroom shelf when it slipped from his grip, falling to the floor and striking him on his right foot, causing immediate swelling and discomfort. The lid on the can broke open, allowing chemicals to spill on the floor. The undesired event in this case was the can striking the worker's foot and floor. The physical harm was an injury and while the extent cannot be determined from the description, the worker was injured. We can also assume that some property damage was involved, since the can broke open. The contact with a source of energy involved the falling can with its kinetic energy, which proved to be above the threshold limit of the worker's body and the can's structure. There is insufficient information to determine whether other injury was inflicted by the can's contents.

Example B: A worker was using a paint brush to clean the surface of machined parts with an organic solvent, preparatory to coating it with a plastic protector. After working in the heavy odor of the solvent's vapors for several hours, he became ill and reported to the first-aid department for permission to go home.

The undesired event in this case was the exposure to the solvent vapors. The physical harm was an illness and, while we may not know how serious it was, it is evident that the contact was with a chemical energy that was above the threshold limit of his body.

Example C: An office worker flipped a cigarette butt into a large metal receptacle that was sitting on a janitor's dolly near the door, as he left the building. The butt smouldered in waste at the bottom of the receptacle, and eventually broke into flames, causing heat that caught nearby window drapes on fire. Fortunately, a roving patrolman spotted the fire through the window and summoned the fire department in time to save the building.

The undesired event in this case was the fire that resulted from the discarded cigarette butt. The fire (and fire-fighting efforts) certainly resulted in some property damage. The contact was with a source of thermal energy above the threshold limit of the property.

Let's review the key points in the definition of an accident (incident).

1. An accident is an undesired event.

2. An accident results in physical harm (illness or injury) and/or property damage.

3. It usually results from a contact with a source of energy above the threshold limit of the body or structure.

FURTHER ANALYSES OF INCIDENTS

Some additional comments about the key factors involved in an accident and certain important facts regarding the nature of accidents will help clarify several important points.

Are accidents unexpected, unplanned, and undesired?

Few people would disagree with the fact that normally a person does not desire to be injured or have his property damaged. We might question the regard of some people for the property of others but, in the final analysis, a person normally does not desire to have an accident. On the other hand, let's ask the question: "Are accidents always unexpected or unplanned?" Do people knowingly take risks or chances even if they don't desire the accident to happen?

It is suggested that the circumstances conducive to an accident's occurrence are frequently recognized in advance and that taking a calculated risk is much more than an infrequent occurrence in industry, in the home or on the highway. Anyone that has any depth of experience in industrial management knows that the causal factors in many accidents were recognized in advance of the undesired event. Many companies have actual procedures to follow to minimize the risk when they knowingly engage in risk taking. As will be stated in other chapters, this is not to indicate that managerial decisions are any different than they are in every day life of individuals in their own homes and in daily activities. A number of safety and loss control writers have expressed the viewpoint that there is no such thing as an accident because management's knowledge prior to the typical industrial accident usually anticipated the event's occurrence but did not alter their actions substantially enough to avoid it. It has actually been suggested by several writers in professional journals that the word accident is inappropriate and should be dropped from the "Professionals" vocabulary.

The following excerpt was taken from a publication of the Hartford Insurance Group and gives substantial support to the argument that accidents are not unexpected, unplanned events.

"Are accidents "unexpected"?

An accident is by dictionary definition "a happening that is not expected, foreseen, or intended." Webster goes on to state that it is "an unfortunate occurrence or mishap." We can't quarrel with the latter, but after reviewing a fair number of 11,000-plus accident investigations conducted by Hartford Steam Boiler last year, we're wondering about "not expected."

Improper maintenance, lack of any maintenance program, no procedure for periodic testing of controls, no protective controls, little or no concern for operator training, absence of any education or training program, no emergency plan or shutdown procedure, no spare parts program, improper installations, improper applications, no maintenance records, no use of logs, and to top it off—little common sense even.

With these conditions prevailing in plant after plant, large and small, how can we by strict definition refer to these "happenings" as being "unexpected."

Many are by invitation."

An analysis of modern accident definitions from such sophisticated industries and organizations as aerospace and the U.S. Army Air Force gives clear insight to the need to replace the commonly accepted definitions of the past with a functional definition to meet the real practical applications of today. That's exactly why we have selected the "functional" definitions we have presented here. It is ridiculous to think the word accident will disappear from the loss control scene.

Jerome Lederer, Director of the Office of Manned Space Flight Safety for NASA at the time of the early moon landings, said this in one of his speeches:

"This nation was built on risk. Personal risk in tackling the wilderness, financial risk in business, risks in exploring the scientific unknown, enormous engineering risks, management risks. We shall continue to take risks of greater magnitude than in the past. But the consequences of failure are becoming less permissible. The political, social, as well as economic and personal risks that now accompany our ventures can have enormous repercussions when failure occurs.

Growing risk factors require a more comprehensive approach to hazard management than our wealth and isolation have permitted in the past."

Mr. Lederer also presents a message that indicates accidents are not always unexpected or unplanned events. In many cases people knowingly take risks, feeling "it won't happen to me" or "the odds are against an accident happening". Our objective as supervisors should be to make the consequences of risk-taking so undesirable and unattractive that people will not want to take them. In order to accomplish our goal, we must develop a sensitivity to the severity of risk-taking, so that we reflect it to our people in decisions that we make and actions that we take.

Are all accidents contacts with an energy source?

All accidents do involve a contact with either a source of energy or a substance in man's environment, but some contacts do not involve a source of energy. For example, a worker fell into a large water tank and (because he does not know how to swim) died from a lack of oxygen under water. The fall placed the worker in contact with water, a substance which interfered with his normal body processes, but it was not a source of energy in this involvement. Since there are some exceptions, we have said in our definition regarding an accident, "it is usually the result of a contact with a source of energy above the threshold limit of the body or structure". While all accidents will be dealt with as undesired events, we will give additional emphasis to considering "the contact", since it is involved in most accidents and since controlling the amount of energy that contacts people and structures is probably the greatest area open to the supervisor for potential loss control.

How important is the property damage accident problem?

The study described below will help the reader understand why accidents that result in property damage should be given a great deal of his attention.

In 1969, a study of industrial accidents was undertaken by one of the authors when he was Director of Engineering Services for the Insurance Company of North America. An analysis was made of 1,753,498 accidents reported by 297 cooperating companies. These companies represented 21 different industrial groups, employing 1,750,000 employees who worked over 3 billion man-hours during the exposure period analyzed.

The study revealed the following ratios of accident reporting:

For every serious or disabling injury (ANSI, Z16.1-1967) reported, there were 9.8 injuries of a less serious nature, an average of 15 serious injuries were reported for each disabling injury by 95 companies utilizing the serious injury index. (ANSI, Z16.1-1967)

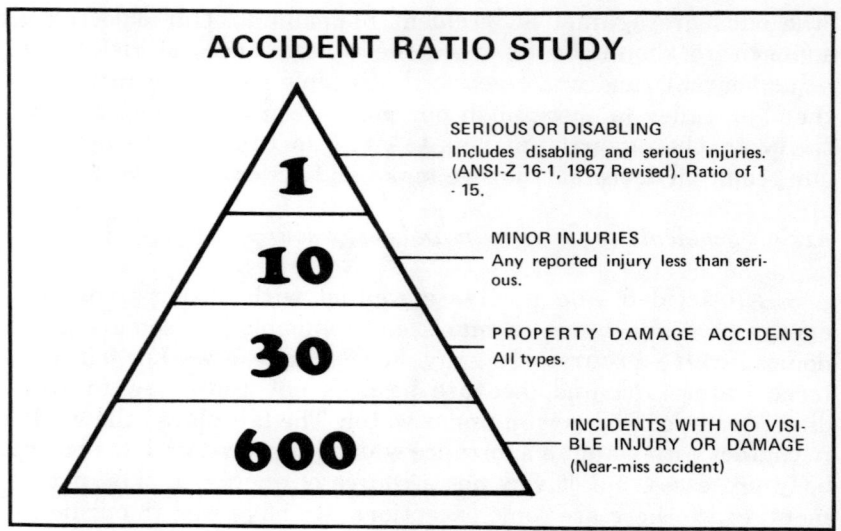

Figure 1.

Forty-seven percent indicated they were investigating all property damage accidents and eighty-four percent stated they were investigating serious and major damage accidents. The final analysis indicated that 30.2 property damage accidents were being reported for each serious or disabling injury.

Part of the study involved 4,000 hours of worker interviewing by trained supervisors on the occurrence of incidents that under slightly different circumstances could have resulted in injury or property damage.

In referring to the 1-10-30-600 ratio, it should be remembered that this represents accidents and incidents reported and not the total number of accidents or incidents that actually occurred.

As we consider the ratio, we observe that 30 property damage accidents were reported for each serious or disabling injury. Property damage accidents cost businessmen billions of dollars annually and yet they are frequently misnamed and referred to as "near-miss" accidents. Ironically, this line of thinking gives recognition to the fact that each property damage situation could probably have resulted in personal injury. This term is a holdover from earlier training and misconceptions that led supervisors to relate the term accident only to injury.

The 1-10-30-600 relationships in the ratio would seem to indicate quite clearly how foolish it is to direct our total effort at the relatively few events terminating in serious or disabling injury when there are 630 property damage or no-loss incidents occurring that provide a much larger basis for more effective control of total accident losses.

The valuable loss control potential that exists in the information available to organizations that expand their investigative efforts to include property damage and near-miss accidents or incidents with no visible injury or damage is substantial. The catastrophic potential for loss that can put the average business-man out of operation overnight is becoming more realistic every day. The need to expand safety and loss control programs to gain information of preactive rather than reactive value is an acute need. The chapters on "Property Damage and Waste Control" and "Incident Recall" were written with this need in mind.

Are illnesses from environmental health exposures and fires accidents?

We know from our earlier reading in this chapter that both of these results of undesired events fully meet our definition of accident. Both usually involve a contact with a source of energy above the threshold limit of the body or structure. In the case of environmental health exposures, the energy source is usually chemical, radiation, acoustical, or biological. In the case of fire, the energy contact is thermal.

These facts need emphasizing, since most supervisors in the past have not looked upon the causes of fire and illness as having a great deal to do with the causes of injury. As we examine the causes of accidents in the remainder of this chapter, we should think in terms of injury, illness and fire.

It is important to look at these important areas in this light, for the causes of fire and occupational illness are identical to those of injury. Again, we are reminded that the causes of all downgrading incidents are usually the same.

WHAT ABOUT NEAR-LOSS INCIDENTS?

These incidents are frequently referred to as "Near-miss accidents" and in most cases the use of this term is quite accurate. Under slightly different circumstances the types of incidents thus described could have resulted in personal injury or property damage. Let's consider one example of this true "near-miss" accident or incident:

> An extra crane operator from Building A was sent over to Building B to operate the crane there, since the regular

operator was ill and had gone home unexpectedly. The crane he was asked to operate was parked very close to another crane under repair, with a maintenance crew working in many positions on its bridge. The operator moved the control of the Building B crane to the left but, unexpectedly, the crane moved quickly to the right, and only by quick reversing of the control did the operator avoid striking the crane under repair.

Several scientific studies have been conducted by researchers that proved conclusively that information obtained from analyzing near-miss accidents (incidents) can be utilized effectively to prevent or control personal injury or property damage type accidents.

The proportion of incidents that occur that could have resulted in injury or damage (refer to ratio study) compared to those that actually do affords the supervisor many opportunities to take preventive action before the same incidents occur again . . . resulting in loss.

THE SOURCES OF LOSS PRODUCING (DOWNGRADING) EVENTS

Management leaders have written thousands of articles through the years on the complex nature of the errors and problems that cause loss producing events (downgrading) incidents in the business world.

A combination of factors or causes come together under just the right circumstances to bring about these undesired events. Seldom, if ever, is there a single cause of a downgrading incident involved with safety, production or quality.

As complex as the problem may sound, tremendous achievements (such as those in aerospace) have proved beyond doubt that it is possible to prevent or control the causes of downgrading incidents. While the enormous resources put into the aerospace program may not be available to everyone, there is well-documented evidence that high levels of success can be achieved by the average businessman. For example, one recent study projected mathematically that the national disabling injury rate could be cut 75% if the average businessman would adopt and promulgate those safety program activities used by leaders in general industry. Available information has led management leaders to accept the following conclusions:

1. The events that downgrade our businesses are caused; they don't just happen.

2. The causes of loss producing events (downgrading) incidents can be determined and controlled.

In order to better understand the circumstances that give rise to the causes of undesired incidents, it would be helpful to consider the four major elements or subsystems in the total business operation that provide their sources. These four elements would include: (a) people, (b) equipment, (c) material and (d) environment.

All four of these elements must relate or inter-act properly with each other, or problems may be created that could result in downgrading incidents. Let's examine each of these elements briefly:

(a) People: This element includes both employees and management. While it has been well established that the human element is involved in a high percentage of incident causes, we must be ever mindful that what employees receive or fail to receive by way of education, motivation and job tools depends on their relationship with management people. The employee is usually the human element directly involved with most accidents, since what he does or fails to do is seen as the immediate causal factor. We must, however, remember the employee-management relationships that greatly influence these employee actions as we determine what the people-related causes really are.

(b) Equipment: By equipment, we mean the tools and machinery the employee works with. Machinery could involve such items as drill presses and lathes, as well as cranes, lift trucks and automobiles. This element or subsystem of our business operations has been a major source of incident causes since the 1900's . . . and a big target for laws involved with mechanical safeguarding and operator training.

In more recent years, the improper design of controls and displays on machinery and equipment has been frequently named as the source or cause of many downgrading incidents involved with safety, quality and production problems. Emphasis on power equipment is not meant to belittle the source of accident causes provided by such simple tools as wrenches, hammers and chisels.

(c) Material: The material people use, work with or make provides another major source of incident causes. The 500,000 products-liability cases in American courts allegedly claim this element is the source of serious injury for this staggering number of people. Material can be sharp, heavy, hot or toxic. In all cases, this element of the business system can be a big source of contacts that result in downgrading incidents.

A recent issue of a well known business magazine allocated several pages and its cover to the causes of cancer. Included in this extended coverage of carcenogenic causing chemical substances was the fact that three new hazardous chemical materials are created by industry every day.

(d) Environment: All those parts of the physical surroundings that include the buildings that house people and the air they breathe would be considered environment. Environment is usually associated with such items as lighting, noise levels and atmospheric conditions. This element or sub-system of the business operation represents the source of causes of an ever-increasing number of diseases and health-related conditions. In addition, environment is also being named as the major source of incident causes associated with absenteeism and poor work quality.

The four major elements or sub-systems in the business operation (people, equipment, material, environment), individually or in combination, provide the source of causes that contribute to a downgrading incident. In every cause evaluation of an accident or other downgrading incident, the supervisor should make sure that he has considered the potential for individual or interrelated involvement of any or all of these four sources of cause. (See fig. 2)

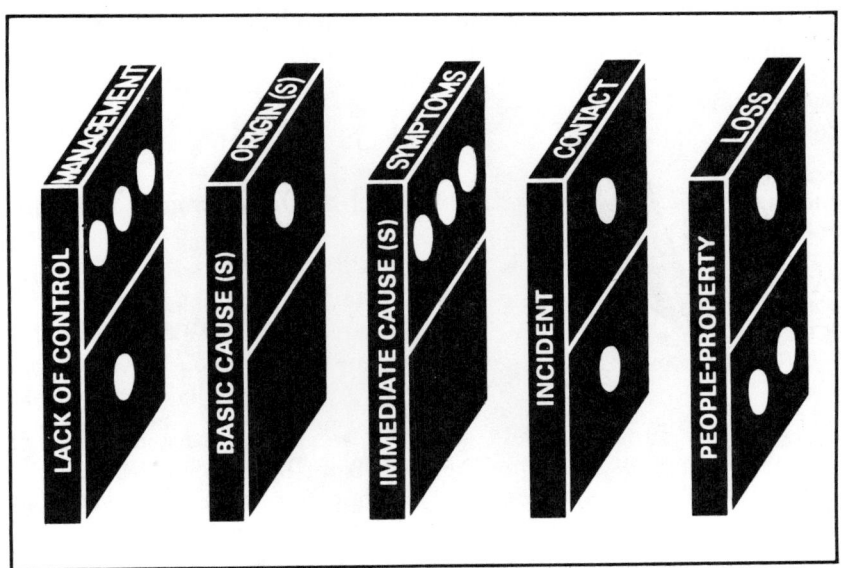

FIGURE 2 — CIRCUMSTANCES THAT LEAD TO LOSS

AN UPDATED DOMINO SEQUENCE

Dominoes are used here to represent modern loss control thinking, since they have been used so widely in the past to convey the principles of accident prevention and loss control.

Full credit is given to the late H. W. Heinrich, whose original domino sequence was a classic in safety thinking and teaching for over 30 years in many countries around the world.

The domino sequence shown above has been updated to reflect the direct management relationship involved with the causes and effects of all incidents that could downgrade a business operation.

1. *Lack of Control – Management*

This first domino in the sequence of events that could lead to a loss producing event is the lack of "control" by management. Since the word "control" would seem to apply to almost everything we will be talking about in this book, we ought to quickly point out the special meaning of "control" on the domino. "Control" as used here refers to one of the four functions of any professional manager:

INADEQUATE PROGRAM

INADEQUATE PROGRAM STANDARDS

FAILURE TO COMPLY WITH STANDARDS

1. Planning
2. Organizing
3. Leading
4. Controlling

All of these functions relate to the work that any member of management does, whether he is a front-line supervisor or the president of an organization. Whether his work is directed at safety, quality, production or costs, he must plan, organize, lead and control to make sure work is done properly.

When a supervisor accomplishes these job functions properly, he can rightfully say he is managing his job professionally. Let's specifically relate these functions to a loss control program.

Obviously, in order to do any work properly, a supervisor must know what work is required of him so that he can manage the doing of it.

Common areas of loss control work for a supervisor could include any of the following:

Making inspections, Conducting group meetings, Indoctrinating new employees, Making investigations, Doing a job analysis, Making a job observation, Reviewing rules and procedures and Giving proper job instruction.

The simple fact that many supervisors are unaware of the total work expected of them in loss control prevents them from doing their job properly, and this alone can cause the first domino to fall, starting the sequence of events that leads to loss.

Where Does the Supervisor Learn to Plan and Organize His Loss Control Work or to Lead his People to Loss Free Performance?

One of the major objectives of this book is to teach modern techniques that will assist professionals in their efforts to aid the supervisor in planning and organizing his work as well as leading his people to produce proper work without undesired losses.

Nearly every chapter in this book contains specific suggestions that can be utilized to make management's job easier and help produce more efficient results. The chapter on "Behavior Motivation" is designed to help you teach the supervisor how to lead his people to improved job performance. Upper management and staff personnel in most organizations give ongoing coaching and personal training to their supervisory personnel. Regardless of the source of help, the first domino will fall for the supervisor who does not plan and organize his work well and apply the necessary motivational skill to lead his people to the proper level of performance.

How Does the Supervisor Know If He Has Done His Loss Control Work Properly?

Proper work requires that the supervisor and his people perform to the standard required in each area of loss control work. This is the real meaning of "control" on the first domino. He does this by knowing the required program standards and measuring his own actual performance and that of his people by those standards. This is what enables the good supervisor to take corrective action before losses occur, rather than constantly reacting after the loss. When he doesn't do this, he can't possibly know how effective his loss control work really is. The first domino could fall because of many deficiencies in the supervisor's own personal loss control efforts, as well as those of the people he supervises. For example, if the company standard for inspection requires that the supervisor inspect an assigned area each month, 12 inspections per year would be the standard by which he should measure his own performance. Performance of less than 12 inspections could permit substandard practices and conditions to exist, setting the stage for the first domino to fall.

One more example: consider a company standard for Job Rules or Procedures, requiring that they be reviewed with each employee annually. If, in practice, the supervisor covers only 75% of his people by the end of the year, he has failed to re-educate 25%, and again the first domino could fall because he is not controlling his own work performance in loss control.

The supervisor who manages professionally: (a) knows his loss control program, (b) knows his loss control program standards,

(c) plans or organizes loss control work necessary to meet standards, (d) leads his people—to desire to achieve standards with him, (e) measures his own performance and that of his people to the standards, (f) evaluates levels of performance, (g) corrects his own performance and the performance of his people. THIS IS CONTROL.

Here are the most common reasons for lack of control by the supervisor. They permit the first domino to fall and the cause-effect sequence to be triggered: (a) An inadequate program and inadequate program knowledge, (b) inadequate program standards and inadequate knowledge of program standards, (c) Failure to perform to standards or to manage employee compliance to standards.

> Special Note: Most of the comments in this chapter relative to the functions of management control have been directed to the front line supervisor, because he is the point of control for upper management. The safety or loss control manager must accept full accountability for professional advisement to upper management on what constitutes an adequate program and adequate program standards. In addition, one of the major functions of safety or loss control personnel should be to measure and remeasure, audit and reaudit in order that management at all levels have professional evaluations of substandard performance by company/organization, division, department and individuals upon which they can act to correct performance differences. While the ultimate responsibility and accountability for performance rests with operating management, staff advisors on this important subject cannot escape the important role they play in providing management with timely professional information upon which they can act. Deficiencies in the program, its standards and standard compliances, regardless of the reason would constitute a lack of control and could cause the front domino to fall. Chapter three indentifies the critical program activities and standards of a modern safety/loss control program.

2. *Basic Cause(s) – Origin(s)*

EXISTENCE OF PERSONAL AND JOB FACTORS

A lack of management control permits the existence of certain basic causes of incidents that downgrade the business operation. These causes have also been referred to as root causes, indirect causes, underlying causes, or real causes, since the substandard practices and conditions (immediate causes) most closely associated with the incident originate directly from them.

Basic causes are frequently classified into two groups:

Personal Factors

Lack of knowledge or skill, improper motivation and physical or mental problems.

Job Factors

Inadequate work standards, inadequate design or maintenance, inadequate purchasing standards, normal wear and tear and abnormal usage.

The basic causes referred to as personal factors explain why people engage in substandard practices.

It is only logical to assume that a person may not follow a proper procedure if he or she has never been told or shown . . . that a crane or lift truck operator, for example, would not have the skill necessary to operate the special equipment efficiently and safely if proper training was lacking . . . that poor quality of work would result from placing a person with faulty vision on a job where good vision was critical . . . or that a worker who was never told the importance of a job would not perform with a high degree of pride in his work.

Likewise, the basic causes referred to as Job Factors explain why substandard conditions are created or exist. Equipment and material will be purchased and structures designed without proper consideration for loss control if adequate standards do not exist and standard compliance is not managed. Machines and equipment will wear out and cause substandard performance and unsafe conditions, if proper maintenance is not effected. Abuse and reuse of material, machines and equipment can cause many substandard conditions that result in waste and inefficient operation, and thus present hazards to people and property.

Basic causes, then, are clearly the origin(s) of substandard acts and conditions, and failure to identify these origins of loss at this stage in the sequence permits this domino to fall, initiating the possibility of further chain reaction.

3. *Immediate Cause(s) – Symptoms*

EXISTENCE OF SUBSTANDARD PRACTICES AND CONDITIONS

OCCURRENCE OF ERRORS

When the basic causes of incidents that could downgrade a business operation exist, they provide the opportunity for the occurrence of substandard practices and conditions (sometimes called errors) that could cause this domino to fall and lead directly to loss.

Definition: A substandard practice or condition (error) is any deviation from an accepted standard or practice. The practice could involve both acts of people and conditions related to physical things.

Several safety spokesmen have advocated dropping the word unsafe and substituting the word error (operational error – management error) in order to more clearly identify management's role in the control of the practices and/or conditions involved with accidents. There is also a vast amount of significant research and error removal information in published form that offers an enormous potential source of information that has direct application to accident prevention and loss control. It is felt that the term substandard not only provides this same value but more appropriately relates to the activities of "management control".

The supervisor who directs his attention to the control of all downgrading incidents will find the term "substandard" in keeping with professional management thinking, since it relates to a "standard" with which a professional recognizes he has a responsibility to manage compliance. In other words, it would automatically cause the good supervisor to raise the questions, "What's the real cause here?" "Did this person have the knowledge, skill and motivation?" "Is there a physical or mental problem that I missed?"

Of course, the use of the words "substandard practice" or "condition" enables the supervisor to apply his knowledge of cause-and-effect relationships to all areas of his management work.

Safety references most frequently refer to immediate causes as unsafe acts and unsafe conditions.

> Definition: "The unsafe act is a violation of an accepted safe procedure which could permit the occurrence of an accident."
>
> Definition: "The unsafe condition is a hazardous physical condition or circumstance which could directly permit the occurrence of an accident."

Unsafe acts or conditions are usually classified according to the ANSI Z16.2-1962 (Revised 1969) code:

UNSAFE PRACTICES

1. Operating without authority
2. Failure to warn or secure
3. Operating at improper speed
4. Making safety devices inoperable
5. Using defective equipment
6. Using equipment improperly
7. Failure to use personal protective equipment
8. Improper loading or placement
9. Improper lifting
10. Taking improper position
11. Servicing equipment in motion
12. Horseplay
13. Drinking or drugs

UNSAFE CONDITIONS

1. Inadequate guards or protection
2. Defective tools, equipment, substances
3. Congestion
4. Inadequate warning system
5. Fire and explosion hazards
6. Substandard housekeeping
7. Hazardous atmospheric conditions: gases, dusts, fumes, vapors
8. Excessive noise
9. Radiation exposures
10. Inadequate illumination or ventilation

Whether we refer to these deviations as substandard practices and conditions, errors or unsafe acts and conditions, there is one important thing common to all. Each and every one is only a symptom of the basic cause that permitted the practices or conditions to exist. When we fail to determine what the basic causes behind the symptoms really are, we fail to keep this domino from falling, and the direct potential for loss exists.

4. *Incident – Contact*

AN UNDESIRED EVENT THAT COULD OR DOES MAKE CONTACT WITH A SOURCE OF ENERGY ABOVE THE THRESHOLD LIMIT OF BODY OR STRUCTURE

Whenever substandard practices and conditions are permitted to exist, the door is always open for the occurrence of the incident that may or may not result in a loss. The incident is "undesired", since the final results of its occurrence are difficult to predict and are most frequently a matter of chance. As we said earlier in this chapter, incidents that result in physical harm or property damage are referred to as accidents and usually involve a contact with a source of energy. The 1-10-30-600 ratio study also pointed out that more incidents occur that do not result in loss than do. It is important to recognize that each incident, whether or not it results in loss, provides an opportunity to obtain information that could prevent or control a similar future incident that could become an accident.

Those incidents called accidents are frequently classified according to their types, as indicated by the ANSI Z16.2-1962 (Rev. 1969) code. More common types are listed below:

1. Struck Against

2. Struck By

3. Fall to Below

4. Fall on Same Level

5. Caught In

6. Caught On

7. Caught Between

8. Contact With

    a. electricity

    b. heat

    c. cold

    d. radiation

    e. caustics

    f. noise

    g. toxic or noxious substances

9. Overexertion (Over-load)

When we do not use information available from the incidents and the accidents to prevent or control future losses, the incident domino can fall again with its chance for major loss.

5. *People – Property – Loss*

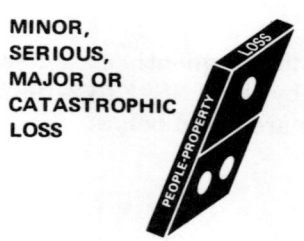

MINOR, SERIOUS, MAJOR OR CATASTROPHIC LOSS

Once the entire sequence has taken place and there is a loss involved with people or property, the results are usually chance events. The element of chance is involved in quality and production losses as well as those involved with safety, health and security.

Losses involved with all areas of the business activity could be considered as minor, serious, major or catastrophic.

The results of accidents can be evaluated in terms of physical harm and property damage, as well as humane effects and economic effects. See Figure 3.

The costs of accidents (excluding fires) that are referred to as uninsured are tremendous. Extensive analyses of property damage costs around the world have led experts to accept the fact that the area of uninsured property damage costs is 5 to 50 times the insured and compensation costs of injuries, while other uninsured areas constitute an additional 1 to 3 times the costs of compensation and medical expenses.

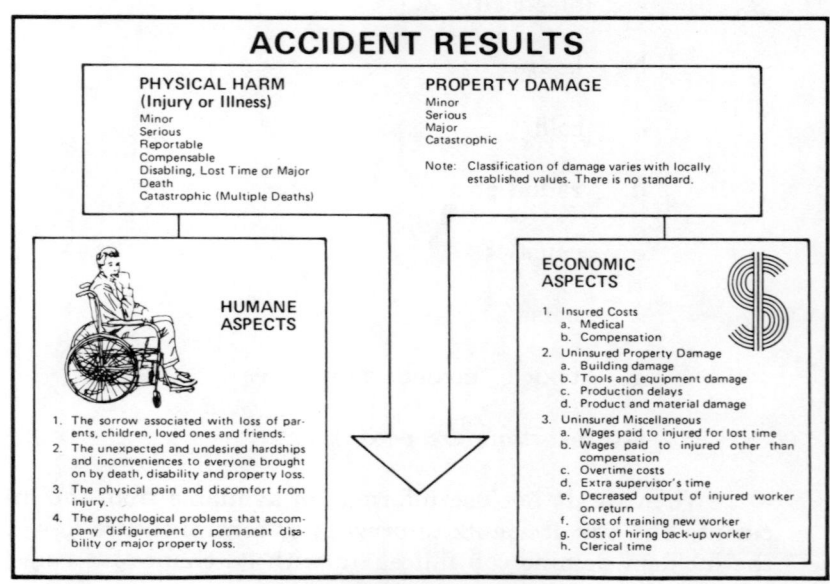

FIGURE 3 – ACCIDENT RESULTS

The reader should keep in mind that the classifications and costs in Figures 4 and 5 relate to losses involving injury, illness and property damage. It staggers the imagination to guess what the costs of fire, products and general liability, off-the-job accidents, air and stream pollution, rehabilitation, alcohol and drug abuse and theft cost the business world each year.

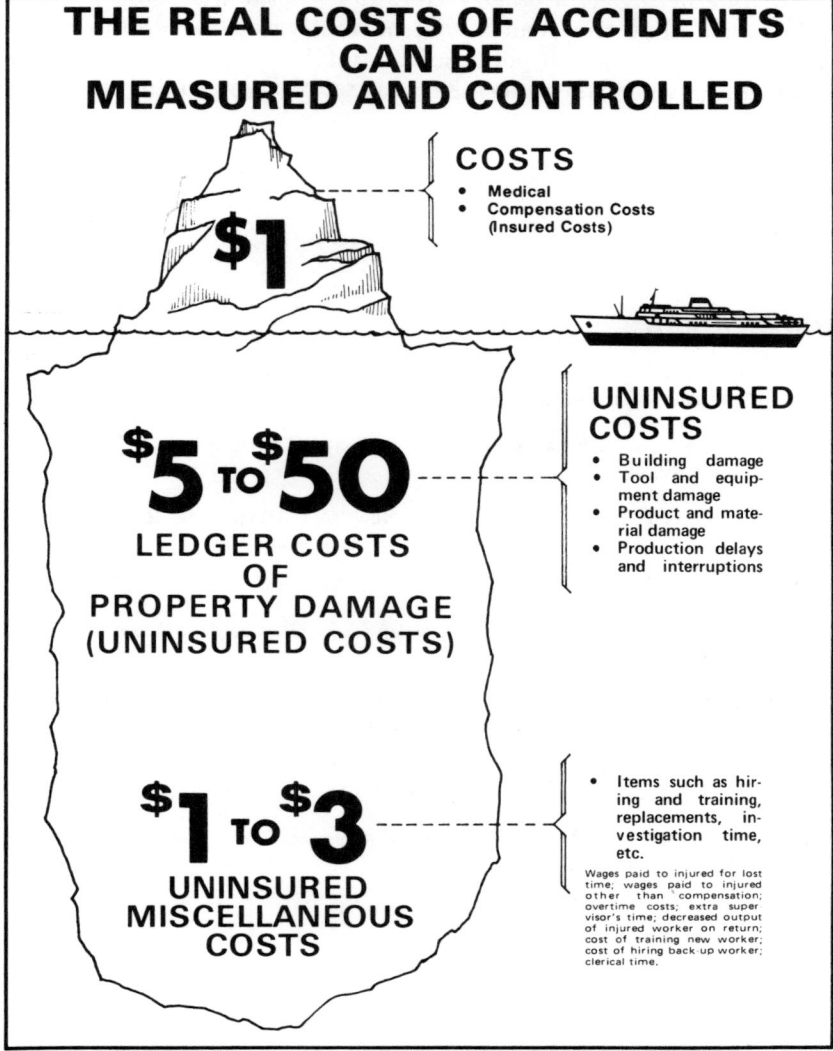

FIGURE 4 — Like the top of an iceberg, the insured costs of accidents are only a small part of the real costs that can be measured and controlled with modern Loss Control techniques.

SUMMARY

An atempt has been made to show the interrelationship of events that lead to a downgrading incident. *Re-emphasis is directed to the fact that the domino effect is not necessarily a direct chain reaction involved with single events. It is rather a reaction involving the potential of multiple events at each stage, with each established causal factor capable of continuing the reaction itself and of interacting with other factors to continue the domino effect.* It is hoped that the reader will be motivated to recognize the broad dimension of the potential for total losses for his business organization when the unnecessary causes of downgrading incidents are permitted to exist.

BIBLIOGRAPHY

1. Bird, Frank E., Jr., and Germain, George L. *Damage Control*. New York: American Management Association, 1966.

2. Bird, Frank E. Jr. "How to Spot Drug Users." *Environmental Control and Safety Management,* April, 1970.

3. Drucker, Peter F. *Managing for Results*. New York: Harper and Row, 1964.

4. Drucker, Peter F. *The Practice of Management*. New York: Harper and Row, 1954.

5. Gilmore, Charles L. *Accident Prevention and Loss Control*. New York: American Management Association, 1970.

6. Haddon, W., Jr., M. D., Clark, D. W., and MacMahon, G. "The Prevention of Accidents", *Preventive Medicine*. Boston: Little, Brown, and Company.

7. Johnson, W. G. *The Management Oversight and Risk Tree* — MORT, Prepared for the U.S. Atomic Energy Commission, U.S. Government Printing Office, 1973.

8. Ryan, G. Anthony, M.D. "The Aetiology of Accidental Injury". *Australian Safety News,* May-June, 1972.

III.
Management Control of Loss

INTRODUCTION

The need to know what a manager is—what work he does, and how he does it in a professional way—is not only a basic need of most safety and loss control people, it is at the base of most of their problems. Peter F. Drucker, author of the book, *The Practice of Management,* and one of the most widely known management consultants in the world, has said, "The ignorance of the function of management, of its work, of its standards and its responsibilities is one of the most serious weaknesses of an industrial society—and it is almost universal." Without exaggeration, one of the biggest needs throughout the world among business organizations that have not attained desired levels of loss control performance is the matter of "managing control" of the work necessary to get the desired job done.

One of the most common practices in industry, when accident records or other indices of loss are going the wrong way, is for an executive of responsible operating managers to share ideas and offer suggestions as to why things have gotten out of control. It is more the exception than the rule that a program coordinator can accurately pinpoint management's performance in all activities of the safety or loss control program to clearly establish the performance deficiencies that logically explain the problems. Invariably, the unsafe practices and conditions that management permits to exist become the focal point of the staff advisor's explanation for the problems at hand. While certainly an important *symptom* of the real problem and point of definition, that is all that this inadequate analysis really is. Behind this gross inability to diagnose management's specific needs in a deficient program is the need for the program coordinator to have a fundamental knowledge of how to manage a simple safety or loss control management system. In effect, Peter Drucker's, "ignorance of the function of management, of its work, of its standards, and its responsibilities", is being demonstrated again.

This chapter describes the characteristics of the professional loss control manager, presents the principles or truths that he should apply in his application of management skills, and outlines a simple loss control management system with which he can maintain control of his program.

While there is a variety of modern management styles and techniques that most certainly have application in any safety or loss control program, the critical need in the majority of programs is a basic system of management control that forms a framework upon which to build. The contents of this chapter are directed toward the attainment of this goal.

The Profession of Loss Control Management

In order to understand more clearly the individual engaged in this specialized work, it seems most appropriate to provide a definition of his work as well as his profession.

Loss Control Definitions

Loss Control is any intentional management action directed at the prevention, reduction or elimination of the pure (non-speculative) risks of business.

Loss Control Management is the application of professional management techniques and skills through those program activities (directed at risk avoidance, loss prevention and loss reduction) specifically intended to minimize losses resulting from the pure (non-speculative) risks of business. Loss Control Management involves the following:

1. the identification of risk exposures
2. the measurement and analysis of exposures
3. the determination of exposures that will respond to treatment by existing or available loss control techniques or activities
4. the selection of appropriate loss control action based on effectiveness and economic feasibility
5. the managing of program implementation in the most effective manner subject to economic restraints.

Generally speaking, the loss control manager will direct his efforts toward the attainment of these objectives:

1. *To upgrade the level of his programming activity.*

   The professional knows that the whole process of loss control management is a dynamic thing requiring timely change.

More important, he knows that he can optimize loss control effectiveness by programming to the levels of leaders in industry.

2. *To broaden the program's loss control utility.*

The professional will strive to simplify management's work as his program broadens and grows. He will accomplish this teaching how to inspect for fire, safety, environmental health and security problems on one loss control inspection. He will design investigation forms with total loss control utility in mind. He will, in effect, build as much loss control utility as possible into every program aid and technique.

3. *To improve loss identification and analysis capability.*

The effective loss control manager will recognize that his goal should be "total loss control" and that to achieve the professional recognition his position deserves in the enterprise, he must strive to optimize his system of loss identification and analysis in order to separate from the total group of loss items the "critical few" that justify the application of loss control within the constraints of available resources. (See Fig. 1)

4. *To optimize the application of professional management skill.*

The loss control specialist recognizes that the thing that will most guarantee achievement of his overall loss control objectives is the application of the same professional management skills that are being applied to production, quality, cost control and other important areas of his company's activity. He is also fully aware that it is this application of professional management skills that will make his approach more scientific, more precise, and above all else, more certain of success.

While the professional loss control manager does his management work of planning, organizing, leading and controlling, he is a *professional* because of several characteristics that make what he does far more scientific today than what his counterparts did many years ago. This is the big reason why results from a loss control or safety program can be assured if professional management techniques are utilized. Conversely, it is probably the most

TOTAL LOSS CONTROL

COULD INCLUDE BUT NOT BE LIMITED TO:

ON - THE - JOB INJURIES

ON - THE - JOB ILLNESSES

OFF - THE - JOB INJURIES

FAMILY INJURIES

MOTOR VEHICLE DAMAGE

MATERIALS HANDLING EQUIPMENT DAMAGE

GENERAL PROPERTY DAMAGE

PRODUCT AND MATERIAL WASTE

SHRINKAGE - THEFT - BURGLARY

VANDLISM

COMPUTER FRAUD LOSS

FIRE RELATED LOSSES

PRODUCTS LIABILITY CLAIMS

GENERAL LIABILITY CLAIMS

AIR AND STREAM POLLUTION COSTS

DRUG ABUSE AND ALCOHOLISM

ABSENTEEISM COSTS

Figure 1.

important reason that many past efforts were, at best, haphazard guessing games.

Following are four major characteristics of a professional loss control manager:

1. He can clearly identify and classify the work he manages through other people.

2. He can measure the work performance of management people who do the work required to produce desired results.

3. He uses a specific vocabulary.

4. He follows certain fundamental truths or principles.

In order to best understand the significance of these characteristics of every professional loss control manager and how they relate to the control of losses, let's examine them individually.

1. *He can clearly identify and classify the work he manages through other people.*

This is the management work that goes into the activities that constitute the loss control program. These activities could include, but not be limited to:

*A. Hiring and Selection
*B. Supervisory Investigation
*C. Supervisory Training
*D. Planned Inspections
*E. Skill Training
*F. First Aid Care
G. Emergency Preparedness
H. General Promotion
*I. Engineering Controls
*J. Purchasing Controls
*K. Proper Job Instruction
*L. New Employee Indoctrination
M. Physical Protection
N. Planned Job Observation
*O. Rules and Practices
*P. Job Analysis/Procedures
*Q. Group Communications
R. Key Point Tipping
S. Loss Analysis
T. Incident Recall
U. Disposal procedures
*V. Protective Equipment

(Asterisks indicate program activities believed to be critical to the attainment of any significant results in loss control.)

Specific management work in each of these activities would be determined by the standards established within the local program.

Like any other management system, the loss control program is a dynamic thing and today's standards may not meet tomorrow's needs. There must be constant growth through the addition of new program activities as well as additions and changes to the standards for each of these areas of management work.

Leaders in industry would be engaged in all the activities listed. Fig. 2 identifies the critical program activities for a hypothetical company and the minimum standards of related work for all supervisors that could form the basis for maintaining a system of management control for that company. Further explanation that follows will show how management's work performance can be measured and evaluated within this company.

2. *He can measure the work performance of the management people who do the work required to produce desired results.*

Perhaps no other characteristic of the professional loss control manager is more important than his ability to measure performance or input into the program. The chapter on Measurement Tools for Management gives an in-depth discussion of the difference between "measurements of consequence" (results) and "measurements of control" (management). While it is unquestionably the results that in the final analysis determine success or failure of a program, it is the manager's ability to measure performance or input on a timely basis that has the greatest influence on those results. Effective measurements of performance enable the loss control manager to provide meaningful evaluations to operating managers so that they can correct their program deficiencies.

The significant value of measurements of control in producing desired results is made clear in a recent comment made by Don Shula, coach of the Miami Dolphins professional football team. Mr. Shula was asked what was the biggest reason for the consistent success of his great Miami team. His answer was quoted in two major business magazines as a fine example of what professional management really is. This famous coach's answer was, "Every player is given a performance grade on every play in every game." This measurement of performance enables Don Shula to correct his game (program) deficiencies and produce desired results by winning games. It has been said that a Coca-Cola bottle also summarizes the value of measuring performance or input into the program—"no deposit-no return".

MINIMUM STANDARDS OF LOSS CONTROL
EVERY MEMBER OF MANAGEMENT WILL ENSURE *THAT:*

JOB INDOCTRINATION
1. Each employee has received the proper job indoctrination prior to the start of his work activity.

RULES AND PRACTICES
2. Each employee has had a complete indoctrination on all rules prior to beginning his job and that he knows and understands them. He will also make sure that a complete annual review of all rules is performed and take such action to ensure that they are enforced.

STANDARD JOB PROCEDURES
3. An approved standard job procedure has been developed for each critical task under his responsibility and that it has been issued to and thoroughly reviewed with each employee involved. He will update these procedures and review them with employees as required but not less than annually.

HAZARD CORRECTION
4. Any alleged unsafe practice of condition reported to him by an employee is promptly placed in the condition report system and followed up promptly. He will conduct and record the results of his inspection of the entire area under his responsibility not less than once every two months and make sure that all critical parts in his area are inspected as required.

PROPER JOB INSTRUCTION
5. Each employee receives proper job instruction (PJI) with every new or different task assigned and that loss control tips are given frequently during routine contacts on a day-to-day basis.

SKILL TRAINING
6. The skill training program required for operators of machinery and equipment is properly given and recorded.

PROTECTIVE EQUIPMENT
7. All employees are fitted for and issued required protective equipment and that this is properly recorded. He will ensure that 100% compliance is maintained.

GROUP MEETINGS
8. Each employee attends a weekly group loss control meeting.

SUPERVISORY INVESTIGATION
9. Every accident resulting in physical harm or property damage and all incidents that could have resulted in loss are immediately investigated and reported on the supervisor's report form.

PERSONAL CONTROL
10. He maintains an accurate knowledge of the degree of compliance he is maintaining with required minimum standards of this program. He will be prepared to discuss his performance with upper management at all times.

Figure 2.

Failure to utilize measurements of control in programs is perhaps the greatest professional deficiency of a majority of safety and loss control managers.

Fig. 3 reveals how the results of measurements of control could provide operating managers with timely evaluations upon which they would take corrective action. Use of performance evaluations enables management to be predictive rather than reactive, to recognize where they should concentrate additional management effort and take action before-the-loss.

3. *He uses a specific vocabulary.*

Every professional manager has a vocabulary that is very specific to the profession or discipline of which he is a part, whether he be a lawyer, physician or loss control manager. Words used in this book are typical of those that characterize the loss control manager and his profession. Words like biomechanic, human factors engineering, planned job observations, incident recall, proper job analysis, disabling injury, etc., may not be readily understood by a manager in another professional discipline. While there is no great need to discuss this in any further depth, a specific vocabulary is one of the four major characteristics of the professional loss control manager.

4. *He follows certain fundamental truths and principles.*[2]

As the loss control manager plans, organizes, leads and controls the important work in his program through other management people, he should make effective use of certain principles, or common truths, followed by professional managers.

A. The Principle of Resistance to Change:

"The greater the departure of any planned change from the accepted ways of the past, the greater the potential resistance on the part of people involved."

This essentially means that as the loss control manager introduces anything new, he should be fully aware that the greater the change he is making from past practice, the greater the management resistance he should expect. Only he can determine whether the change should

LOSS CONTROL RATING

"It is seldom that a tool provides a double edge sword of opportunity that we have in our loss control effort."

T.R. Pledger
President

DEPARTMENT	1 Weekly Meetings	2 Supervisor Investigations	3 Standard Job Procedures	4 Jobsite Inspections	5 Comprehensive Inspections	6 Protective Equipment	7 Skill Training	8 Employment Procedures	9 Loss Improvement Project	10 Pre-use Check	11 General Promotion	12 Rules	Performance this Quarter	Performance Year to Date	Honor Status
TELEPHONE GROUP															
Burnup & Sims, Fla.	90	96	88	92	80	85	90	85	86	89	90	93	88.6	86	::
Deviney	85	89	84	80	80	78	80	76	90	88	92	90	84.5	85	*
Dysard	87	90	93	89	92	90	96	89	97	92	93	92	91.6	92	:::
Fitton & Pittman	96	89	90	87	92	96	93	87	89	87	91	89	90.5	91	::
Burnup & Sims-Texas	92	97	89	92	93	89	97	94	96	90	93	96	93.1	92	:::
ELECTRICAL GROUP															
Burnup & Sims, CATU	86	90	83	76	91	86	90	86	89	86	90	92	87	88	::
Burnup & Sims, Electrical	75	69	72	80	63	76	84	87	79	82	86	79	77.6	80	*
Georgia Electric	85	90	87	89	86	73	88	79	84	90	83	82	84.6	83	*
Line Dismantling	96	92	89	97	88	99	97	86	97	91	98	96	93.8	94	:::
Southeastern Printing															
West Coast Line	69	73	84	79	82	67	84	76	83	72	69	81	76.8	79	
UTILITY GROUP															
BSI, Inc.	93	92	98	97	94	99	96	92	99	98	96	91	96	93	:::
Ford-Wehmeyer, Inc.	79	80	78	81	83	76	82	81	90	79	76	82	82	85	*
Greenbank	86	89	91	87	93	88	92	94	90	87	91	96	90.3	91	:::
Smith - Sweger	97	94	91	96	93	89	97	94	87	96	91	99	93.6	90	:::
TOTAL - CORPORATION															

PEOPLE ...... PROFITS ...... PROPERTY

Figure 3.

be taken in one, two or three steps to minimize management's resistance. The professional manager will constantly bridge the past to the present change, making the change more logical, and easier to deal with and understand. By being aware of this principle, he also knows that the bigger the change, the more he should plan and organize to minimize the resistance that can be anticipated. For this reason, the principle is usually associated with the planning function of management.

B. The Principle of Definition:

"A logical and proper decision can be made only when the basic or real problem is first defined."

This principle is involved in almost every activity of a loss control program. Whether the manager is dealing with a hazard, an accident cause, or other item of concern, he must constantly be reminded that effective remedial results are possible only when the real or basic problem is clearly defined. While not always possible, he should always attempt to get at the "point of definition" for best results.

C. The Principle of Reciprocated Interest:

"People tend to be motivated to accomplish results you want, to the extent you show interest in the results they want to achieve."

This simply means that the loss control manager can best sell any program or idea when he clearly establishes a bridge or connection of values between what he wants and something wanted by the operating manager or worker that he is trying to sell. This principle is discussed more fully in the chapter on "Economics in Loss Control". It is the key to persuasive communications with management at all levels. Successful application of this principle, more than any other, will gain the top management support every safety and loss control manager seeks. It has been summarized as, "You scratch my back, and I'll scratch yours."

D. The Principle of The Critical Few:

"In any given group or array, a relatively small number of items will tend to give rise to the largest proportion of results."

Vilfredo Pareto, an Italian economist and sociologist, was the first to set this concept down as an economic principle in 1906. It has since been referred to as "Pareto's law." While his view of this principle was rather narrow and based almost exclusively on economics, Joseph M. Juran, a well-known consultant and Fellow of the American Management Association, voiced the opinion in 1954 that the Pareto principle has application far beyond that of economics. Juran described the principle as "universal for management planning and control". He referred to the principle in terms such as the "vital few" and "trivial many"—terms that are used widely today in business. As startling as it might seem to some, this principle indicates that only one problem out of four is worth more than a loss control manager's fleeting glance. The other three can be dealt with summarily, since they are rarely of significant consequence. The loss control manager who recognizes this well-documented but often overlooked economic law takes a giant step in directing his limited time and efforts where they will do his enterprise the most good. It also permits him to let his subordinates handle the majority of problems in complete detail.

This principle can relate to the causes and costs of loss, anatomical location of injury, occupational and job involvement, pieces of equipment, material or operators. Its application is so wide and its value so great in terms of time and effort conservation that it is considered one of the most important management principles.[6]

E. The Principle of Recognition:

"Motivation to accomplish results tends to increase as people are given recognition for their contribution to those results."

The successful safety or loss control manager is the individual who can associate every facet of his program with a member of upper management who assisted him in promoting it. Application of personal and behavioral recognition serves as a powerful motivating force in reinforcing management and employee efforts. Recognition feeds the individual's strong desire to feel he is contributing, regardless of his stature or position.

F. The Principle of Future Characteristics:

"The past performance of an organization or unit tends to foreshadow its future characteristics."

This principle tells us that we can study past performance in safety, quality or production and assume that, with no change in those factors that would alter results, present and future results will continue to reflect similar trends. The value here is to use past performance as a guide to professional management action that can prevent history from repeating itself.

G. The Principle of Multiple Causes:

"Problems and loss producing events are seldom, if ever, the result of a single cause.

More frequently than not, there are several causes involved with every incident. Since this principle will be emphasized several times in this book, its meaning should be reasonably clear. It would be good to keep in mind that, while we must try to identify every possible cause of a problem, we should give the greatest amount of attention to those causes that possess the greatest potential of loss severity and the greatest probability of recurrence.

H. The Principle of Application:

"The more often a loss control manager communicates a message, the more certain he can be that it is understood and will be retained."

This truth clearly emphasizes the diverse ways and means that the loss control manager must devise to repeat his message. This is not meant to imply that communications should not be different, interesting and exciting messages and experiences. They must be to be effective. But most important of all, they must be repeated over and over. The reverse truth is also worth emphasizing. "The less we communicate the prevention and control message, the more certain we can be that losses will continue, because people didn't know, didn't understand and didn't receive the message."

I. The Principle of Point of Control:

"The greatest potential for control tends to exist at the point where the action takes place."

The front-line supervisor is a key in any successful program since he is the bridge from management to the worker. He interprets management policy. Workers see the viewpoint and attitude of the entire organization through the actions, deeds and words of their immediate supervisor. With this important statement made, it may seem contradictory to indicate that while the front-line supervisor is *a* key man in the program, he is not *the* key man. In order of importance to the success of the program, he could be the least important of all management levels. This "point of control" supervisor sees the safety and loss control program exactly as his "boss" wants him to see it. He is merely a mirror reflection of upper management attitude toward the subject. The "key" man in any program (whether it be loss control, quality, production or costs) is the highest level of operating management. The second key figure in any program is usually the second highest ranking executive. These "key rankings" flow down to the frontline supervisor, who is a mighty important cog in the wheel of success because he is "the point of control". While he isn't *the* key, he is *a* key, and an important one. In order to utilize this principle, it is necessary to operate a program that recognizes where the "keys" to success are located—and they are above the front-line level. Upper management in many organizations delegate their responsibility for safety and loss control to the fellow they say is the key, and add further frustration to this bottom-of-the-ladder supervisor whose hands are tied by lack of direction and motivation.

Chapters in this book such as "Economics in Loss Control" and "Property Damage and Waste Control" give enormous insight into motivating upper management involvement and commitment that results in a cascading effect to bring out the real "point of control" value in the front-line supervisor who makes the program go.

J. The Principle of Reporting Relationships:

"The higher the level to which a loss control manager reports, the more management cooperation he obtains."

The safety or loss control manager in most companies and organizations is an authority on the validity of this widely accepted management truth. In most companies and organizations he answers to a third-level member of this executive team. This reporting relationship in most organizations establishes the loss control manager/coordinator at the general foreman/supervisor level of management. In a relatively small number of organizations, his reporting relationship places him at a department head/superintendent level. Only in a relatively few organizations, particularly corporate level positions with large corporations or government departments, does he reach a true managerial or fourth level strata in management. While rapidly increasing numerically, an even fewer number of loss control managers have achieved vice president level. As will be said several times in other chapters, this position is usually held by individuals who have staff advisory responsibilities for control of property as well as personal protection. Property damage control, so important in motor fleet, air and rail transportation, is a factor common to staff responsibility of these high-ranking loss control executives.

A key question, of course, is whether or not the individual in most safety and loss control positions with a relatively low reporting relationship deserves anything higher. This is one of the most important messages of this book. As individuals recognize that broader loss control application, with its enormous potential for cost reduction and profit improvement, is the career development path, reporting relationships tend to evolve upward.

Regardless of the need for the individual to prove his worth and earn recognition, there is no doubt that the higher a loss control manager reports, the more management cooperation he obtains.

K. The Principle of Management Results:

"A manager tends to secure most effective results through and with others by performing the management work of planning, organizing, leading and controlling."

Perhaps no other principle provides more insight into making a program effective than this great truth. Several chapters of this book deal with planning and organizing specific and broad aspects of the loss control program. While this chapter deals essentially with the controlling function of the professional loss control manager, it should be emphasized that control is only possible when people have been led to effect the control system. The Director of Manufacturing of the Deere and Company recently addressed a safety conference with these words: "Authority deals with people's bodies while leadership deals with their minds. Authority is not as necessary as leadership."

Everything that a loss control manager does in all areas and phases of his work is eventually through and with other people, usually management personnel. Teamwork, so important to effective program management, is best achieved when supervisors know we want their ideas, their thoughts and reactions. Cooperation is stimulated most when we perpetuate by our actions the affirmation that each supervisor/manager is completely responsible for the safety/loss control program in his area of responsibility and we are there to serve as advisors and consultants.

Management Commitment-A Key to Management Control

While a number of other chapters will hopefully contribute to the reader's knowledge regarding this vital subject, it is deemed essential to this chapter's message to highlight several points on gaining upper management commitment. Figure 4 lists the critical program elements necessary to achieve prompt effective results in loss control program implementation. These items are those most commonly dealt with in a systematic manner when the help of an outside consultant is sought to expedite program results. Item number one, top management endorsement, would probably appear in this high position on any safety or loss control manager's list. And rightfully it should be. While a policy statement or

> **CRITICAL PROGRAM ELEMENTS NECESSARY TO ACHIEVE PROMPT EFFECTIVE RESULTS IN LOSS CONTROL PROGRAM IMPLEMENTATION**
>
> 1. Top management endorsement and continuous active leadership.
> 2. The establishment of minimum program standards.
> 3. Adequate management training at all levels.
> 4. The establishment of objectives to achieve minimum standards.
> 5. Development of necessary forms for information system.
> 6. Adequate training and advisement for program coordinator.
> 7. Establishment of effective reporting relationships for program coordinator.
> 8. A management system to measure and evaluate performance to standards at all levels.
> 9. In-depth evaluation of existing program.
> 10. Establishment of base line indices of performance for future comparisons.
> 11. Organization of a total management participation system.

Figure 4.

letter of endorsement is an extremely important document and point of reference, the continuous active leadership and on-going involvement of upper management, and all levels of management, is largely obtained by developing positive attitudes toward loss control through a planned program of participation. While many program coordinators talk about getting management involved, few promulgate a participative system to manage that involvement. Much too often the participation and involvement of upper management in program planning and organization is limited to very infrequent attendance at a dry policy meeting on safety committee meeting. Such participation is usually shared by a very small number of people with an assignment for a relatively long period of time. This type of assignment is hardly conducive to positive attitude building of the entire management team, or even the few participants. A well-designed system that guarantees 100 percent effective involvement of everyone on the management team in an exciting experience of participative team action that can produce significant results. While there are many approaches to this type of activity, it is vital that its design obtains the kind of participation that leads to positive attitude building and reinforcement.

A special appendix at the end of this chapter contains a description of one corporation's loss control management team activity.

Critical to establishing and maintaining management control is the establishment of minimum program standards (See Fig. 2)

The establishment of clearly defined standards provides the base for management performance that can be measured and evaluated, giving upper levels of management a barometer or indicator to correct substandard performance.

The chapter on "Measurement Tools for Management" discusses techniques for measuring *management control* in detail.

A Management Control System

The acronym ISMEC helps the student of loss control management remember the key words shown in Fig. 5 that summarize the activities involved in the function of "management control".

Identification of work
Standards for work performance
Measuring performance to standards
Evaluating performance in quantified form
Correcting substandard performance and recognizing desired levels

---

Figure 5.

A simple graphic outline of a very practical system to control management performance in a loss control program is shown in Figure 6.

This graphic portrayal of a simple management control system shows that a loss control manager should have a well-organized program that could include, but not be limited to, the activities shown. In order to guide management's control of losses, standards must be promulgated for each work activity, with compliance to each standard measured and evaluated. By providing upper management at all levels with a timely evaluation similar to the one shown in Figure 6, substandard performance can be corrected before-the-loss occurrence. Program control actions are predictive rather than reactive.

From time to time it is necessary to upgrade the program standards; e.g., a rule regarding the reporting of all injuries may be upgraded to include property damage accidents. This upgrading of standards is shown in the "maintenance loop" of the management control system in Figure 7.

Since any loss control effort is a dynamic thing that evolves over a period of time, new activities may be added to programs

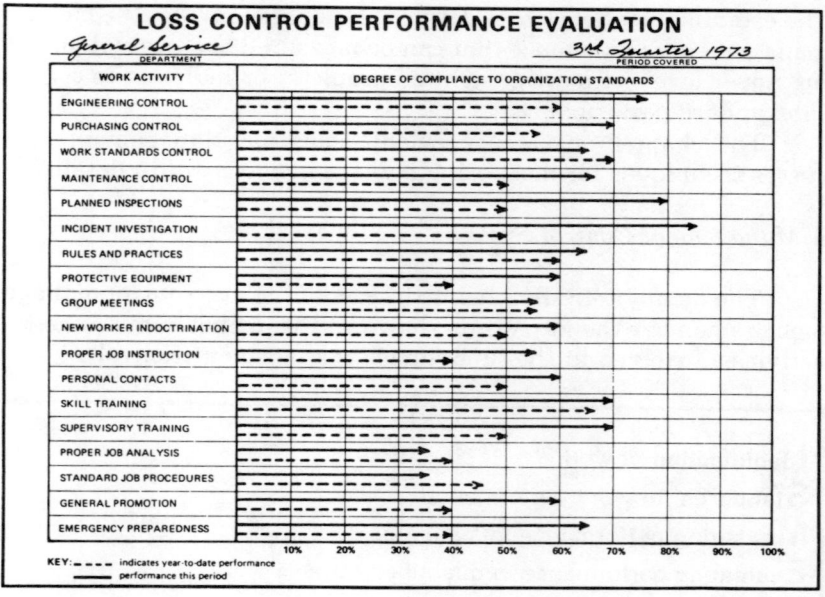

Figure 6.

when attitudes and other factors indicate a timeliness for change. When this is accomplished, all stages in the ISMEC system must be considered. A new standard must be promulgated; compliance with the standard must be measured; and a display revealing quantified evaluations distributed to management so that substandard performance can be corrected and acceptable performance can be recognized appropriately.

Managing An Imperfect System

It is not difficult to understand that the total prevention of all loss-producing events is neither economically feasible nor practical. It has been stated that the Apollo program proved beyond doubt that with unlimited budget, a near-perfect system is possible, though not practical for the average company or organization. The loss control manager must strive to optimize loss control effectiveness within the constraints of his budget and available resources. For all practical purposes, his program must be carefully planned to evolve over a period of time at the fastest practical rate—subject to the same business considerations involved in production, quality and cost improvement.

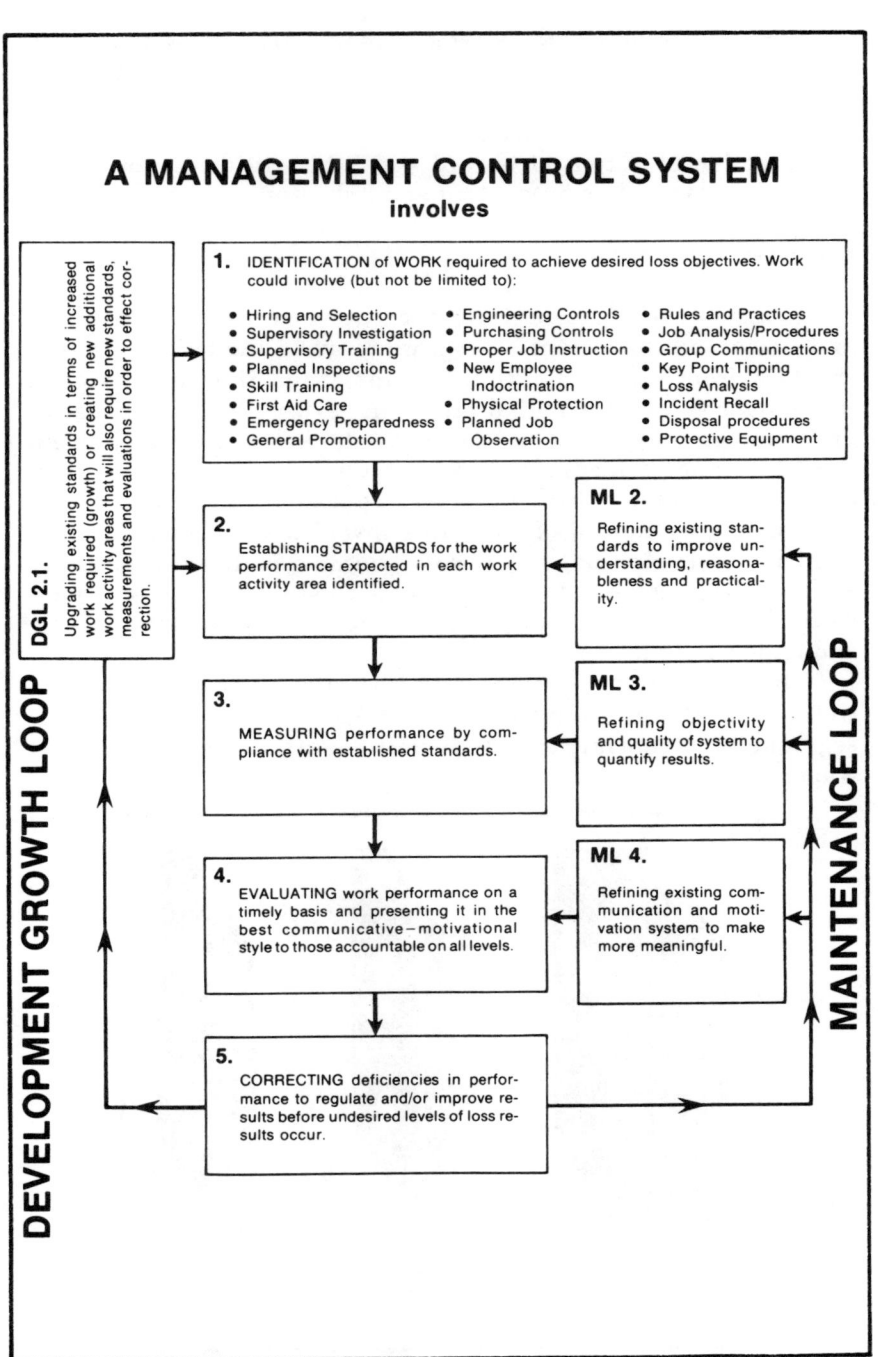

Figure 7.

The professional loss control manager recognizes that, with an imperfect system, he can only optimize program effectiveness by directing management skill at three stages or levels of loss control. Figure 8 shows the dominoes arranged to indicate three stages or levels of control toward which he will direct his efforts in risk avoidance, loss prevention and reduction.

The Pre-Contact Stage

This is the stage or level that includes everything we do to develop and implement a program to avoid the risk or prevent the loss from occurring. It is at this stage that we design a well-balanced loss prevention program; devise effective standards; and manage compliance to those standards in order to prevent the basic and immediate causes of undesired events, as well as check and correct them when they appear.

Managing control of standards involved with such activities as hiring and selection, supervisory training, planned inspections, skill training, general promotion, proper job instruction and standard job procedures can keep the first three dominoes from falling, and triggering the others.

By managing program activities effectively, the loss control manager takes a giant step toward the prevention of the largest segment of potential loss-producing events at the pre-contact stage.

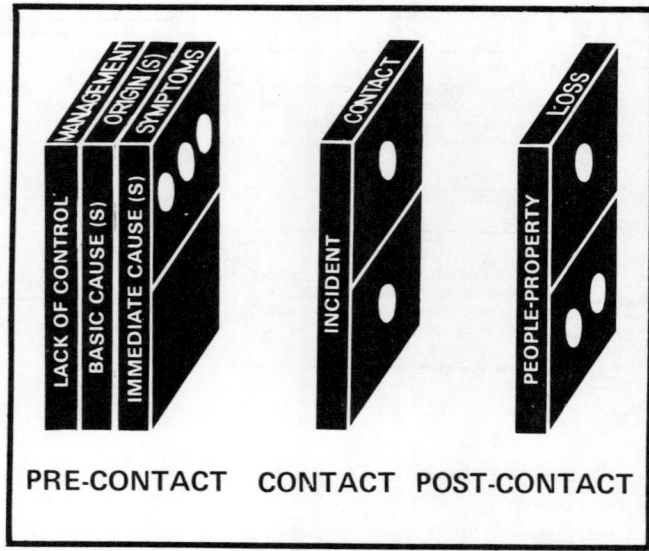

Figure 8. Three Stages of Control

The Contact Stage

This is the point at which the incident occurs that may or may not result in loss, depending on the amount of energy exchanged. A program should be designed to encourage the reporting of no-loss incidents or near-accidents, in order to take action *before* repetitive incidents occur that could result in losses. The chapter on Incident Recall gives valuable insight into ways to obtain this useful information. It is at this stage that compliance with protective equipment standards will reduce the effects of certain potentially harmful contacts and (hopefully) prevent or minimize loss. There are numerous measures or strategies that can be employed at this stage of control in order to minimize the amount of energy exchanged. The measures that follow could be employed to minimize loss involved with injury, illness or property damage (including fire) at the contact stage of control:

A. Eliminate a potentially harmful energy source by substitution or use of alternative source (e.g., the manager could recommend the use of electrical motors instead of shafts and belts in powered machinery; or he could suggest a solvent with a higher threshold limit value or flash point than the one proposed).

B. Reduce the amount of energy used or released (e.g., the manager could recommend the reduction of the temperature of a hot water system to reduce the danger of scalds to personnel in shower rooms; or he could suggest the slowing of vehicle speed between buildings in a plant by periodic bumper pads in the road).

C. Separate the energy by time or space from persons or property that could be exposed (e.g., he could recommend the barricading and locking of a safety space around stored hazardous material; suggest placing electric power lines outside of a building in a less accessible location; or move an expensive electronic unit off the floor to a protected overhead position).

D. Place a barricade or barrier between the source of energy and the people or property potentially exposed (e.g., personal protective equipment, a bumper guard on a column, or insulation on a noise-emitting machine).

E. Modify the contact surfaces of materials or structures to reduce injurious effects to people or property (e.g., recommend placement of shock-absorbing materials on a low ceiling point of a stairway or on a sharp corner of a building edge near a walkway, to minimize risk of head injury or body bruises).

F. Strengthen the body of the workers or the structure of equipment and buildings to support the energy exchange (e.g., the manager should suggest weight control and physical conditioning for such people as railroad conductors, pole climbers or maintenance personnel, to prevent ankle injuries while climbing "up and down" or "on and off"; or reinforcement of the beds of railroad cars and truck beds, or floors of buildings, to resist dropped loads during handling).

Obviously all these measures tend to reduce the loss potential once a contact occurs and are not intended to prevent the occurrence of the contacts themselves. It should be well understood that prevention or elimination at the pre-contact stage should always be the manager's first objective. Realistically, for reasons mentioned earlier, this cannot always be accomplished completely and the "contact" loss control measures give further assurance that losses will be reduced. They are "in addition to" measures, not substitutes for prevention efforts.

The Post-Contact Stage

No program manager wants harmful contacts or incidents that result in loss. Realistically, however, he knows that, although their occurrence can be reduced to a very low frequency, an imperfect system or organization will have the potential for some loss; and he should be prepared (by his planning and organization) to effect as much control as possible over the degree of loss that may occur.

There are several significant things that can be done at this post-contact stage to minimize loss, once a harmful contact has occured. While these efforts are after-the-fact, they could mean the difference between injury and death, or damage and total loss. It should be emphasized again that management's work at each of these stages of loss should be managed to comply with the organization's related standards. Loss control work at the post-contact stage includes:

A. Preparation for prompt emergency care for the ill or injured.

The logic of securing or rendering prompt emergency care, as an effective measure to reduce death and disability in the business environment, is supported by many occupational medicine specialists. There is no way of knowing how many lives might have been saved last year, or permanent disabilities averted, if prompt, proper care had been more readily available. When we consider that one in every four disabilities involves some permanent loss of body part, the importance of this vital subject becomes more evident. Expert consultants returning from Viet Nam have publicly asserted that, if seriously injured, their chances of survival would be much better in the zone of combat than in some business operations they have seen. Excellence of prompt emergency care proved to be the major factor in the phenomenal reduction of the death rate for battle casualties who reached medical facilities, from 4.5% in World War II to less than 2% in Viet Nam. The size of the death and disability problem in industry around the world justifies improved preparation by everyone on the management team as a post-contact control measure. The program should start with the supervisor, by making certain that he has had at least a good basic course in first aid, and that his follow-up refresher training is maintained. The reason for this is self-evident when we consider that the supervisor's team immediately looks to him for guidance in every emergency, and particularly when one of their own group needs help. Their respect for their team captain should not be diminished by his lack of knowledge about what to do. Once the supervisor has basic first aid training, he can better prepare his whole team to cope with emergencies involving illness or injury, in keeping with the recommended standards of his organization's program. The good supervisor will encourage all members of his team to take a basic course in first aid, but will give special encouragement in this regard to the key man/men on the team, who are usually looked to as the natural leader/s in emergencies. The fact that comprehensive studies completed by York University in Toronto have indicated a positive correlation between first-aid training and safe behavior is strong additional reason for the supervisor to stimulate interest within his group to take first-aid training. He will also utilize adequate time in group meetings and personal contacts to teach emergency care on an ongoing basis, as well as encourage the early reporting of minor injuries and illnesses.

While we have discussed emergency care as a post-contact stage loss control measure, its newly recognized value in developing safe behavior justifies it as a pre-contact measure to motivate the prevention of accidents.

B. Preparation for fire and explosion emergencies.

Since the subject of fire loss control is discussed in detail in another chapter, it is sufficient to say here that the supervisor who has complete control will have his people trained and ready to cope with these emergencies, to prevent injury to themselves, and to minimize property damage loss. Such training should include knowledge and experience in evacuation and escape, the use of fire-fighting equipment and protective devices (such as gas masks, extinguishers and fire blankets), the sounding of alarms, the shut-off of critical controls, and related skills.

C. Prompt reparative action for certain property damage.

Just as prompt first aid and emergency care have important values in averting more serious loss, there are numerous occasions where prompt maintenance or repair of equipment or other items of property can avert greater loss. This is why the alert supervisor will look for any evidence of property damage or premature wear-out of equipment through normal use, in his routine informal work activities as well as during planned inspections. Small cracks in machines, equipment, or building structures can frequently be repaired easily at certain points in their progress. Beyond those points, repair is impossible and major loss cannot be averted. One maintenance supervisor reported that a promptly-done $20.00 welding job could have averted the loss of a big press valued at $65,000, if the operator had reported the crack when he first observed it. There is no question that maintenance and other staff personnel share an important responsibility to maintain and keep the facilities in proper condition, but the front-line supervisor should be made to realize that he has an inescapable key responsibility for everything in his area. Managing critical parts inspections, as described earlier, is one of the better ways to detect minor loss before it becomes major.

D. Salvage and waste control measures.

Millions of dollars worth of salvageable items are thrown away in the business world each year because workers feel they are no longer usable or have any value. Waste cans and scrap hampers are the best places to look for items that could be repaired, reused in another fashion for another purpose, or simply sold when identified and repainted, for extra value. Even small items like gloves and protective equipment can frequently be repaired, sanitized and reused, if good "controls" are exercised. The most control, perhaps, is that the supervisor manage effective "control" of any existing standards of work that presently exist.

SUMMARY

While overall program results depend on the combined efforts of everyone, it is the application of professional loss control management skills that provide the clear roadmap of action that counts most. The loss control manager must plan, organize, lead and control the work involved in each of his program activities for which he has staff responsibility. His application of the professional management principles and skills described earlier in this chapter assist him in providing all levels of management with the tools necessary to achieve their desired goals in loss control.

BIBLIOGRAPHY

1. *Accident Facts.* Chicago: National Safety Council, 1975.

2. Allen, Louis A. *The Management Profession.* New York: McGraw-Hill Book Company, 1964.

3. Bird, Frank E., Jr., and Germain, George L. *Damage Control.* New York: American Management Association, 1966.

4. Bird, Frank E., Jr. "How to Spot Drug Users." *Environmental Control and Safety Management,* April, 1970.

5. _____. *Management Guide to Loss Control.* Atlanta: Institute Press, 1974.

6. Bittel, L. R. *The Nine Master Keys of Management.* New York: McGraw-Hill Book Company, 1972.

7. Blake, R. P. *Industrial Safety,* 3rd Edition. Englewood Cliffs: Prentice Hall, Inc., 1963.

8. Drucker, Peter F. *Managing for Results.* New York: Harper and Row, 1964.

9. _____. *The Practice of Management*. New York: Harper and Row, 1954.

10. Gilmore, Charles L. *Accident Prevention and Loss Control*. New York: American Management Association, 1970.

11. Haddon, W., Jr., M. D., Clark, D. W., and MacMahon, G. "The Prevention of Accidents", *Preventive Medicine*. Boston: Little, Brown, and Company.

12. Heinrich, H. W., *Industrial Accident Prevention*, 4th Edition. New York: McGraw-Hill Book Company, Inc., 1959.

13. Johnson, W. G. *The Management Oversight and Risk Tree*—MORT, Prepared for the U.S. Atomic Energy Commission, U.S. Government Printing Office, 1973.

14. Ryan, G. Anthony, M. D. "The Aetiology of Accidental Injury." *Australian Safety News,* May—June, 1972.

Appendix A

THE
LOSS CONTROL MANAGEMENT
TEAM

An effective tool to accomplish....total management involvement, improved loss control attitudes, formal supervisory training, special program promotion, planned inspections...

THE LOSS CONTROL MANAGEMENT TEAM

WHAT IS IT?

The Loss Control Management Team is an appointed group of management people from all levels responsible for promoting, advertising, and encouraging the company's loss control program. The team may consist of as few as three members, or as many as ten. Every member of the middle and upper management team from the Senior Officer to the district level manager would be assigned to serve on the team for two months at least once every two years.

The Chairman (a senior member of management) and two assistants chosen from the team help the Loss Control Coordinator promote a theme for the two-month period based on the critical loss control problems in the company. After the theme promotion has been developed by this sub-committee, it is presented to the entire team at its first meeting. Then, the team members, representing all operating locations of the company, carry the group's message to every supervisor and employee in their respective groups and see that the program goals are achieved.

The team chairman is responsible for making a smooth transition when the next team takes over the safety promotion program.

This type of organization should provide important management involvement in the loss control program, thus insuring its complete success.

WHY NOT CALL IT A "SAFETY COMMITTEE"?

"Safety Committee" is a technical term generally used to describe a group of people, usually representing management and labor, which meets regularly to discuss unsafe practices and conditions found in the company and ways to eliminate them. This committee is frequently one of the requirements of the labor contract, and its members serve for relatively long periods. While this type of safety committee is quite common and is becoming more common because of negotiated labor contracts, and in some instances, state law, its value in promoting good safety attitudes is highly questionable. Survey after survey indicate that 50% feel that these groups actually make safety efforts more difficult.

Since the emphasis of this group is negative ("what's wrong with the company?") and its approach "after-the-fact," the best it can do is make corrections of physical hazards. Groups organized in this pattern do not involve themselves in efforts of safety education and motivation. For this reason they contribute little to the improvement of the safety attitudes and awareness of the workers.

The Loss Control Management Team is designed to be a management-oriented tool; relegation to the lower status of the classic safety committee would greatly lessen its value. Where the traditional safety committee is a must, it is strongly recommended that the "Loss Control Management Team" be established as an entirely different group to fulfill its purpose as the source of education concerning safety, health, property, conservation, fire and security.

WHAT ARE ITS OBJECTIVES; WHAT WILL IT ACCOMPLISH?

1. TOTAL MANAGEMENT INVOLVEMENT IN THE LOSS CONTROL PROGRAM

    Total management involvement is essential to the success of any loss control program. The team concept provides for every management individual to be a member of the team or participate at least once every two years. The responsiblities assigned to the team members will provide more than ample activity to let them know they have an important role in the loss control program.

2. THE DEVELOPMENT OF POSITIVE MANAGEMENT ATTITUDES

    The first responsibility of the chairman and his planning committee is to assess the attitudes toward loss control of every team member and to plan methods of improving these attitudes during the two-month program. As they help to improve team members' attitudes, the chairman and his assistants should find that their own positive attitudes have been reinforced. Techniques will be described later which will help the chairman in his analysis of attitudes.

3. THE FORMAL TRAINING FOR MANAGERS IN THEIR RESPONSIBILITIES TO THE LOSS CONTROL PROGRAM AND IN LOSS CONTROL MANAGEMENT TECHNIQUES

    Part of every meeting should be devoted to formal training so that every manager's knowledge of loss control is kept up-to-date.

4. THE COMPANY-WIDE PROMOTION OF A CRITICAL--LOSS CONTROL PROBLEM WHICH HAS BEEN SELECTED BY AN ANALYSIS OF LOSSES OR PROGRAM NEEDS

    A good promotion program will develop in all employees an acute awareness of loss control problems and their remedies, thus preventing, controlling, and reducing losses and related costs. Notices, slogans, posters, pamphlets, and talks--any of these may be used to publicize the program theme. A contest can be a real incentive for employee participation. Each team representative will be responsible for bringing the message to his own group.

5. A THOROUGH GENERAL COMPANY INSPECTION WITH A FOLLOW-UP OF CORRECTIONS AND COMPLIMENTS AS NEEDED

The assignment of team members to coordinate inspection responsibility for their own group or one other than their own is assigned by the chairman. In addition, they would be responsible for the follow-up of items uncovered in their own groups.

HOW DOES IT WORK AND IN WHAT ORDER?

(a) TEAM ASSIGNMENTS

To establish a team, the number one operating executive (usually a vice president) assigns a different key executive to the chairmanship for two months each over a two year period. He also names two assistant chairmen, usually from middle management. Then, each chairman selected should receive a copy of the listing with a letter describing the Loss Control Management Team and requesting that each chairman complete a personal report outlining results at the end of his program. This letter should indicate that the Loss Control Coordinator will meet with each chairman well in advance of his term to assist with planning. A second letter is sent from the President to all assistant chairmen explaining the importance of the program and urging their support.

-3-

All team members should receive advance notice of the time and place of meetings they must attend.

(b) PLANNING MEETINGS

At least one month prior to the assigned period, the team chairman should hold their first planning session with the loss control coordinator. This first meeting should be private so that in subsequent meetings the chairman will have sufficient knowledge of the total program to fulfill his leadership role. The loss control coordinator should be prepared to discuss the entire program at this first meeting. At this point, the plans should be developed for carrying out the team training required for the presentation.

The critical loss control problem selected for promotion during this period should be discussed with emphasis given to the best means of communicating and implementing it most effectively.

The loss control coordinator, at this time, should explain a simple test to administer at the first team meeting which will inventory existing management understanding of attitudes toward loss control. The tests take only a few minutes to complete, and, for best results, they should be handed in unsigned. The leaders should be able to discern individual attitudes as meetings progress.

Methods of inspections should be discussed, and plans made for correction of problems uncovered during inspections. The loss control coordinator must impress upon the chairman the need for strong leadership by himself and his assistant chairmen. Finally, a time for the second planning session (which will include the assistant chairmen) is set.

At subsequent meetings with the assistant chairmen, dates and times of the four team meetings are selected so <u>that a notice can be sent out by the team chairman.</u> The critical loss control problem is discussed, and the chairman assigns some responsibilities to his assistants. One assistant usually handles promotion aspects of the critical problem and miscellaneous responsibilities, while the other organizes the inspections.

WHAT ELSE IS IMPORTANT TO SUCCESS?

(a) The most important ingredient for success is the active involvement of the chairman and his assistants in all phases of team activity. Experience has shown that the more active all leaders are in all areas of the program, the better the results will be.

-4-

The really big challenge for the Loss Control Coordinator is to make sure that the chairman and his assistants feel they are making a significant contribution. The personal satisfaction that can come to these important management leaders with full realization that a meaningful job has been accomplished will be a giant step to winning them to the cause of loss control for a long time to come. Experience has proven that one of the biggest mistakes that can be made is to try to make the job easy by keeping them out of the mainstream of activity. On the contrary, using them wisely where others will see and know their active role will be a dynamic loss control tool that has no substitute.

One technique frequently used (in addition to their leadership at team meetings) is to have the chairman and his two assistants draw three locations by random selection for their personal close inspections. Each may join with a team member assigned to inspect one of the locations drawn but may decide to conduct a visit independently.

In conjunction with this inspection effort, other activities should be considered. The chairman and his assistants may select certain areas to be photographed for highlighting. They should make a personal review of the most critical hazards uncovered by team members. These activities will provide an opportunity for team leaders to talk with employees and supervisors to gauge the effectiveness of the program. They

-5-

can then discuss their personal observations at upper-level management meetings, as well as at subsequent team meetings. Another technique is to select two locations by random selection for the chairman and his assistants to appraise as an audit task force. By pre-arrangement with the manager of each of these locations, they first meet with each manager to discuss and to evaluate what each feels his management group is doing for the loss control program. Using pre-designed forms they note his reactions to their questions and then field survey his area of responsibility to determine if the level of loss control performance in his area actually conforms to his beliefs. The results of the audits are forwarded to the managers concerned with copies forwarded to their managers and the loss control office.

This special assignment isn't too much if handled properly. Since participation on the team may be as infrequent as every two years, this activity is also regarded as an opportunity to see what's going on in an area not directly concerned with the busy executives' every-day business life.

Getting into the company and playing a direct role in the action part of the program gives the chairman and his assistants a much deeper feeling of contributing as they deal with team members at regular meetings.

Regardless of techniques used, an important key to a successful loss control management team effort is getting these important management people out into the mainstream of action where they can be seen and heard representing the cause of loss control.

(b) Meetings planned in advance make interesting meetings. The better the planning, the better the meetings. It is always best to have a short planning session prior to each meeting of the team. The first meeting is a lively one, because it involves introduction of new ideas and thoughts, but the last three can become boring if not planned effectively. Unless team members are made to understand clearly that follow-up reports at these meetings must be specific and accurate, information given will be vague and general. Firm direction by the chairman is important in this matter. Interest in the last meeting can be improved by having the president or another guest speaker come in for a few minutes as a highlight. The sincerely expressed thanks of the chairman and proper presentation of training cards and awards will help make the last session more meaningful.

(c) The chairman must insist on perfect attendance by everyone. Unless he makes the importance of perfect attendance completely clear at the first meeting, there will be an amazing number of good excuses given by many members. Some companies give a small incentive such as a pen and pencil set for perfect attendance. This practice is justified because certain members of the team may be travelling distances and have heavy work schedules ; thus, attendance will sometimes require very special effort. The practice of calling the team members' manager whenever someone is absent is an extremely strong motivator to keep attendance up.

(d) Interest and follow-up by the top executive is absolutely necessary for best results. Team assignments require a little extra effort from everyone. This is much more acceptable and, in turn, enjoyable when all know the importance placed on their effort by the number one executive. His interest can be demonstrated at operating meetings and whenever he is called into discussion on the subject. His attitude toward this effort will inevitably be reflected in those of the team members.

(e) Effective staff support is fundamental. This program requires some significant behind-the-scenes time and effort of the loss control staff - and accounting personnel, but the results to be gained are so tremendous that this effort deserves top priority. In reality, the loss control management team will soon become the backbone of the entire loss control program.

EXHIBIT A

LOSS CONTROL MANAGEMENT
1976 —FOR— 1977

ASSIGNMENT PERIOD	CHAIRMAN	ASSISTANT CHAIRMEN
MAY – JUNE 1976	T. P. JONES	H. A. THOMPSON L. GALLOWAY
JULY – AUGUST 1976	W. F. TAYLOR	L. B. LAMMEY M. A. GRUBB
SEPT – OCT. 1976	L. B. DONAS	P. R. ANTRIM E. E. WILKINSON
NOV – DEC 1976	C. C. ELLIS	J. A. MOTEER C. F. ALEXANDER
JAN – FEB 1977	M. W. ALBINSON	J. E. MILLER G. M. MORGAN
MARCH – APRIL 1977	R. B. STEVENS	A. N. HOWE J. C. PETERS
MAY – JUNE 1977	T. P. JONES	E. E. WILKINSON M. A. GRUBB
JULY – AUG 1977	W. F. TAYLOR	J. A. MOTEER H. A. THOMPSON
SEPT – OCT 1977	L. B. DONAS	L. GALLOWAY C. F. ALEXANDER
NOV – DEC 1977	C. C. ELLIS	L. B. LAMMEY P. R. ANTRIM
JAN – FEB 1978	M. W. ALBINSON	A. N. HOWE J. E. MILLER
MARCH – APRIL 1978	R. B. STEVENS	G. M. MORGAN J. C. PETERS

-8-

EXHIBIT B

Mr. Tom P. Jones
Vice President
Short Hills Location
XYZ Company

Dear Tom:

In order that we may further improve the loss control performance of our company, I am assigning one of our managers to the chairmanship of a new group we will refer to as our "Loss Control Management Team". This group will meet at least four times during its two-month period of charge to accomplish a number of goals:

1. Total management involvement in the loss control program.
2. The improvement of management attitudes.
3. Formal training of our managers.
4. Company promotion of a major critical problem.
5. Coordination of company inspections.

Four meetings of two-hour duration should be planned during the term of your assignment.

You should contact the loss control coordinator at least one month prior to your assigned period in order that your program achieves the desired results through proper planning.

I am also requesting that you assign each of the managers reporting to you to one of the teams. Please use your judgment in distributing their assignment over the two-year period. Please return these assignments to me within 30 days with a copy forwarded to Paul Brown, our loss control coordinator.

Further explanation will be forthcoming before your scheduled assignment. I'm sure I can count on you for your full cooperation on this vital subject.

Sincerely,

Robert E. Jones
President

REJ:smh

Enclosure: 1 schedule

-9-

EXHIBIT C

Mr. H. A. Thompson
Manager
Deep Valley Location
XYZ Company

Dear Hap:

In order that we further improve the loss control performance of our company operations, I am assigning two of our department heads to assist each chairman of a new group we will refer to as our "Loss Control Management Team". This group will meet at least four times during its two-month period of charge to accomplish a number of goals:

1. Total management involvement in the loss control program.

2. The improvement of management attitudes toward safety.

3. Formal training of our managers.

4. Company promotion of a major safety theme.

5. Coordination of company-wide inspections.

Four meetings of two-hour duration should be planned during the term of your assignment.

You will be contacted by the chairman of your team at least one month prior to the assigned period in order that your program achieve the desired results through proper planning.

I am also requesting that you share necessary responsibilities with each of your supervisors.

Further explanation will be forthcoming before your scheduled assignment. I'm sure I can count on you for your full cooperation on this vital subject.

Sincerely,

Robert E. Jones
Robert E. Jones
President

REJ:smh

Enclosure: 1 schedule

-10-

EXHIBIT D

LOSS CONTROL MANAGEMENT TEAM

AGENDA - - - - FIRST MEETING

1 P.M.	Greeting by Chairman Get-acquainted period--introductions	T. P. Jones
1:15 PM	Introduction of person who will present training phase 1	T. P. Jones
1:55 PM	5 minute break--coffee can be served	
2:00 PM	(A) Reading of company policy on loss control or special communication from Pres.	T. P. Jones
	(B) Brief explanation of the purpose and importance of the team.	T. P. Jones
	(C) Firm position on attendance-- mention of training certificate and award for perfect attendance.	T. P. Jones
2:09 PM	1. Area to be inspected	H. A. Thompson
	2. Assignments of each team member (Pass out assignment sheet)	
	3. Explanation of how to prepare for and make the inspection.	
	4. Explanation for recording results-- use of forms.	
	5. Explanation for reporting follow-up at subsequent meetings with final report at last meeting.	
	6. Offer of assistance in case of problems.	
2:24 PM	Introduction of Assistant Chairman, L. Galloway, who will explain the critical problem or special program promotion.	T. P. Jones
2:25 PM	Critical problem explanation	L. Galloway
	1. What the problem is.	
	2. Why it has been selected.	
	3. What everyone's responsibility is to effectively promote it.	
	4. Hand-out of materials.	
	5. Questions.	

-11-

EXHIBIT D

2:40 PM Tie in comments by chairman on
         inspection or special problem assign-
         ments.   T. P. Jones

2:45 PM Administration of short quiz--no names
         required--will be discussed at future
         meetings.   T. P. Jones

2:55 PM Closing comments--final challenge
         announcement of time and place of
         next meeting. (best to hand out in
         writing)

3:00 PM Meeting adjournment.

EXHIBIT E

LOSS CONTROL MANAGEMENT TEAM
AGENDA--2ND AND 3RD MEETINGS

Time	Item	Presenter
1 P.M.	Call to order by chairman--introduction of person who will present next phase of training.	T. P. Jones
1:05 PM	Loss Control training session.	As assigned
1:55 PM	5 minute break--coffee can be served	
2:00 PM	Check on attendance--have assistant make calls if necessary--refer to its importance again.	T. P. Jones
2:10 PM	Inspection reports	H. A. Thompson

    1. Determine that all inspections have been made.

    2. Get commitments for those incomplete.

    3. Determine that everyone has findings on his own group.

    4. Challenge members to do conscientious job of follow-up report specifics at next two meetings with final report at last meeting.

Time	Item	Presenter
2:20 PM	Critical problem follow-up	L. Galloway

    1. Determination that everyone has communicated.

    2. Direction as needed.

    3. Restress importance of critical problem and challenge effective follow-up (try to have any examples of actual incidents related to critical problem for discussion).

Time	Item	Presenter
2:35 PM	Discussion of first three questions on quiz (discussion of final two questions at 3rd meeting). Make pre-arrangements for two team members to be ready to give their opinions on question 3. Try to have general idea of what their answers will be.	T. P. Jones
2:55 PM	Final announcements--questions	T. P. Jones
3:00 PM	Adjournment.	

-13-

EXHIBIT F

LOSS CONTROL MANAGEMENT TEAM

AGENDA--FINAL MEETING

Time	Item	Responsible
1 PM	Call to order by chairman--introduction of person who will present final phase of training.	T. P. Jones
1:05 PM	Loss control training session	As assigned
1:55 PM	5 minute break--coffee can be served.	
2:00 PM	Check on attendance.	T. P. Jones
2:05 PM	Final inspection reports	H. A. Thompson
	1. Determine that all final reports are complete.	
	2. Have brief final report from representative of each group.	
	3. Express personal thanks.	
2:25 PM	Critical problem wind-up	L. Galloway
	1. Brief discussion of results attained and success testimony by members of group (make pre-arrangements with 2 or 3 members for brief report).	
	2. Express personal thanks.	
2:45 PM	Introduction of guest speaker (President or other top executive) 5 minute presentation on importance of proper management attitude and action.	T. P. Jones
2:47 PM	5 minute presentation.	Guest Speaker
2:52 PM	Training certificates and perfect attendance awards by chairman and assistants. Final thank you.	T. P. Jones
3:00 PM	Adjournment.	

-14-

EXHIBIT G

SUGGESTED SHORT ATTITUDE SURVEY

Please circle the most correct answer for each of the questions with choices.

Give a brief answer to other questions as required.

1. How many days has our company operated without a major loss?

    500 - 200 - 100 - 50 - 10

2. What is the present lost time injury frequency for our company?

    50 - 25 - 15 - 10 - 5 - 3 - 1

3. What was the last $3,000. property damage accident experienced within our company?

4. Who is responsibil for loss control?

5. Who is at fault when an employee violates a company rule?

6. How important do you think our management considers loss control?

IV.
Property Damage and Waste Control

INTRODUCTION

Many safety and loss control leaders have said that the next logical step from an injury-oriented safety program to an effective loss management system is the control of property damage. While the program direction of a multiplicity of major corporations and government organizations indicates the validity of this feeling, selected published statements lend appropriate emphasis to its truth. John A. Fletcher and Hugh M. Douglas in their book, *Total Environmental Control,* summarize this general point with the comment, "Yes, the safety movement is an evolution; an evolution moving from injury prevention–to total accident control–to total environmental control."[7]

"The Robens Report on Safety and Health at Work in the United Kingdom" precipitated a comprehensive law there, known as The Health and Safety at Work Etc. Act 1974. In that internationally known report, Lord Robens made this significant comment: "The essential input of these approaches (damage control and total loss control) is that the employer who wants to prevent injuries in the future, to reduce loss and damage, and to increase efficiency, must look systematically at the total pattern of accident happenings–whether or not they caused injury or damage."[13]

The following words also appeared in a 1974 comprehensive report of the National Committee of Inquiry in Australia, entitled, "Compensation and Rehabilitation in Australia": "Safety measures will not succeed if they concentrate only on the one major injury and the 29 other injuries (H. W. Heinrich reference). Such an approach chips at the apex and ignores the broad base of the pyramid. Total loss control is concerned with injury prevention, total accident control, fire prevention, industrial health and hygiene and pollution."[11]

Whether or not one accepts the tenet that Property Damage Control is the natural bridge from an injury-oriented loss control program is not the important issue. The problem of Property Damage and Waste Control is of such significant magnitude that it must inevitably be dealt with in any business organization that desires to optimize its profit potential. A representative of the Anglo-American Corporation of South Africa recently stated in a report, "The question is not whether or not it is desirable, but how it can be effectively and practically applied."

It is specifically to this latter point that this chapter is addressed. Every effort will be made to clearly define practical steps that can be taken by almost any occupational establishment to set up a program to control this major source of loss. Since the book, *Damage Control,* originally published in 1966 by the American Management Association, presents a great deal of related information, there will be little attempt to repeat the important basics contained in that publication. Sufficient reference to that work will lend continuity of thought and will share updated knowledge to augment that which is available elsewhere.[1]

THE SIZE AND SCOPE OF THE PROBLEM

While there are extensive statistics concerning the frequency and directly-related costs of accidental injury, no single known source at this time offers summary costs of property damage in the work establishment. Some insight into the magnitude of the problem can logically be deduced by evaluating the statistics presented in the "Accident Facts" publication of the National Safety Council. A recent issue reveals that 200,000,000 man-days were lost by persons with non-disabling injuries and by persons with no injuries who stopped to help the injured persons or discuss the accidents. Common sense would tell anyone who has ever worked in industry that there are many more work stoppages from property damage accidents than from those terminating in injuries.

Following this same line of thought, it staggers the imagination to estimate what the actual total cost of property damage may be to the employer if we consider that the "other costs" related to injuries reported in the "Accident Facts" publication are presented as $6,800,000,000. This includes the money value of time lost by workers other than those with disabling injuries, who are directly or indirectly involved in accidents. Also included is the time required to investigate the accidents, write up reports, etc. It is logical to assume that "other costs" of property damage would be much greater than those involved with injuries. Several examples from specific industries are presented here to emphasize this logic.

- An automotive truck manufacturing plant manager reported the direct costs of property damage in the previous 12 months to be $4,605,000. This figure was approximately 10 times the cost of the workers' entire compensation for their injury losses.

- An oil company reported general property damage costs of $5,740,700, compared to $60,500 for injuries. Motor vehicle damage costs for this company were listed separately and represented $276,800 in *additional* property damage losses.
- A mining company's records show the reporting of 892 property damage accidents for the year, costing $615,750, compared to 157 reported injuries, costing $37,562.
- A steel plant's records indicate reported property damage costs for 12 months at $928,544, with its workers' total compensation costs for injuries less than $100,000.
- A heavy machinery manufacturing corporation executive reported the total costs of accidents (including general property damage) to exceed 10 million dollars during the year analyzed. While not revealing the costs of injuries, he indicated they were *less than 20%* of the total accident costs.

These examples have been used with the knowledge that the companies involved had outstanding property damage reporting systems and general safety programs acknowledged to be comparable to leaders in their industries.

It's interesting to note that in one recent year there were 148,857,145 freight damage claims paid by class 1 railroads alone, according to an editorial in *Materials Handling Engineering Magazine*. In addition, records of the Federal Railroad Administration indicate that over 5,000 railroad derailments of sufficient magnitude to require reporting occur annually, with costs exceeding $100,000,000.[6]

An executive of a New York department store reported that from 30 to 40 percent of all furniture gets damaged *somewhere* between the time it is made and the time it rests in your living room.

Other bits and pieces of information are available from professional associations, insurance companies, and from the Transportation Department of the Federal Government, to indicate that motor fleet, marine, rail and air transportation cargo damage alone substantially exceeds the total costs of injuries in our country.

The magnitude of this industry-wide problem of property damage and waste is reinforced over and over by the reports and acknowledgements of the hundreds of supervisory personnel contacted annually in management training conferences.

A special exercise utilized in training programs with the management personnel of 12 different companies during this past year involved a 15-minute group discussion to identify items personally observed to be damaged or wasted frequently. It was surprising to find that the number of specific items identified by every group was consistently limited in its size only by the time allotted. Training records of these conferences show that the supervisory personnel of each of the 12 companies readily identified, in the 15-minute period allotted, over 125 items involving property damage or waste that were substantially reducing their own companies' profitability. (See Figure 1 for typical pictures of property damage.) It is believed that the quantity of individual statistical data fully supports the estimate that the costs of property damage range from five to 50 times the costs of injuries in our nation. (See iceberg illustration, Figure 2) and that the total annual costs of property damage to industry exceed 20 billion dollars annually within the United States and approximately 2 billion in Canada.

The total costs of waste in industry are considered even more staggering than the enormous figures on property damage. For example, the U.S. Department of Commerce reported recently that Americans utilize 32 million barrels of oil daily. American business, in its industrial, commercial, and transportation activities, accounts for 70 percent of the total energy consumption and use consumption. The President has asked all Americans to cut back 5%, but many businesses have found that they can do more. Experiences being shared during this one area of waste control indicate that waste is often the result of long-unquestioned plant practices that, upon analysis, can be shown to have no effect on employee performance or product dependability. Energy used for heating represents about 18 percent of the nation's energy consumption. The national Bureau of Standards estimates that approximately 40 percent of the energy used for heating is wasted. NBS research indicates further that energy requirements for cooling can be reduced 30 percent with little sacrifice to comfort, and that energy utilized for illumination can be effectively reduced 15 percent in most existing buildings by turning off lights when not needed.[14]

Of course, there is unbelievable waste in the utilization of many sources of energy, such as compressed air, gas, and raw water. Comments on energy control and other waste-related areas will be made throughout this chapter.

Perhaps it is sufficient to indicate at this point that there seems to be little doubt that the costs of waste in industry at least equal, and probably exceed, the costs of property damage.

WORKABLE DEFINITIONS — A STARTING POINT

Most accident definitions in the past have included the words "property damage", even though the practical application of the acci-

TYPICAL EXAMPLES OF PROPERTY DAMAGE

(Reprinted from book Damage Control)

Figure 1.

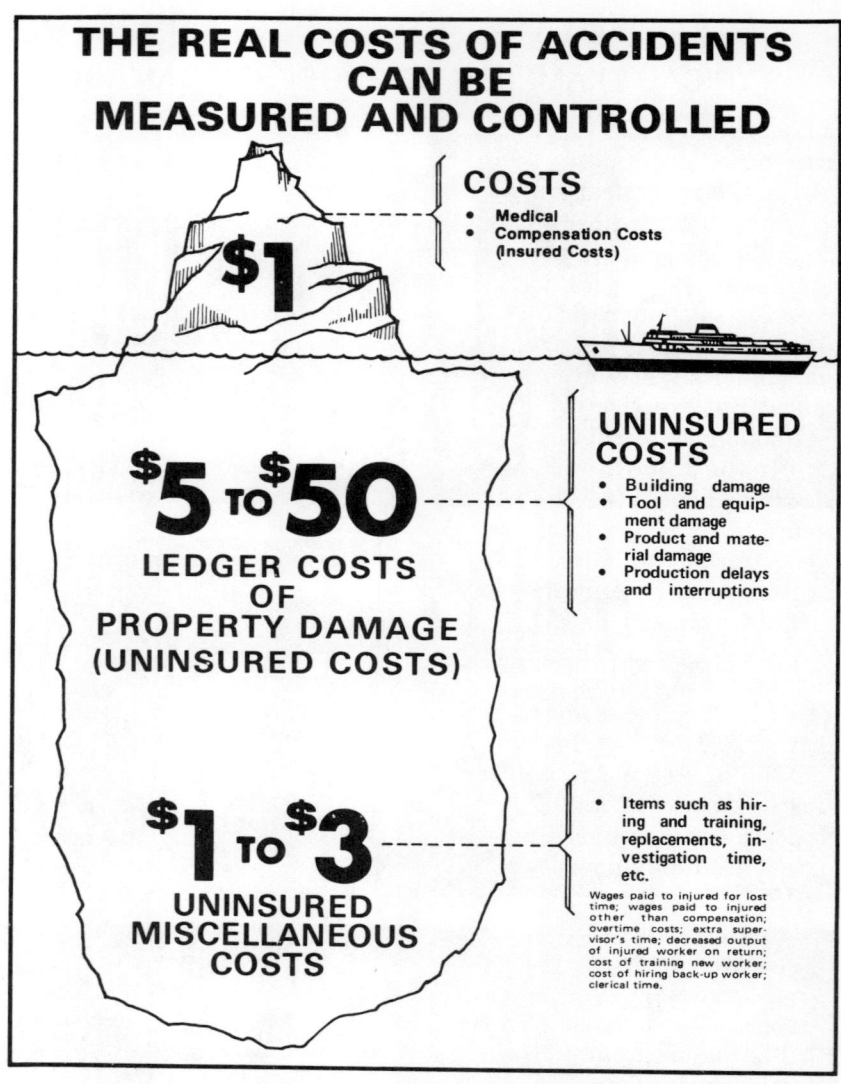

Figure 2.

dent prevention control discipline in general industry frequently excluded property damage considerations. A typical accident definition of the past decade would be represented by this extensively used one:

"An accident is an unintended or unplanned happening that may or may not result in personal injury, property damage, work process stoppage or interference, or any combination of these conditions under such circumstances that personal injury might have resulted".[1]

It's also interesting that some safety program coordinators, even today, will include the words "property damage" in management training program definitions but will refer to them improperly as near-accidents or near-miss accidents in practice. The term "no-injury accident" has also been used extensively in referring to property damage accidents other than those terminating in injury. Two important facts are revealed by past terminology and program practice. On the one hand, management will accept the logic of a connection between certain events terminating in property damage that under slightly different circumstances would have resulted in injury. But practice indicates that this definition is either not understood or not accepted.

In the words of the familiar proverb, "Actions speak louder than words". Training and experience have taught us that undesired events terminating in controllable property damage need not have the potential for injury. However, to make the big transition to a total accident control program or a loss control program that includes property damage, the majority of program coordinators must cross a bridge that connects these two. In effect, management's resistance to change can be greatly reduced at early program introduction stages by making sure that property damage references and examples in program promotion are carefully selected to reflect the connecting bridge to personal injury. Enlightened management will accept and even insist upon broadening the base of property damage inclusion in the program as their education and experience break down past barriers.

The functional and modern definition of an "accident" presented in Chapter Two is a good practical definition to use at all stages of program development:

"An accident is an undesired event that results in physical harm to a person or damage to property."

The modern thinker would also connect this definition with the thought that an accident is usually the result of a contact with a source of energy (i.e., kinetic, electrical, chemical, thermal, etc.) above the threshold of the body or *structure*. It is also understood that accidents interrupt work or activity. The value of energy exchange considerations were discussed in detail in Chapter Two and are important to the solution of many property damage problems. The reader is encouraged to refer back to this Chapter for reinforcement of his knowledge of this concept.

Several program coordinators have found it helpful to provide management with a guidepost in addition to accident definitions for property damage accounting purposes. Two such examples are provided here:

"Any property damage incident that involves *personal injury* and all other damage incidents that require *repair* or *replacement,* other than those resulting from normal wear and tear, shall be considered as accidental and must be included in the reporting program."

"Any damage incident that is considered outside the standards established or desired for fair wear and tear by the most knowledgeable persons will be considered accidental and included in the reporting system".

Perhaps the most significant lesson learned through experience is that making decisions on fair wear and tear versus property damage is not the big problem that many expect it will be. This issue is usually amplified during program introduction as one of several demonstrating management's resistance to change. If coordinators use the same management counsel and advice in property damage determination that they use in establishing the potential of occupational relationship to injury, there will be relatively few, if any problems.

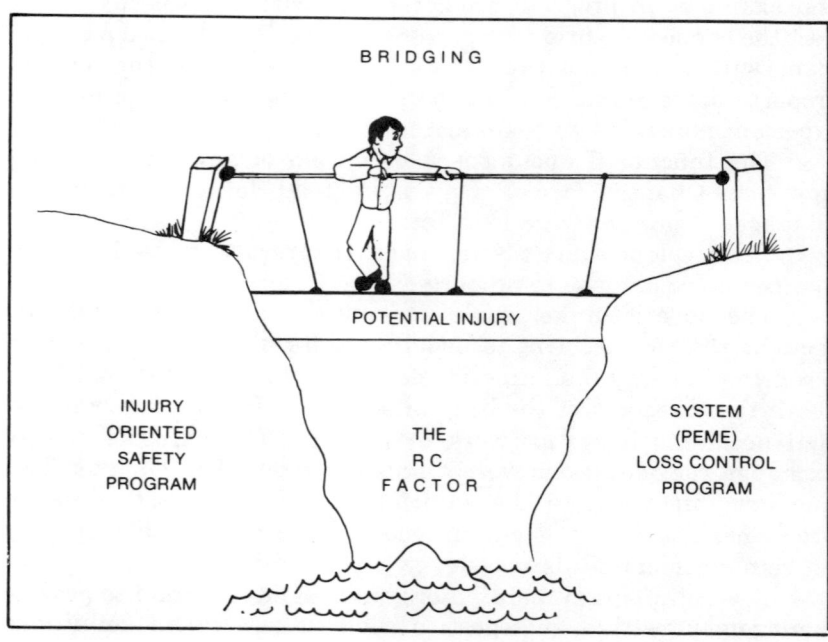

Numerous feature articles in business magazines and several best-selling books have indicated that the greatest opportunity for profit today is in the business of helping management deal with, or even benefit from, inflation. Programs and services that help management save time, and money, and protect assets, are more acutely needed today than ever before. For this reason, it may not be necessary to be as concerned about the bridge between the injury-oriented program and the total accident or loss control effort. While this will be the exception rather than the rule, it is important to recognize that the probablity is greater today than ever before. The coordinator's knowledge of management attitude is greater today than ever before. The coordinator's knowledge of management attitude is always the best guide in program development.

ROADBLOCKS TO PROGRAM IMPLEMENTATION

In discussing programs directed at the control of property damage with management people in many conferences, both here and abroad, certain specific roadblocks or common impediments to smooth effective program implementation deserve comment. Not only have these roadblocks impaired program implementation, but they have invariably been major causal contributors in those instances where programs have failed.

Let us briefly consider these common roadblocks, in order to better understand how to avoid problems and streamline efforts to successful program results in Damage Control.

1. *Injury-oriented habits are difficult to change.*

   This is the big reason for the repeated emphasis on building a mental bridge in the early stages of program implementation that connects property damage to personal injury in all aspects and phases of the program. We should keep in mind that this mental bridge will be equally helpful with representatives of any labor organization as well as with management personnel in promoting all phases of the program.

2. *Management's failure to recognize the problem.*

   More frequently than not, there is a strong initial sentiment by management personnel that there is no problem, or that the problem has been exaggerated. Experience has proven that there are many reasons that this additional mental roadblock exists. Few management people like snoopers or witch-hunters. And no one likes to be shamed or made to feel he is not performing properly. Many management personnel visualize the whole program as another big hoopla that will only end up in a mass of unnecessary paperwork. Anyone with a knowledge of modern motivational techniques would quickly grasp the value of keeping the program on a positive note at all stages. Credit and recognition for problem-solving should be utilized consistently. Programs should be designed to keep people involved and contributing. While well-documented investigative and audit efforts involved with repair centers and disposal areas are usually necessary on a continuing basis, they can also be handled in a manner that encourages cooperation rather than contempt. The value of a camera properly used can be a major factor in gaining management's acceptance of the problem and its magnitude. (This is discussed in greater depth later.)

3. *Acceptance of damage reported as representative of damage occurrence.*

There is usually a naivete on the part of upper management as well as staff and loss control personnel about the efficiency of the accident-reporting system. This is unquestionably a carry-over from injury-oriented program attitudes, where staff personnel typically accept the number of injuries reported as representative of the number of injuries occurring. Research has proven that most accidents, including those terminating in injury, are *not* reported. The skeptic need only stand with a physician during a physical examination to observe the number of small scars on the arms and legs of workers, and then attempt to correlate them with reports of the injuries that produced them. The majority of property damage accidents will *not* be reported, even in an atmosphere of significant motivation and educational effort. This condition continues long after a total accident program is underway. In the absence of a good investigative audit by repair centers and point-of-control areas where damage or waste is ultimately handled, substantially less than 50 percent of the total damage occurrence will be voluntarily reported. In the absence of good investigative and regular audit activities, this naivete about the level of accident reporting can mislead people to say, "We really don't have a problem. Only a minor amount of damage is occurring." It is important to stress that, while recognizing this roadblock, we should not infer that the accident reporting is unimportant or that it shouldn't be emphasized. On the contrary, emphasis on accident reporting produces several important results, not the least being the awareness of management's interest and concern about the problem.

4. *Weak reporting relationship of staff advisor.*

The degree of cooperation of all levels of management with a property damage and waste control program is directly related to the level of upper management to which the program coordinator reports. The weak reporting relationship of many safety and loss control specialists is one of the big reasons that management cooperation is not what it could potentially (and deservedly) be. While the professional management principle related to this truth (principle of reporting relationships) applies to all aspects of a safety and loss control program, it has even greater relevancy to property damage control implementation. As will be said several times in this chapter, operating management personnel see advisement on the control of property damage as a maintenance or engineering responsibility. While management resistance to change during property damage control program implementation can be expected to be greater than with other types of problem-solution, it can be minimized if reporting relationships of the staff advisor are at a high level. There appears to be a definite correlation between the level of management that is achieved by safety and loss control people and the dimension of the loss problem on which they give counsel; i.e., most Vice-Presidents of Safety and Loss Control are in motor fleet and railroad operations, *and have* responsibility for staff consultation on property damage, as well as personnel injury. Obviously, one's title gives inference to the level of reporting relationships. While there is no easy advice to give on how to strengthen this important factor bearing upon management cooperation, the inclusion of property damage and waste control areas offers an avenue to bring about stronger reporting relationships.

Harry Philo, a prominent trial lawyer, spoke at an annual meeting of the A.S.S.E. over eight years ago, offering this advice:

"If I were 20 years old today, and if I had the desire to become vice-president of a major corporation, I would become a safety engineer. Most major corporations today have vice-presidents in charge of marketing, purchasing, sales, engineering and recently lawyers have made it. The next group to make it will be the safety engineers.

If I were the board chairman of an American corporation today, the first thing I would do is get someone of vice-presidential status in charge of safety. The corporations that don't do this are going to have an unfavorable profit and loss picture."[12]

5. *Inadequate overall capability of staff advisors.*

Just as weak reporting relationships may hamper the spirit of cooperation received, the lack of overall capability of the staff safety or loss control coordinator could present another roadblock to achieving desired program objectives. The need for professional development within this group of important specialists is being emphasized broadly on a continual basis. The individual lacking total proper capability cannot establish credibility and confidence in a program of this magnitude. The significant potential to reduce operating costs and increase profit substantially through management control of this important area of loss concern justifies selecting an individual of unquestionable capability to coordinate the program.

THREE SIGNIFICANT STUDIES

There are three studies that deserve reference in this chapter because of the significant implications for those interested in the control of property damage. The first study was made when one of the authors was Supervisor of Safety for the Lukens Steel Company a number of years ago. The study involved an analysis of well over 90,000 recorded accidents, spanning a seven-year period of experience from 1959 to 1966. The following report on this study is quoted directly from the book, *Damage Control,* published in 1966 by the American Management Association.

"The year-end costs of property damage for the base year (1959) were estimated to be $325,545 per million hourly-rated man-hours worked. Partial reporting immediately preceding 1959 indicated further that this base-year level of damage occurrence and cost was conservative.

But this conservative figure was accepted as a definite and finite point of reference against which to measure change.

During the years which followed, the all-accident control program included analysis of over 75,000 property damage accidents, the great majority of which did not involve personal injury. During this same period of time, more than 15,000 injuries were reported and analyzed, of which 145 were classified as disabling. These facts—based on well over 90,000 incidents and spanning a seven-year period—indicate the following as conservative ratios:"

Reference in the text is made to the ratio shown in Figure 3, which reveals one disabling injury represented in the statistics for every 100 minor injuries and 500 property damage accidents. These figures were rounded off for ease of reference. The reference closes with these words, "Notice the tremendous magnitude of the base of the triangle—that is, property damage accidents. Their number is five times that of injury-type accidents for the same period; their cost is phenomenal; and, of course, they imply a high injury potential. The cost of property damage accidents in 1965 was $137,832 per million hourly-rated man-hours—a reduction of $187,713 per million hourly-rated man-hours for the year, as compared with the base year of 1959."

The second study was also made by one of the authors when Director of Engineering Services for the Insurance Company of North America. Having made a previous extensive study within one heavy industry, he was interested in similar relationships within a large cross-section of general industry. The following information is quoted directly from the book, *Management Guide to Loss Control,* published by the Institute Press, Atlanta, Georgia, in 1974. It is also contained in Chapter Two of this book and is reprinted here for ease of reference and subject continuity.

"The study described will help the reader understand why accidents that result in property damage should be given a great deal of attention. In 1969, a study of industrial accidents was undertaken by the Director of Engineering Services for the Insurance Company of North America. An analysis was made of 1,753,498 accidents reported by 297 cooperating companies. These companies represented 21 different industrial groups, employing 1,750,000 employees, who worked over 3 billion man-hours during the exposure period analyzed.

ACCIDENT RATIO STUDIES

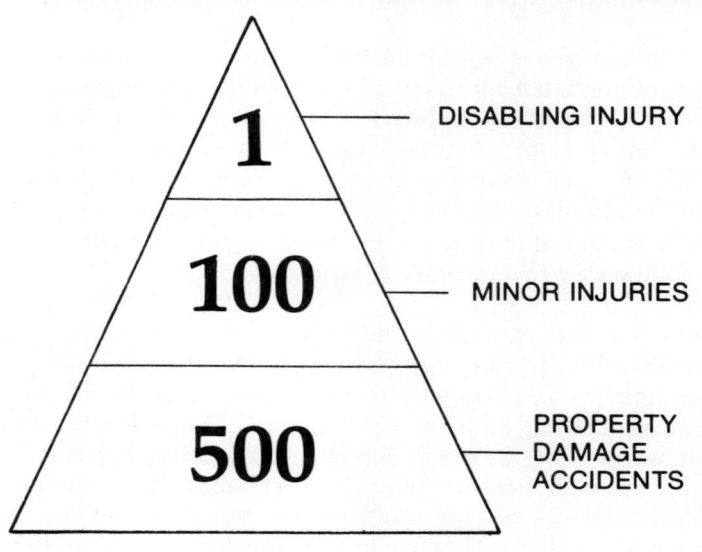

1965 Luken's Study

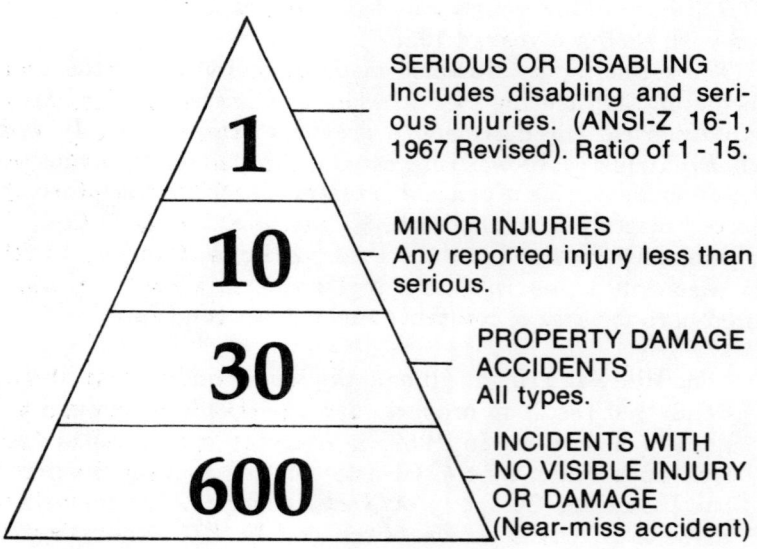

1969 General Industry Study

Figure 3.

The study revealed the following ratios of accident reporting: For every serious or disabling injury (ANSI, Z16. 1-1967) reported, there were 918 injuries of a less serious nature. 15 serious injuries were reported for each disabling injury by 95 companies utilizing the serious injury index. (ANSI, Z16. 1-1967)

Forty-seven percent indicated they were investigating all property damage accidents and eighty-four percent stated they were investigating serious and major damage accidents. The final analysis indicated that 30.2 property damage accidents were being reported for each serious or disabling injury.

Part of the study involved 4,000 hours of worker interviewing by trained supervisors on the occurrence of incidents that under slightly different circumstances could have resulted in injury or property damage.

In referring to the 1-10-30-600 ratio, it should be remembered that this represents accidents and incidents reported, and not the total number of accidents or incidents that actually occurred."

The 1-10-30-600 relationships in the ratio would seem to indicate quite clearly how foolish it is to direct our total effort at the relatively few events terminating in serious or diabling injury when there are 630 property damage or no-loss incidents occurring that provide a much larger basis for more effective control of total accident losses.

Since one of the authors was involved with both of these studies, there are several important points to make about a comparison of findings and summary conclusions. First of all, they should not be compared, since the base of comparison was different in each study. In the Lukens study, the disabling injury was used as a point of comparison, while the base of comparison in the second study was serious or disabling injuries. A second reaction is that one should not assume that the ratio of accident relationships is the same for his company or organization.

It isn't really important what the ratio is for a specific operation; the important point that can generally be assumed is that many more property damage and near-miss accidents occur than those terminating in loss. The value of this truth is twofold. There are many opportunities occurring that do not result in injury but that possess an enormous potential for predictive injury or damage preventive efforts, as the case may be, and the enormous cost reduction potential is there to be tapped.

The third study of property damage accidents deemed important to any in-depth analysis of the problem also relates to the occurrence of accident and incident results. This study was also made by one of the authors during his career as supervisor of safety at the Lukens Steel Company in Coatesville, Pennsylvania. It included three thirty-day programs of observations by members of the safety department. The industrial Engineering Department worked jointly with safety personnel because of their interest in materials handling efficiency. Participants trained in the whole program, including use of data collection forms and schedules, were established in teams for 24-continuous-hour periods. Observations were made of the same plant locations (involved with crane handling) on three different occasions of 30 days each, over a three year period. Data collected were based on the purposeful observations by trained management people of 8,669 different crane lifts. The following were key findings:

- There were 184 accidents with the potential for personal injury and/or property damage (near-miss accidents).
- Nine accidents resulted in property damage.
- A small number of operators were involved in the majority of damage accidents.
- Groundmen followed established safe job procedures less than 33 percent of the time.
- Certain handling devices were obviously safer to use than others.
- A damage study of the 30-day observation periods was compared with the damage incident rate for these areas during the 30 days prior to each of the studies. There was an average of 25 percent fewer damage incidents during the study periods.

Of the many beneficial suggestions for improved accident and cost control which came out of these studies, several observations seem especially noteworthy.

1. The "critical few" operators involved emphasize the need to apply this management concept broadly in our accident prevention efforts. In property damage control, look for the critical few operators, as well as items, and zero your controls on them.

2. Having safe job procedures in written form does not in itself bring about safe behavior. There is a need for *periodic*

planned observations to see that procedures are being followed and for *regular updating* of established procedures.

3. While the number of accidents involved in the study may not have been large enough to be statistically rated, it is interesting that there were 20 near-miss accidents for each property damage incident. The comprehensive study at the Insurance Company of North America revealed the identical relationship.

4. Of very special interest was the 25 percent reduction of property damage incidents during the period of time when observers were *openly paying attention* to this problem.

There is sufficient evidence in various similar studies to indicate that safety awareness is increased substantially, with resultant accident reduction, when people know you are interested and paying attention to a problem that involves them. Logically, a well-organized property damage and waste control program should produce a substantial loss reduction, by virtue of this classic employee "positive behavior" that results when attention is paid to a problem.

A DUAL APPROACH TO PROPERTY DAMAGE AND WASTE CONTROL

The approach that has the greatest appeal to the typical executive is usually the one that brings the quickest effective results with the least expenditure of money. It is invariably appealing to management people that prompt reduction of property damage and waste can result in a high level of confidence by the utilization of the loss improvement project system described later. It is not at all beyond reasonable expectations that this type of system can bring about appreciable loss reduction within 6 months or less.

The grave warning is that such approaches do not usually bring about the type of management control that lends itself to long-range program continuity. It can be said without reservation that there is no instant solution to such important basic causes of property damage losses as inadequate knowledge or skill, improper motivation, inadequate design and work standards. In effect, to achieve the immediate and long-range results so desirable to management, an approach that maintains continuous long-range control must be utilized simultaneously. The long-range program that will integrate control of property damage and waste into every facet of the safety or loss control program will take several years, without a doubt. This does not mean that quality planning and organizing cannot streamline program implementation. It will, and that's the purpose of this chapter. Both approaches have their place, and both are needed for effective immediate and long-range results. Let's look at each.

A. CONTROL THROUGH LOSS IMPROVEMENT PROJECT SYSTEM

The Loss Improvement Project System is an effective way to focus problem-solving attention on the "critical few" items of damage costing an organization the majority of its losses. Results through this technique can produce the prompt substantial results that management looks for quickly when they realize that a major problem exists.

This LIP (Loss Improvement Project) System, as we refer to it, is based on the widely-accepted statistical premise that, in almost every kind of economic-statistical situation, approximately 25 percent of the items account for 75 percent of the total value of the whole; i.e., 25 percent of the items manufactured by the ABC Company account for 75 percent of the company's total profit. This concept is based on the studies of an Italian economist, who set the principle down as an economic law in 1906. It is also widely known as "Pareto's Law".

It is sufficient to say that management authorities through the years have constantly substantiated "Pareto's" early findings. Perhaps no one emphasized this more than the well-known independent consultant and Fellow of the American Management Association, Joseph M. Juran, who in 1954 described this principle in his lectures and writings as "universal for management planning and control". Juran described the concept in terms that continue to be used today, "vital few" and "trivial many".[4] Other terms, such as "main issue" or "critical few", are also found in various management applications. We will refer to the "critical few" items of property damage as the "critical items".

There are two major ways to identify critical items damaged:
1. Identification by management directive.
2. Identification by systematic professional audit.

Either method can be utilized alone or, better, they can be combined.

The first technique requires the chief operating executive to write a letter (through appropriate management channels) requesting that department head level managers utilize the knowledge of their supervisory group to identify their own critical items of damage and waste. This request would naturally follow suitable "mind preparation" so that everyone would be thoroughly familiar with the purpose and objective of the exercise. The staff safety/loss control program coordinator would naturally be a critical figure in this pre-program kick-off preparation, along with the top official. This kind of approach has great appeal to top management officials because it enables them to permit their managerial staff to identify their own problems in their own way. The management resistance-to-change factor is immediately reduced, compared to what it might have been if outsiders identified the items for them. This total management involvement also emphasizes the need for everyone's cooperation and, of course, everyone is made aware that top management is interested in this project.

The Loss Improvement Project Team (LIP)

Once the lists have been channeled to the chief operating executive, a subsequent letter from him will outline the program of control to be described, emphasizing the important interaction and role of the safety or loss control organization as the consulting agency. This letter would essentially request the organization of Loss Improvement Project Teams and suggest a frequency of follow-up reports, emphasizing that a periodic summary of results would be channeled to everyone.

By virtue of this latter directive, the program coordinator (or a designate) would guide management in the selection of a chairman for a Loss Improvement Project Team for each critical item identified. It should be noted that some of the items submitted may not be "critical items" statistically, and guidance would be offered on this matter.

A department head is usually the most practical choice for team chairman, since he is close enough to the "action" to contribute innovative and creative thinking about the problem, and yet is at a high enough level in his reporting relationships to make certain important decisions. The team should be made up of people with the greatest knowledge of the problem or its potential solution. Purchasing personnel and supplier representatives, repair center supervisors and engineers could be considered as valuable sources of information and potential team members. The team should be large enough to provide information upon which to act, but small enough to move ahead efficiently. It should be kept in mind that those with valuable information need not be regular team members. Experts can be invited to one or more meetings for their advice or sharing of information. Perhaps one of the most important factors involves the selection of the chairman. We have indicated the value of a department head, but an equally important decision involves the "which one?" question. Basic concepts in professional management offer this sound logical advice: the person who makes the best chairman is the manager on whose budget the "critical item" appears. This individual has the accountability to control these costs. Inadequate recognition of the problem and its magnitude will usually be the fundamental reason for past "lack of control" by these managers. It seems appropriate to re-emphasize that these losses were "ledger" costs on this manager's budget—and were referred to in the past as "indirect" costs. They are, in effect, *real*—direct—"ledger" costs that merely needed identification.

Prior to the actual operation of the LIP teams, it may be necessary to give the team chairmen or leaders a short training experience in team leadership and problem-solving. It is also important to indicate that the same individual could be chairman of several LIP teams. In the interest of time conservation, several or all meetings could be held by the leader consecutively on the same day.

While the success of a LIP team does not rise or fall on the use of forms, experience has proven that a minimum number of good forms will prove of great value in the management of the program. Their utilization serves to strengthen and maintain effective communication for everyone and provides an easy-access document for efficient statistical analysis of results.

A brief description of the forms (Figures 4, 5, 6 and 7) and their intended utilization will also give additional insight into the operation of the LIP team.

The forms shown in this chapter are intended to be illustrative of the type that could be helpful in maintaining a good management system. Modification to the specific needs of a given industry or organization would be necessary.

General Information Form (Figure 4)

This form should be completed at the earliest possible time. The chief operating official may request that this completed form on each "critical item" be returned (with other information requested) in response to his second letter directing formation of LIP teams. It will serve as a starting point for team operation. Its completion requires causal analysis and the objective setting that sets the stage for a management-by-objectives approach to each loss improvement project. This form is a good source reference for any new LIP chairman or advisor, when promotions or changes in assignment are made. An updated form should be completed every six months to a year on those projects that are ongoing, in order to keep this source document current.

LOSS IMPROVEMENT PROJECT REPORT
GENERAL INFORMATION FORM

1. Project No.	2. Subject:	3. Report No.	4. Date:
12	GRINDER DAMAGE	1	Feb. 15, 1970

5. Project Leader:	6. Project Adviser:	7. Frequency of Progress Reports:
D. J. Brown	RAY SLIDER	QUARTERLY

8. Purpose of Project:
To determine the cause of spindle fracture and stator burn-outs since the cost constitutes over 78% of grinding repair costs and is believed to be substantially outside the perimeters of fair wear and tear.

9. Most Frequently Damaged Item or Injured Body Part:	10. Main Accident Type Involved:
Spindles and Stators	Unknown

11. Departments or Areas Involved: All grinding areas, Purchasing, Maintenance shop, Maintenance cost control	12. Supervisory, Staff, or Technical Personnel Involved: D. J. Brown Paul Martin B. I. Lowery Park Dague
13. Equipment and Related Products Involved: B & D Portable Hand Grinder Model 20	**14. Occupations, Jobs, and Crafts Involved:** Grinder
15. Maintenance Centers Involved: General Maintenance Shop	**16. Sources of Additional Information:** Black & Decker representative R. Bunter

17. What Conditions or Practices Apparently Resulted in the Losses:
Not accurately known at present. Based on early studies it is believed that striking the locking nut by the grinder in order to losen it is causing crystalization that results in fracture under load. It is also felt that improper grinder practice is resulting in burn-outs since formal job observations apparently bring about substantial reductions during the periods of observations.

18. Project Objectives with Target Date for Progress and Control:
To reduce grinding maintenance cost 15% by July 25, 1970 and 35% by December 15, 1970 through engineering and/or grinder behavior controls.

19. Methods for Achieving Objectives:
1. Design new wrench to provide proper torque on nut - May 15, target date
2. Establish positive controls for operator grinding behavior - June 1, target date.

20. Additional Comments or Recommendations:
The above methods for achieving objectives are based on early studies of this problem but are believed to possess excellent potential for control.

SEE REVERSE SIDE FOR COST ANALYSIS

Figure 4.

Cost Analysis Form (Figure 5)

It is absolutely necessary to establish a valid measurement of results of the work of the LIP team for each critical item of damage. Upper management might well ask when told that the damage was decreasing, "How many pieces of equipment are operating, and for how many hours?" Just as with an injury frequency rate, the costs

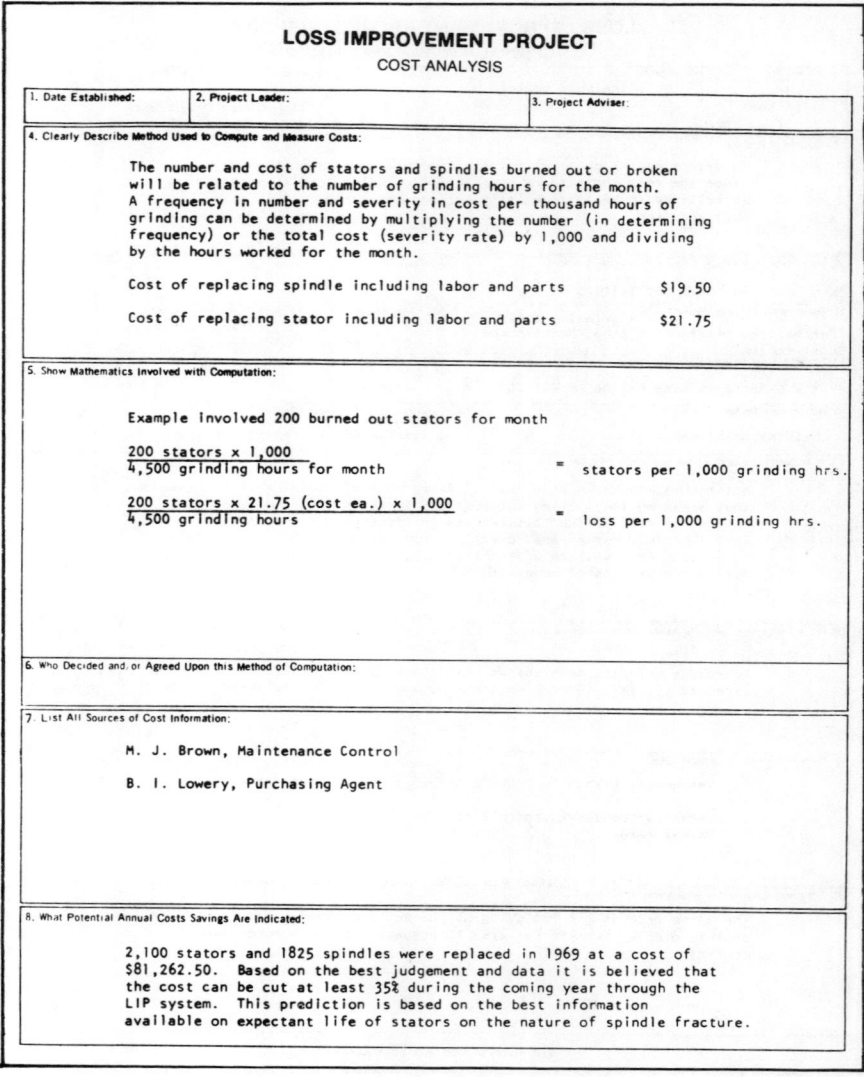

Figure 5.

involved with any critical item must be related to an accepted base of exposure hours or units accepted as best for the purpose by members of the LIP team. The application of the indices of measurement defined on individual cost analysis forms will be used to compute results on each item and on the overall program. The input for this form should also be obtained as soon as possible after the team is formed.

Team Progress Form (Figure 6)

Routine progress reports from teams or committees can take many forms. They are usually in an informal newsletter style. This simple form is suggested to serve as a constant reminder of certain critical information that should be included in follow-up reports. There is also psychological value in the thought that it is easier to submit a form than to write an independent report.

Summary Form (Figure 7)

This form summarizes the results of all LIP team activity. It should be designed to be as simple as possible. Words can be abbreviated. Every loss improvement project should be given a different number, which is retained throughout its life. Additional information on any project is always easy to obtain through this numbering system. A form in this general style is usually forwarded to members of upper management, at least quarterly.

A few additional general comments about the operation of the loss improvement project system should be of value. The safety/loss control program coordinator (or his designated representative would be advisor to each LIP team. In addition to other responsibilities, the

---

LOSS IMPROVEMENT PROJECT REPORT
TEAM PROGRESS FORM

Project No 12   Subject Grinder Damage   Report No. 3
Date July 1, 1970   Project Leader D. J. Brown   Project Adviser R. Slider

Report should include 1. What has been accomplished since last report? 2. How was it accomplished and who did it? 3. What new recommendations and/or improvements have been made? 4. What are the results?

Jack Bumer has received six band wrenches now in use in building three. These wrenches have been in use for about 2 months. The grinders were properly instructed in their use and all open end wrenches were collected. According to the daily log in the repair shop, replacement of spindles has dropped off at least 35%. Since actual man-hours are not known at this time this figure is a conservative estimate. Based on this early response, D. J. Brown has ordered another 12 band wrenches to equip all grinders in the shop.

Job Observations were made of 5 grinders in June by O. J. Tober each day for 10 work days. They were generally found to be following their T.J.A. Of particular interest, however, was the fact that burned out stators during the month of June dropped down $500. per thousand grinding hours. There seems to be a good possibility that standard grinder practice is not being followed in the absence of a grinder foreman. D. J. Brown has requested formal job observations on 5 grinders in building #1 and frequent informal observations on all grinders during the next two weeks to diagnose this problem.

Figure 6.

	LOSS IMPROVEMENT PROJECT REPORT					
	SUMMARY FORM					
Period:	Project Leader			Loss Control Coordinator:		Year
No.	Project	Project Start Date	Primary Action Taken		Fiscal Year Results	Total Project Results
GYTD	Total # Projects		FYTD Results		Period Results	

Figure 7.

advisor must insure that all forms are completed and channeled as required. He should also maintain a master file system on *all* reports. He would prepare the summary form for distribution as required. While he may not be the person writing or completing the progress reports, he should serve as "conscience" to the team chairman and be a motivational catalyst to keep the team's effort progressive. All of this activity should, of course, be promulgated in the role of an advisor to the team and to the team leader.

The second method for establishing "critical property damage and waste items" is through a systematic audit by safety or loss control professionals. As stated earlier, the combination of both techniques is necessary in order to best identify the "critical few", and thus optimize the accuracy of program results. While a majority of top executives will feel more comfortable getting the information by directive, they should be made aware that many "critical items" will not be voluntarily identified. This does not necessarily indicate that management people will be dishonest. It does mean that the whole picture is not always revealed. One reason may be intentional cover-up, but the biggest reason is a lack of awareness that some of the problems *are* property damage and waste. Regardless of the reason, it is best to conduct the systematic audit to uncover as many items of damage and waste as possible (in order to more precisely determine the "critical few") without the bias of the people with vested interests in the problem. For most accurate results, the systematic audit should be very well planned, including thorough explanations of the program to all levels of management. Special efforts should be made to assure the free exchange of all related information. Frequently, there is subtle resistance by the lower echelons of management to revealing information that could reflect negatively on the boss. This loyalty and the resultant "resistance to change" should be recognized and advance planning should include efforts to minimize it.

A final important stage of preparation would be the systematic listing (the form shown in Figure 8 could be helpful) of every point-of-control front-line supervisor in direct charge of any repair shop, area or location where anything damaged would ultimately come for repair, replacement, salvage or disposal. This listing must be very carefully done to make sure that no areas or sources of damage information are omitted. Usually, an individual will think he has identified all areas—only to name several more upon questioning. People need to be reminded that many things are repaired in the field, as well as in the shop, and that some damaged items are sent outside for repair.

In addition to the names of the front-line supervisors for each of these locations, the names of the department heads for each of these

LOSS IMPROVEMENT PROJECT REPORT		
CRITICAL ITEM IDENTIFICATION FORM		
SOURCES OF INFORMATION		
1. LOCATION	2. FRONT-LINE SUPERVISOR OR INFORMATION SOURCE	3. SUPERVISOR OR PERSON LISTED IN COLUMN 2.

Figure 8.

point-of-control people should also be placed on the list. In effect, we now have the locations to audit and the names of the persons who probably have the greatest knowledge of everything moving in and through the area.

The next step is to arrange a meeting with each of the front-line supervisors, in the proximity of the shop or area involved. If it is at all possible, it is of proven significant value to have the department head present during the audit. It is very important that a time best suited to everyone be selected for the meeting. Every effort should be made to have the group recognize the importance of taking the time necessary to do this audit properly.

The next vital exercise involves an interview that could be called "accident recall". The program coordinator (or his designated representative will ask the point-of-control supervisor to recall any items of damage or waste that have come into his area of responsibility on a more-than-infrequent basis. It is usually good to tour the area with the supervisor and to ask as many questions about the nature of activities in the area as possible. (The form shown in Figure 9 might be helpful to accummulate information, or a lined pad could also be used.) Each item identified should be listed, with information on unit cost and the estimated number damaged or wasted per month. It is

LOSS IMPROVEMENT PROJECT REPORT
CRITICAL ITEM INVENTORY

LOCATION	COMPANY	INFORMATION SOURCE/PERSON	SUPERVISOR OF NO. 3
ITEMS DAMAGED - WASTED	UNIT COSTS	ESTIMATED NO. DAMAGED/WASTED — MONTHLY / ANNUALLY	ESTIMATED ANNUAL COSTS	OTHER SOURCES COST INFORMATION

Figure 9.

not unusual for a front-line supervisor to be quite precise about numbers and costs. After all, he probably orders replacement parts, and his observations on numbers of items coming through his area should be more accurate than anyone else's. This same exercise must be carried out in each shop or area; if it is properly conducted with patience and a positive appreciative attitude (using the form in Figure 9), the interviewer can obtain data of great value, from which a majority of the "critical few" items can be selected.

The presence of the department head (or his assistant) serves several valuable purposes. He offers an additional source of related information at a responsible management level, and exerts a significant motivational influence to assure the cooperation of the front-line supervisor. The possibility that the supervisor may "clam up" on vital information which he might otherwise have shared does not outweigh the overall value of this manager's presence. The information the department head gains in this exercise may also be of great value in his work as a future LIP team leader.

Once "critical items" have been identified through this technique, their total cost can be summarized as the estimated cost of items believed to be the "critical few items damaged or wasted". When this exercise is done with the care it deserves, the resulting costs are substantial enough to serve as a powerful motivator to upper management to move the program ahead. Critical items so identified can now be assigned to loss improvement project team leaders for analysis and action. The same system and forms described earlier for use with LIP teams can be utilized. The establishment of actual costs for each "critical item" will be accomplished as part of the team activity.

It must also be remembered that while the two methods described should uncover at least 60 to 70 percent of the critical items, there remains a need to use other investigative techniques (described later) to identify the remainder.

Critical items will frequently be those with small individual costs but large total group costs. (Figure 10 shows this important fact, also discerned in the seven-year study of 90,000 property damage accidents.) The front-line supervisor needs a great deal of help in problem-solving and remedial action areas on many of these items. It isn't difficult to see the enormous time-saving value of the LIP team in this regard.

B. *Maintaining Long-range Damage Cost Control*

As we said earlier, to accomplish the long-range damage cost control goal, property damage control must become an integrated

SEVEN YEAR RELATIONSHIP SURVEY OF 90,000 PROPERTY DAMAGE ACCIDENTS	
COST	NUMBER
$1,000 OR MORE	1
$ 300 UP TO $1,000	50
$ 51 UP TO $300	150
$ 50 OR LESS	300

Figure 10.

part of every facet of the safety or loss control program. As integral parts of the process of total program development, these critical program elements may be helpful to keep in mind during the development and achievement of longer-range program objectives:

1. *Management Training*

    The integration of property damage control into every facet of regular management training in the areas of accident prevention or loss control must be so complete that management people do not think of accidents as injuries, but rather as the undeserved events they really are, that result in personal injury *or* property damage. This concept must be so completely integrated that in any thought given to any activity in the program, property damage is automatically considered to be as real a concern as personal injury.

2. *Supervisory Investigation*

    The small changes necessary to alter an injury-oriented accident investigation form to one that enables the supervisor to investigate a property damage accident are essential (See Figure 11). Upper management involvement should be designed into the property damage control program, just as it has histori-

SUPERVISOR'S INVESTIGATION REPORT

1. COMPANY OR BRANCH	2. DEPARTMENT		
Western Box Company	Sheet and Paint		
3. EXACT LOCATION	**4. DATE OF OCCURRENCE**	**5. TIME ☐ AM ☒ PM**	**6. DATE REPORTED**
Shipping Dept. Southside at Column 20 B	4-8-74	3:20 PM	4-8-74

INJURY OR ILLNESS		PROPERTY DAMAGE		OTHER INCIDENT	
7. INJURED'S NAME: Robert K. Berry		13. PROPERTY DAMAGED: 15 ft. ladder		19. PERSON REPORTING INCIDENT	
8. OCCUPATION: painter	9. PART OF BODY AFFECTED?: Lt. upper leg	14. ESTIMATED COSTS: $ 22.00	15. ACTUAL COSTS: $ 28.50	20. OCCUPATION	21. COST (IF APPLICABLE) $
10. NATURE OF INJURY/ILLNESS: Compound fracture		16. NATURE OF DAMAGE: broken rung		22. NATURE OF INCIDENT	
11. OBJECT/EQUIPMENT/SUBSTANCE/ INFLICTING INJURY/ILLNESS: ladder on ground		17. OBJECT/EQUIPMENT/SUBSTANCE/ INFLICTING DAMAGE: injured worker in fall		23. OBJECT/EQUIPMENT/SUBSTANCE/ RELATED	
12. PERSON WITH MOST CONTROL OF ITEM 11.: Robert K. Berry		18. PERSON WITH MOST CONTROL OF ITEM 17.: Robert K. Berry		24. PERSON WITH MOST CONTROL OF ITEM 23.	

DESCRIPTION

25. DESCRIBE CLEARLY HOW THE INCIDENT OCCURRED.

Bob was reaching out with a paint brush while standing on the third rung from the top of a 20' extension ladder section to paint an additional area on the steam pipe located on the south wall of the #2 shipping dept. He lost his balance as the bottom of the ladder started slipping and fell approximately 10' to the floor, landing on top of the bottom section of the ladder.

ANALYSIS

26. WHAT ACTS, FAILURES TO ACT AND/OR CONDITIONS CONTRIBUTED MOST DIRECTLY TO THIS INCIDENT?

It seems apparent that Bob was attempting to reach a bit further than the normal extension of his arm. John Hertz a witness, indicates the legs of the top section of the ladder were positioned about 3' from the vertical to the pipe. Not only was Bob using an improper ladder without safety feet but it was positioned unsafely. In addition the ladder was not tied off to the pipe and Bob was not wearing or using his safety belt.

27. WHAT ARE THE BASIC OR FUNDAMENTAL REASONS FOR THE EXISTENCE OF THESE ACTS AND/OR CONDITIONS?

A check indicates that Bob's last ladder safety rule review was over 2 yrs. ago. There should have been a 6' nylon rope tied on the top of his ladder. While he was equipped with a standard heavy duty safety belt that hung in his locker, he had not been issued one of the new lighter safety belts that are to be worn at all times by all maintenance personnel. The upper section of the ladder was not stamped with the required caption "Do not use this section alone." The area around Bob's ladder was quite disorderly and may have contributed to the seriousness of the injury.

EVALUATION: 28. LOSS SEVERITY POTENTIAL ☒ Major ☐ Serious ☐ Minor — 29. PROBABLE RECURRENCE RATE ☐ Frequent ☒ Occasional ☐ Rare

PREVENTION

30. WHAT ACTION HAS OR WILL BE TAKEN TO PREVENT RECURRENCE? NUMBER ALL ITEMS IN SEQUENCE

1. A monthly tick system for rule review by me has been established with the shipping clerks for all workers.
2. All ladders will be equipped with required 6' nylon rope and checked to see that the tops of extensions have proper notices stamped on them.
3. Weekly ladder inspections will be instituted immediately.
4. Purchasing indicated shipment on all new safety belts to be completed in 10 days. Extra belts to be maintained in store room.
5. Three other painters will be issued new belts as soon as they arrive.
6. I will design a simple ladder safety check sheet to use in personal daily contacts. It will include housekeeping.

31. CIRCLE NUMBER AND GIVE DATE OF INTERMEDIATE ACTION; "x" OUT NUMBER (AND GIVE DATE) WHEN COMPLETED.

	1.	2.	3.	4.	5.	6.	7.	8.
INTERMEDIATE:				(4) 4-8-74				
COMPLETED:	X 4-8-74							

32. INVESTIGATED BY	33. DATE	34. REVIEWED BY	35. DATE
John Smallwood	4-8-74	Sam Booker	4-10-74

Note: These sample forms may appear somewhat different from those used in your company's safety program. They are included as a reference and are only intended to be representative of those generally used.

Figure 11.

cally been with disabling or serious injuries. In order to justify increased involvement by upper management, the establishment of a major property damage severity category by dollar value becomes important. Many companies use the value of $1,000 or more as a break-off point to classify major property damage accidents. One food processing company uses $500 or more as a major property damage category, and one aviation manufacturer uses $50,000. A key to selection is that point at which management places enough import to justify additional investigative attention by key personnel. Once this classification has been designated, major property damage announcements can be utilized to promote control (see Figure 12), and upper management hearings or follow-up reviews (see Figure 13) can demonstrate support and provide more executive leadership and involvement. High priority must be given to the establishment, early in the overall program development, of a major property damage frequency rate and a major property damage severity rate. (These indices for measuring the frequency and severity of major property damage accidents are discussed in detail in Chapter Six, "Measurement Tools for Management".)

Use of these tools will be a major step in building and reinforcing strong upper management support, as their increasing involvement and participation reveals to them continual evidence of control need. This is exactly what the application of these tools will accomplish. In order to utilize these indices, it is essential for the program coordinator to follow up closely all details of investigation and costs related to this select group of accidents. (See Figure 14 for sample Cost Analysis Form.)

It seems appropriate to indicate that the Loss Improvement Project Team effort is really an additional investigative tool. The experienced program administrator will realize that, at some point, it becomes an impractical exercise to require investigation of *all* property damage discovered (i.e., dents in fenders picked up in a garage or repair shop). It is far more practical to handle some damage items with small individual but great accumulative costs as critical items through use of the LIP team concept. This is not meant to de-emphasize the need to investigate all accidents. We recognize that most accidents will not be reported when they happen, and there must be practical approaches to handle solutions of special problems.

Several times in this chapter, the comment has been made that most accidents are not reported. It may help us to understand how to improve property damage reporting by examining

MANAGEMENT REVIEW
MAJOR LOSS

1. Date of Incident	2. Date of Review	3. Location	4. Department or Division

5. NATURE AND EXTENT OF LOSS TO PERSONS OR PROPERTY (INCLUDE COSTS WHEN PROPERTY IS INVOLVED)

6. PERSONAL INFORMATION (WHEN APPLICABLE)

7. BRIEF DESCRIPTION OF INCIDENT

8. SPECIAL INFORMATION FOR ORGANIZATION-WIDE ATTENTION

9. REVIEW MEMBERS

Signature of Review Chairman   Date

LOSS CONTROL PROFITS EVERYONE — THE FAMILY, THE EMPLOYEE, THE ORGANIZATION

Figure 12.

LOSS CONTROL

MAJOR LOSS ANNOUNCEMENT

☐ 1. PERSONAL INJURY ☐ 2. PROPERTY DAMAGE ☐ 3. OTHER INCIDENT

4. DATE AND TIME OF INCIDENT	5. PLANT/DIVISION	6. LOCATION

7. APPARENT NATURE AND EXTENT OF INJURY OR OTHER LOSS TO PERSONS OR PROPERTY. (INCLUDE COSTS WHEN PROPERTY IS INVOLVED.)

8. PERSONAL INFORMATION

9. BRIEF DESCRIPTION OF INCIDENT

10. APPARENT CAUSES

11. IMMEDIATE SUPERVISOR   12. DEPARTMENT HEAD

LOSS CONTROL PROFITS EVERYONE – THE EMPLOYEE, THE COMPANY, THE CUSTOMER.

Figure 13.

MAJOR PROPERTY DAMAGE COST ANALYSIS

_____ _____
Department Accident Date

(1) DESCRIPTION OF DAMAGE AND OTHER POTENTIAL RELATED LOSSES: _____
_____
_____

(2) IMMEDIATE REMEDIAL COSTS INVOLVED: (REPLACEMENT AND/OR REPAIR AND LABOR INCLUDING NECESSARY OVERTIME) _____
_____
_____

TOTAL IMMEDIATE REMEDIAL COSTS _____

- -

LIST OTHER COSTS BELOW AS AVAILABILITY PERMITS:

(3) UNPRODUCTIVE LABOR COSTS: _____
_____
_____

(4) INTERRUPTED MACHINE AND/OR EQUIPMENT TIME LOSS: _____
_____
_____

(5) CONTINUED REPAIR AND/OR REPLACEMENT COSTS: _____
_____
_____

(6) MISCELLANEOUS COSTS (REHANDLING, REHEATING, RESCHEDULING, DELAYS, DEMURRAGE, PENALTY, ETC.): _____
_____
_____

INVESTIGATOR _____ GRAND TOTAL ALL COSTS _____
(USE REVERSE SIDE IF NECESSARY)

Figure 14.

why they are not reported. Here are the most common reasons employees have given for not reporting damage:
- Fear of discipline.
- Nobody cares anyway.
- It's not my property.
- The company can afford it.
- They waste thousands anyway.
- I didn't know.
- Everybody does it.
- They can't prove anything.
- It's too little to fuss over.
- What's the use?
- Why rock the boat?

3. *Rules and Practices*

Another basic program essential is the need to require everyone to report all property damage accidents when they occur, as well as those conditions or practices that could cause either property damage or personal injury. Again, simple changes are all that are needed in existing rules related to injury-type accidents. The two rules below have been adopted by hundreds of companies:

- "Report immediately to your foreman or superior any condition or practice you think might cause injury to employees or damage to equipment."
- "Whenever you or the equipment you operate is involved in personal injury or damage to property, regardless of how minor, you must immediately report it to your foreman or supervisor; get first aid promptly in the company dispensary."

As all rules and safe practices are promulgated, whether they be job, craft, or departmental rules, they should also reflect thought given to guideposts on behavior which prevents property damage.

As a last resort, properly administered discipline is an essential part of any good safety or loss control program, and its presence in a system is considered to be a critical ingredient. One of the significant related factors in effective property damage reporting and the remedial efforts that can follow, is the important need to use discipline more in established cases where accidents were not reported than in cases where accident information is volunteered. This application of discipline as a motivator can be of such value that we frequently suggest a policy of automatic discipline for individuals who were involved in property damage and failed to report it. A dramatic comparison of accident reporting before and after a program of strict equitable enforcement of the reporting rule is explained in greater detail in the book, DAMAGE CONTROL. (Figure 15 reveals the results of this extensive 17-year analysis of accidents reported by one group of materials-handling operators.)

The values that can come through causal analysis and remedial action (including a dramatic increase in the number of accidents reported) are quite evident.

4. *Engineering Controls*

No opportunity to minimize property damage is more important than the integration of the safety or loss control specialist into the conception, pre-design and design stages of new construction and the redesign of operating facilities. It is here that prime opportunity exists to control the exchange of energy involved in the damage accident. Anyone who has ever been involved in the industrial management process knows how much more difficult it is to obtain an additional expenditure to eliminate, segregate

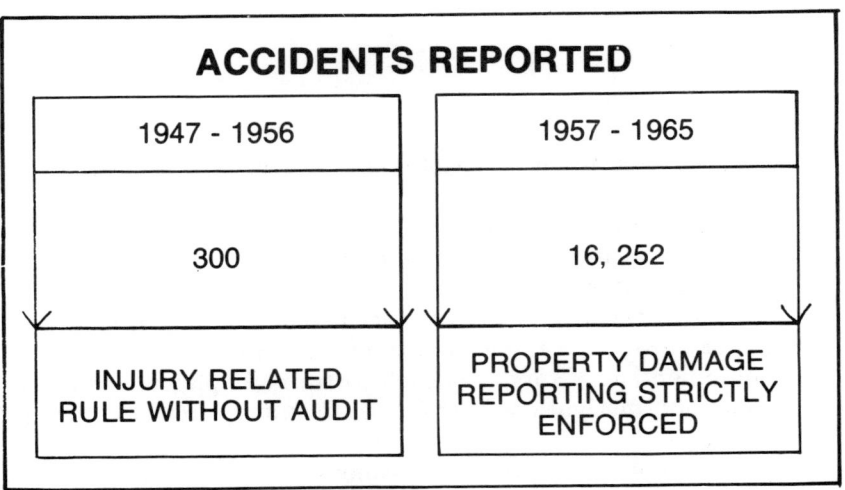

Figure 15.

or protect some building product line structure once construction has begun—and especially after it has been completed.

The total accident control viewpoint must become a habit at this stage and take its important place with the independent concerns for injury and illness. These items are promotion-interrelated, and many staff people miss a golden opportunity to gain what they need in one area of concern because they failed to utilize the multi-value approach in their advisement on design changes and considerations.

Listed below for convenience' sake are eight damage-reduction strategies (described in similar form in Chapter Two) that can be considered when the loss potential is detected at pre-design or design stages.

- A. Prevent the initial marshalling of the form of energy.
- B. Reduce the amount of energy marshalled.
- C. Prevent the release of energy.
- D. Modify the rate or spatial distribution of release of energy from its source.
- E. Separate in space or time the energy being released from the susceptible structure.
- F. Separate the energy being released from the susceptible structure by interposition of a material barrier.
- G. Modify the contact surface, subsurface, or basic structure which can be impacted.
- H. Strengthen the structure which might be damaged by the energy transfer.

5. *Skill Training*

A major percentage of the property damage of most companies will result from accidental contacts and the resulting transfer of energy from materials-handling equipment. This equipment will account for 60 to 85 percent of the damage in most industrial environments.

While not a panacea in the control of property damage, few factors are more important than adequate skill training of equipment operators, with follow-through planned observations and motivational activities, including property damage control.

Listed below are several items believed to be important considerations to optimize the effectiveness of skill training and its follow-up in damage control.

A. Adequate formal training should be given to every equipment operator before assigning him to his job.
B. A special form indicating critical items to review with a new or transferred operator should be completed by the supervisor when first assigning any operator.
C. A complete training manual (designed to include the development of awareness of the importance of property damage control as well as the techniques involved) should be utilized in training, and a copy should be given to each operator.
D. Operator safety rules and related job procedures should be part of the training manual, but should also be issued separately for individual references as independent items.
E. Operators should be licensed or certified as part of their training program, with cards and identification badges or decals worn in a prominent position. The holding of a license should be contingent upon a good operating record.
F. Desired performance of operators following training should be properly recognized, with recorded commendations made when appropriate. Consistent violations of rules and procedures should be handled in accordance with the company's enforcement procedures.
G. A minimum of six planned job observations of each operator should be made annually, with forms completed and proper discussions with operators conducted.
H. Rules and job procedures should be reviewed frequently, but not less than annually, in a formal manner with all operators. Oral or written testing should be an important part of this important exercise, to learn what the operator knows.
I. Provision for recognition of operators who meet desired standards of performance is considered helpful by most supervisors of equipment operators. It is suggested that a total evaluation of personal injury and property damage records, commendations, violations, observations, and testing results, etc. be considered in determining those eligible for such recognition.

6. *General Promotion*

   While not generally considered to be a critical element in any program, promotion can play an important role in the development of safety and conservation awareness that leads to damage and waste control.

   Leaders in the safety and loss control field have long maintained that the safety-conscious worker, working in a hazardous environment will have fewer accidents than the worker lacking safety awareness working under relatively hazard-free conditions.

   While it is difficult to prove the merits of the multiplicity of promotional and educational devices that would go under the caption of "general promotion," there is little doubt that, effectively used, they can create an atmosphere of interest and awareness, proclaiming that a safety and conservation program exists and that it should be a major concern of everyone. Well-prepared property damage control messages should on occasion be the major subject of newsletters to equipment operators, newspaper and magazine articles, employee publications, bulletin board notices and special promotional campaigns. People *and* property conservation should be constant companions in all general promotion activities.

7. *Regular Repair Center Audits*

   Any safety program coordinator knows the value of the dispensary or first aid treatment area as a control point for measuring the results of his injury-oriented program. Injuries and illnesses reported to this area provide statistics that enable him to measure the effectiveness of injury investigations and the frequency of occurrence of those events his program is directed at controlling. Likewise, repair centers become one of the primary "first aid" locations for damaged equipment and material. The information that can be obtained from these important areas forms a point of control for the entire property damage program. Checks of these locations should be regular and systematic. A program should be designed to include an effective yet simple paperwork system for personnel in these locations to regularly report all damaged items.

   Many people on different jobs are also excellent sources of helpful information. Maintenance personnel, equipment operators, purchasing, storeroom and engineering personnel are frequently named as special sources of good information.

Once locations and other sources of information are well known, a system of routine contacts should be established to provide a continuous flow of useful information, which can be utilized as an important point of reference.

Waste and Property Damage Control Inter-related

Not infrequently, investigative efforts related to property damage will indicate the need to consider these two areas of loss potential as one, in remedial solution of the problem.

One company found that its control efforts to minimize damage to the costly tires of heavy equipment were not resulting in any significant loss reduction. The LIP team action directed at this problem resulted in a very successful tire conservation program, by having these pneumatic tires filled with a synthetic foam that enabled them to recap damaged tires several times more often than could be done safely when tires were filled with compressed air. While the initial cost of the plastic foam-filled tire is considerably greater for this company, a cost effectiveness study has indicated appreciable savings in tire loss. Thus, what started as a damage control project ended in the control of waste.

It is interesting that in looking at damage to solid rubber tires on mobile equipment, the LIP team also experimented with synthetic foam-filled tires. A cost effectiveness study indicated that while tire loss was not improved in this case, there was an appreciable injury cost reduction in the decrease of costly back problems previously caused by the transmission of shock through the solid tires. Thus, again, the close connection between personnel injury and property damage is seen.

While waste control can and should extend substantially beyond its relationship to property damage, it is a natural companion. The typical pattern of program enlargement and development is for a company to move into a total accident control program and then to evolve naturally into waste control as well.

One company has established a system to identify and classify all air, water and compressed gas leaks, by size and dollar loss, on a monthly basis. Security personnel have been trained to detect on their regular rounds and record on special forms all leaks found. While leak size determination is estimated, and therefore of questionable accuracy, substantial savings in energy is being reported by this company.

By utilizing this system, and expressing leaks in terms of costs, repairs are prompt and effective, resulting in reduced operating costs and safer conditions. (See Figure 16 for charts utilized in costing by this company.)

Not only does waste control tie in closely with property damage control, but it makes an important contribution to an urgent national need, and can have additional important business benefits that include, but are not limited to:

- Reducing costs in a period of spiraling product prices.
- Preventing plant shut-downs during fuel or other product shortages.
- Keeping business competitive, at home and abroad.

While this chapter is primarily concerned with control of property damage, it is our hope that many will see the value of enlarging the scope of this important area of cost reduction to include waste as an important new target in their loss control programs.

FINAL ADVICE ON PROGRAM IMPLEMENTATION

Perhaps no better advice could be given to the individual beginning to enlarge his injury-oriented program to include the property damage accident than. . . ."use a carefully-planned step-by-step approach".

Briefly described below are some of the reasons for the implementation of this important dictum:

1. It minimizes resistance to change. Management will accept the logic of individual steps (i.e., the fact that damaged equipment should be reported). Usually, little resistance is offered when obviously big values outweigh a small amount of increased effort on everyone's part. On the other hand, many programs fail because the apparent values to be received do not outweigh the enormous effort that seems necessary.

2. It permits systematic problem solving. The best way to gain increased management support and backing is to provide specific "success stories". Success begets sucess, and the surest way to achieve overall goals and objectives is to attack individual problems systematically and thoroughly. Attempting to do too much too fast opens the gateway to failure.

STEAM LOSS

LEAK SIZE DIA. IN.	POUNDS-LOST 160 P.S.I.G.	YEARLY COST $1.30/1000 LBS.
1/32	57,480	$ 74.64
1/16	231,360	300.72
1/8	895,800	1,164.48
1/4	3,657,840	4,755.12
3/8	8,211,480	10,674.96
1/2	14,631,360	19,020.72
3/4	32,845,920	42,699.84

WATER LOSS

LEAK SIZE DIA. IN.	GALLONS-LOST 60 P.S.I.G.	YEARLY COST $0.48/1000 GAL.
1/32	71,880	$ 34.56
1/16	289,320	138.96
1/8	1,119,720	537.48
1/4	4,572,240	2,194.56
3/8	10,264,320	4,926.91
1/2	18,289,200	8,788.60
3/4	41,057,280	19,707.84

AIR LOSS

LEAK SIZE DIA. IN.	CU. FT-LOST 100 P.S.I.G. (NOZZLE COEFF. OF 0.65)		YEARLY COST $0.16/1000 CU.FT.	
1/32		546,816		$ 87.48
1/16		2,187,274		349.96
1/8		8,882,520		1,421.20
1/4		35,050,080		5,608.00
3/8		80,062,680		12,810.02
1/2		140,200,320		22,432.00
3/4	320,250,720	320,250,720	51,240.08	51,240.08

Figure 16.

3. It builds awareness of multiple program values. Each step of the program should be sold on its own merits. There are many, many values and benefits to the program, but they will only be realized to the fullest when opportunity has been provided to let management see and experience them in a way that allows for full and proper evaluation. If a failure occurs, it should only mean strengthening that step of the effort, not losing the entire program. *Total* values can only be realized as management has had the opportunity to build its awareness of *individual* program values.

4. It provides in-depth understanding of the program. Many safety managers have had difficulty in selling a total accident control program because they tried to sell too much at one time. By utilizing the step-by-step approach, two major things are accomplished. First, what one is proposing is much easier to sell and, secondly, management will better understand what is being sold each step of the way. The end result is a complete in-depth understanding of the program once total accident control is working. A checklist of program needs is furnished (See Figure 18.) as a guide to step-by-step program implementation.

CHAPTER SUMMARY AND PROGRAM BENEFITS

The following selected comments made here and abroad effectively summarize the key points in this chapter:

> "You cannot tell which of the accidents will result in an injury. All you can know is that in roughly one out of ten somebody is going to get hurt. All that is certain about the others is that they cost money in the form of damage or delay. Therefore, the root of safety lies not in just investigating the accident that causes injury, but in investigating all accidents. It will also pay handsomely by reducing damage. It is an industrial variation on the theme that if you look after the pennies—which is the damage—the pounds—which are the injuries—will look after themselves."[5]

"Surprisingly few people seem to be able to appreciate that what goes under the name accident prevention is, in the main, injury prevention; and that all so-called accident reports are actually reports on a selected sample of accidents resulting in injury to men. It is, of course, understandable that accidents termina-

CHECK LIST OF PROGRAM NEEDS

____ Letter of endorsement on program from top operating executive.

____ Company loss control policy signed by chief executive.

____ Company loss control policy signed by all members of upper management (PACT)

____ Expansion of safety rule requiring reporting of injuries regardless of how minor to include property damage.

____ Expansion of rule encouraging the reporting of hazards that could cause personal injury to include property damage.

____ Company policy regarding discipline for violation of company rules.

____ Letter from chief operating executive endorsing automatic discipline for apparent willful failure to report property damage.

____ Signatures of all employees of receipt and understanding of company's revised general safety rules including property damage aspects.

____ Permanent posting of rules regarding the reporting of hazards and accidents that include property damage aspects.

____ Labor relations department approval on all program aspects from the start.

____ Union leader notification of all steps of the program.

____ Development of benefits to labor organization for use in discussions of property damage control program with labor people.

____ Development of Standard Practice for investigation of accidents (including P.D.) for management people.

____ Consider inclusion of property damage aspects as all safety rules are re-written and reissued.

____ Letter of endorsement from each major division executive for use in booklets, publications, newsletter references, training, etc.

____ Inclusion of items on property damage in newsletters on safety and loss control with emphasis on near-injury aspects as well as costs.

____ Establishment of severity in $$ for a major property damage loss.

____ Establishment of severity in $$ for a serious property damage loss.

____ Re-design of major injury announcements to include potential for announcing major P.D. accidents.

____ Policy by top operating officer on holding of major accident reviews by department heads and division managers.

____ Develop system to accumulate all related costs on major P.D. accidents.

____ Establish major property damage rate comparable to D.I. rate.

Figure 17. (Continued on p. 136)

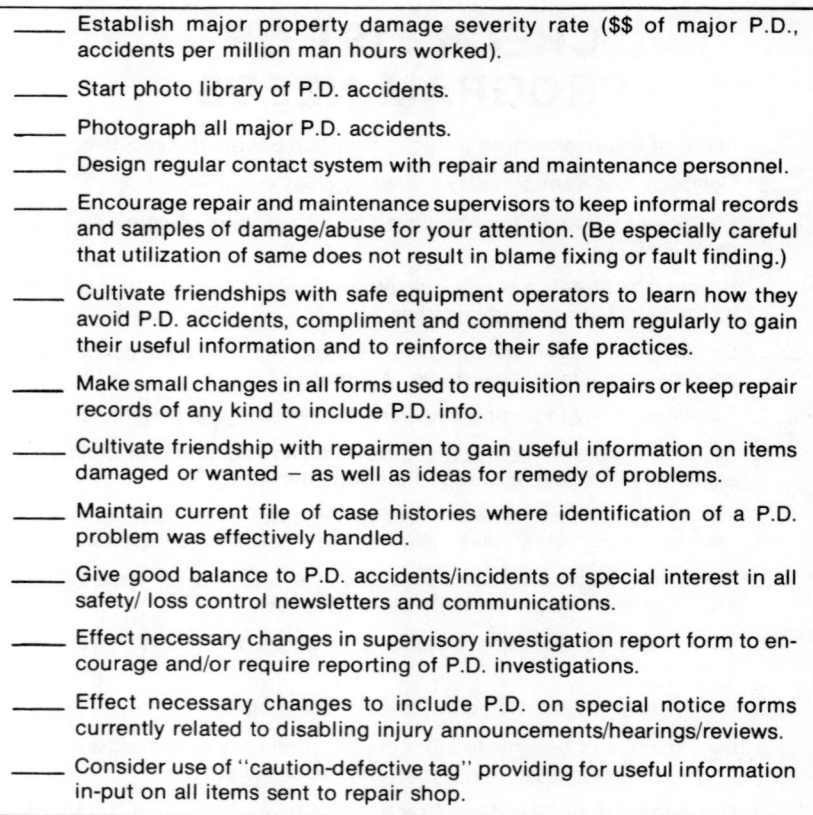

Figure 17.

ting in injury to life and limb should assume special prominence in human minds. Yet, paradoxical though it may sound, this exclusive concentration on injury-causing accidents is also the chief obstacle to further big steps forward toward their prevention. In the light of all this, it is scarcely an exaggeration to designate damage control as a "modern key to safety", in tune with the technological advances of the second half of the 20th century."[9]

"The statistics of accidents prove that as less accidents take place, fewer people get hurt. Prevention of damage to equipment also serves to prevent hazard to life and limb. Safety is indivisible: promotion of safety at home contributes to safety in the mill;

prevention of accidents to equipment helps prevent injury to people. Steelmen know, furthermore, that with procedures improved and equipment better protected, the individual works more effectively and profitably—and with a feeling of greater security. From every vantage point, the Damage Control Program looks like money in the bank."[8]

"So far, a sound foundation for safety within the industry has been laid. Damage Control, to get at the real truth about accidents, is a valuable method of exposing the range and depth of study necessary to achieve effective accident investigation. One of my greatest problems has been the man who could see nothing but a profit and loss account. He did not even know that safe working was efficient and that disregard for safety was not even profitable—and never has been. This is a fact that damage control, directed to the establishment of the truth about accidents, demonstrates beyond argument."[10]

With these thoughts in mind, it isn't difficult to understand that a program directed at the control of property damage and waste could result in these benefits:

- Safer plant environment.
- Fewer serious injuries.
- Reduced production delays.
- Equipment value awareness.
- Reduced operating costs.

BIBLIOGRAPHY

1. Bird, F. E. Jr., and Germain, G. L., *Damage Control.* New York: American Management Association, 1966.
2. Bird, F. E., Jr., "Damage Control." *National Safety News,* September, 1966.
3. Bird, F. E. Jr., *Management Guide to Loss Control.* Atlanta: Institute Press, 1974.
4. Bittel, L. R., *The Nine Master Keys of Management.* New York: McGraw-Hill Book Company, 1972.
5. British Iron & Steel Federation. Ed., "Accidents or Injuries." *Safety Magazine,* Number 18 (1962).
6. Department of Transportation, Federal Railroad Administration. *Summary and Analysis of Accidents on Railroads in the United States,* Accident Bulletin, n.d.
7. Fletcher, John A., and Douglas, Hugh M., *Total Environmental Control.* Toronto: National Profile Limited, 1970.
8. Jackson, Merrick, "Getting at the High and Hidden." *Steelways,* January, 1963.
9. Laner, Dr., S., "Personal Opinion." *Safety Magazine,* Number   (1962).

10. McCullough, T. W., C. B., OBE, "In My Opinion." *Safety Magazine,* Number 21 (May, 1963).
11. National Committee of Inquiry, *Compensation and Rehabilitation in Australia.* Canberra: Australian Government Publishing Service, Volume 2 (1974).
12. National Safety Council, "New York Meeting Stresses Challenges of Tomorrow." *National Safety News,* Vol. 95, No. 6, June 1967. Nuberg, A. B.Sc. "Reducing Damage Accidents." *Australian Safety News,* Nov.-Dec. 1971.
13. Sutherland, Hugh, "Damage Control—A Plain Man's Guide to a Revolutionary Technique." *Occupational Safety & Health,* 1974.
14. U.S. Department of Commerce, *How to Start an Energy Management Program.* Office of Energy Programs. Washington, D.C.: Government Printing Office, October, 1973.
15. Woodhouse, Hon. Mr. Justice, "The Woodhouse Report." *Australian Safety News,* January-February, 1975.

V.
Economics in Loss Control

INTRODUCTION

The need to utilize the economic, as well as humane, aspects of accident prevention and loss control in motivating management has been advocated by leaders within these disciplines for many years. In practice, however, there has not been as much application as theory would indicate we ought to expect. See Figure 1.

Failure to utilize costs more frequently and effectively has been more a lack of knowledge and confidence in the best modes of application rather than failure to accept the general tenet that "money motivates management" as well as people in general.

The primary objective of this Chapter is to present this subject in a manner that will bring more definition and understanding to the proper utilization of costs. Of course, it is hoped that the end result will have been to more clearly focus management's attention on the significant values of an effective safety or loss control program.

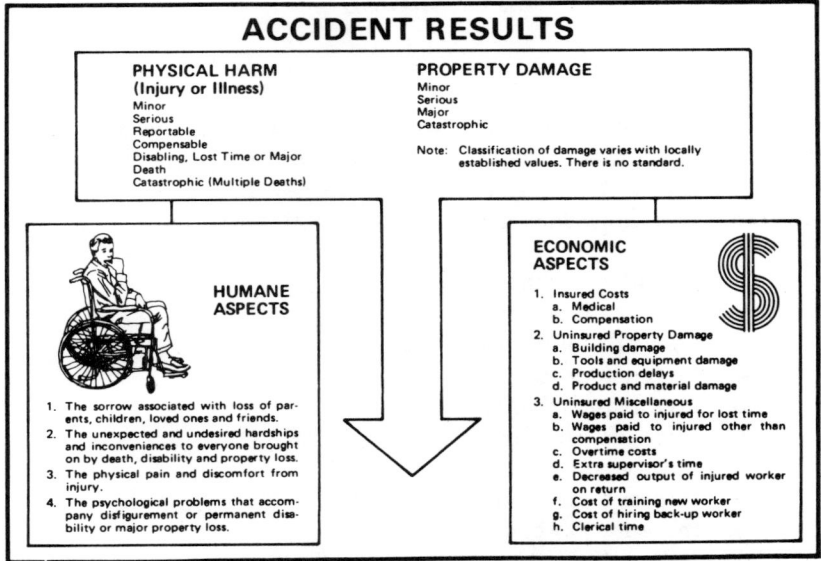

Figure 1.

139

HOW TO SELL TOP MANAGEMENT

There is no question asked more frequently of speakers around the world than, "how do you sell top management on safety or loss control?" Both of the authors have responded to this question dozens, if not hundreds, of times through the years, and both have heard answers given to this question an equal number of times by others.

The multiplicity of answers that we have heard through the years is only exceeded by the number of different ways that we have personally attempted to provide an answer that seemed both satisfactory and useable to the asker. Until recently, neither of us has felt the confidence in our answers to this universal question that we feel today. Perhaps our comfort in using this answer is the reinforcing influence provided by over 2,000 safety or loss control management people that have participated in its development.

Approximately two and a half years ago, we started utilizing a simple individual opinion poll in the many conferences in which we have participated, both here and abroad. During this period of time, we have accumulated 2,352 recorded opinions on the Motivational Scale Exercise, which follows. You may find it interesting to record your own responses and see how you compare with an overwhelming majority of other professionals in eight countries of the world.

MOTIVATIONAL SCALE EXERCISE

Instructions:

Number the items below in the order you believe that upper management places importance on them. *Do not* place them in the order you believe they ought to be, but rather, in the order of importance that you know they actually are in upper management's mind.

1. provide personal satisfaction _____
2. improve labor relations _____
3. enhance public relations _____
4. increase production rate _____
5. give legislative compliance _____
6. improve product quality _____
7. reduce injury rate _____
8. improve operating costs _____
9. increase job pride _____
10. reduce liability potential _____
11. improve customer relations _____

It's most interesting that, around the world, safety and loss control management people believe that upper management places the greatest importance on "improving operating costs" (92%), "increasing production rate (87%), and "improving product quality (84%), as the top three items. It is also interesting that, while many agree that they often hear the word "quality" mentioned ahead of production by upper management personnel, a large majority believe that production actually is a higher priority objective.

The percentages for the order of most other items were not impressive. Forty-seven per cent of all opinions placed "reduce injury rate" between 6th and 8th position, with the 7th choice receiving the most votes. A large majority (96%) did place "provide personal satisfaction" at the very bottom, with "enhance public relations" (94%) close behind.

While many interesting thoughts can be generated by looking at the order of these items, the most important question is, "what does this expression of opinion really mean?" We believe the message is quite clear. Whatever program element a safety or loss control manager wants to sell, he must promote it on the basis of what it will do to accomplish the items high on the motivational scale of the manager whose interest he is attempting to motivate. Those things that will make a major contribution to improving operating costs will have the greatest impact. This is why we call it a "motivational scale for reciprocated interest". It is a direct application of a principle of professional management described in Chapter 3. It should be recognized that the motivational scale could vary from executive to executive; and a program coordinator's goal should be to know the motivational scale of the executive most important to him. See Fig. 2.

At this point, please consider the following thoughts, lest you inappropriately condemn management because of a misunderstanding of the intended message of this Chapter.

... the average worker who gripes about safety conditions in the plant usually does not have his home constructed of non-flammable materials, or equipped with a sprinkler system when flammable construction materials have been used.

... the average worker usually does not have adequate fire extinguishers in his home, or a smoke or heat detection system in the absence of a sprinkler system.

... the average worker usually does not use an approved safety container to store gasoline for his power lawn mower.

The absence of these and many other safety and fire protection items, in the homes of people who love those closest to them, is part of the reason why accident rates are much higher off-the-job and in the home than they are on-the-job. The main point, of course, is that the

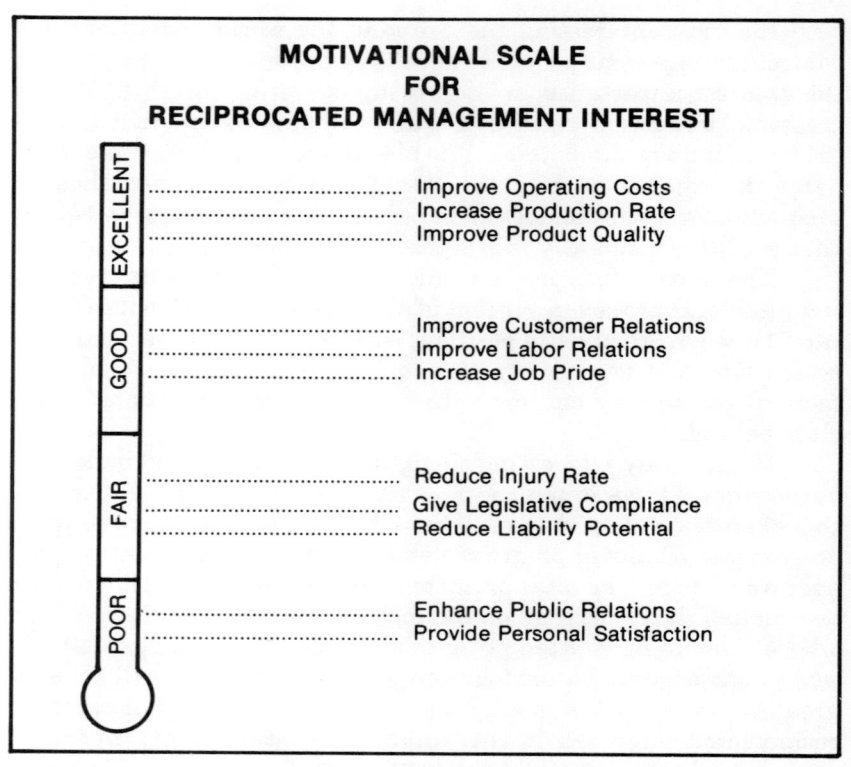

Figure 2.

average person is also motivated by *costs*, and frequently places at risk the lives of those he/she loves most because of the costs of items vital to their safety. It is safe to say that people, in general, place "costs" high on their scale of relative values, and "safety" relatively low.

To motivate a manager or the average individual, we must know their motivational scale and use it to gain their reciprocated interest for our cause. See Fig. 2. Effective use of costs unquestionably is one of the most powerful forces available to motivate man, whether he be the president of a company or the neighbor next door.

With this thought in mind, the following ten principles of economic application have been selected to express what they believe to be representative thinking of current leaders on the use of costs in a safety or loss control program.

1. The Principle of Economic Association
2. The Principle of the Critical Few
3. The Principle of Reciprocated Interest
4. The Principle of Future Characteristics
5. The Principle of Application
6. The Principle of Economic Priorities
7. The Principle of Vested Interest
8. The Principle of Substantial Evidence
9. The Principle of Adequate Evidence
10. The Principle of Dimensional Value

Each of these principles is discussed in this chapter.

The Principle of Economic Association

A manager will usually pay more attention to statistical or general information when expressed or associated with cost terminology.

Management's keen interest in costs is understood when we consider a fact stated by Peter F. Drucker, the well-known management consultant and author. "Management can only justify its existence and its authority by the economic results it produces."[4]

Applications of the principle can be found in many safety programs. Around the world, more and more companies are focusing attention on the effect of uncontrolled accident losses on the profitability of their company. For example, for many years the St. Regis Corporation has expressed the accident statistics given to its management group in terms of an "injury cost rate" rather than a disabling injury rate. Their enviable safety record seems to indicate the value of focusing management attention on the economic side, as well as the humane side, of accident control.

The Employer's Insurance of Wausau advocates a similar approach to its insureds, in the form of a costing-budgeting-accountability approach. A plant's immediate past injury loss history is translated to a cost per man-hour figure, using national average costs. A reasonable cost per man-hour objective for the new year is established by a unique process of injury cost budgeting, and management is held accountable at all levels to achieve the new cost objective.

National average injury costs are utilized in the computing methodology of both the St. Regis and Wausau techniques.

A number of companies have established their own average costs for various categories of injury loss, including: the first aid case,

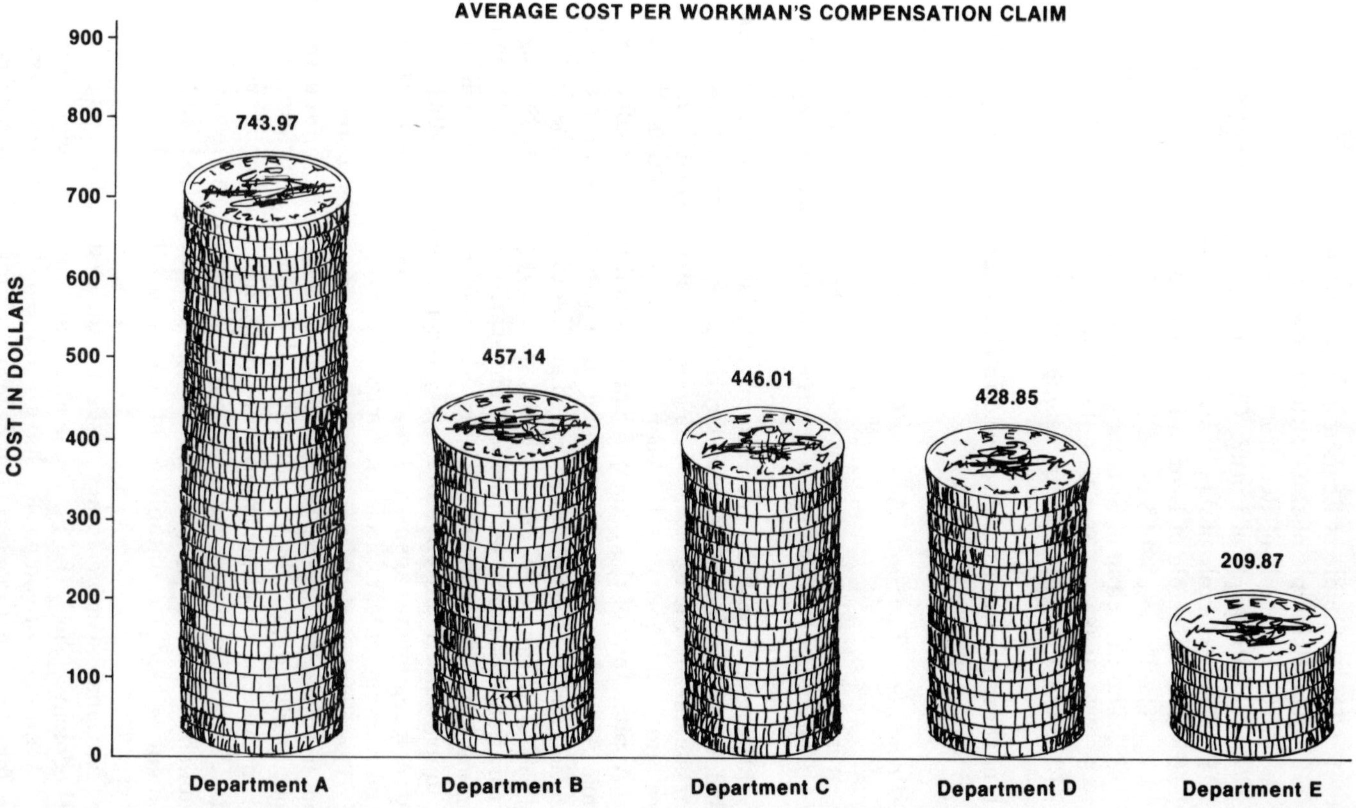

AVERAGE COST PER WORKMAN'S COMPENSATION CLAIM

Department	Cost
Department A	743.97
Department B	457.14
Department C	446.01
Department D	428.85
Department E	209.87

COST IN DOLLARS

Figure 3.

the doctor's care case, temporary total, permanent partial, death and permanent total. On a timely basis, they publish reports showing the current costs of injuries by department, division and organization. Average costs are utilized to keep comparative costing figures up to date with adjustments made when actual costs become available. An increasing number of companies are charging compensation costs back to individual divisions within companies, to bring about a more accurate awareness of accident costs. Several large corporations have required that management of their individual companies establish injury cost reduction budgets with supporting objectives. Periodic reports reflect how well these MBO systems on injury cost reduction are achieving their goals

A good number of companies publish reports by division and department on the average cost per worker's compensation claim. (See Fig. 3)

The accident costing system of the Sandia Laboratories of the Sandia Corporation is still another interesting approach that focuses attention on the economics of accident control. The cost accounting organization of this company compiled data on their various categories of accidental injuries. Costs included in-house expenses incurred by the medical organization, wages for production time lost, administrative expense of preparing accident reports and performing accident investigations, as well as compensation costs. Average costs for a recent two-year period for various types of injuries are shown in Fig. 4, to point out differences and allowances made for categories of employees. Full recognition is given in the Sandia program for losses incurred in non-job-related, as well as job-related, injuries.

In addition to the costs outlined, this company also established estimates of time involved by medical, supervision and miscellaneous

Job-Related		Supervisors		Staff		Graded
Minor	$	80 ea	$	70 ea	$	65 ea
Medical		135		115		100
Vehicle (no-injury)		220		210		200
Disabling 850		+115 /day		+75 /day		+40 /day
Death		98,000		73,000		52,000
Nonjob-Related						
Nondisabling		0		0		0
Disabling 117		+115 /day		+75 /day		+40 /day
Death		73,000		48,000		27,000

Figure 4.

Job-Related	Hours	Days
Minor	2	0.25
Medical	5 1/2	0.68
Vehicle	4	0.50
Disabling	16	2.00+
Death	160	20.00+

+: no. of work days lost by injured employee.

Nonjob-Related		
Disabling	8	1+
Death	24	3+

Figure 5.

related administrative activities required by investigating procedures. Time charges were as shown in Fig. 5.

The worth of people has historically been a point of discussion that seemed impossible to categorize. However, Dr. Lawrence R. Zeitlin, staff consultant for BFS Psychological Associates, has analyzed the replacement costs of recruiting, hiring, training, and re-establishing the level of effectiveness and performance which existed when the replaced individual was lost. The study revealed the following rule-of-thumb for these specific indirect costs (Fig. 6):

JOB CLASSIFICATION	INDIRECT COSTS
Semi-skilled personnel	1/2 Year's Salary
Foremen, Supervisors and Lower Managerial Ranks	1 Year's Salary
Policy and Decision-making Level Executives	2-3 Year's Salary
Top Level Executives, such as President or Head of Major Division	So high as to make estimate meaningless

Figure 6.

Many companies relate their accident losses to the sale of products necessary to offset each $100 accident with profits. For example,

Detroit must construct 1/3 of a popular low-priced sedan; a supermarket must ring up 500 twenty-dollar sales; a restaurant must serve 2,000 two-dollar breakfasts; a department store must sell 1,000 pairs of socks; a publisher must sell 3,000 newspapers; and an office supply retailer must sell 200,000 paper clips.

We have presented some of the many ways that companies and organizations associate costs with program results. We must be aware that the principle of economic association can be applied to the broad spectrum of loss problems including, but certainly not limited to, off-the-job and family accidents, products liability, fire, general damage, theft, vandalism, waste, and drug abuse.

This area of economic application is possibly the most fertile area for motivation in our quest for proper attitude development. The typical operating manager is a pure realist when it comes to allocating his attention. He counts realistic, practical business facts. No other activity in our safety and loss control effort can so readily supply this essential catalyst for action.

The Principle of the Critical Few

In any given group of occurrences, a relatively small number will tend to give use to the largest proportion of results (costs).

As indicated in Chapter 3, this principle is based on the widely accepted statistical premise that in almost every kind of economic-statistical situation, approximately 25 percent of the items account for 75 percent of the total value. This concept is based on the observations of an Italian economist who set the principle down as an economic Law in 1906. In his honor, it is now widely known as "Pareto's Law".

These introductory comments are being repeated to lend continuity of thought to this section. Management authorities through the years have repeatedly substantiated "Pareto's" findings. Perhaps no one emphasized this more than the well-known independent consultant and fellow of the American Management Association, Joseph M. Juran, who in 1954 described this principle in his lectures and writings as "universal for management planning and control". Juran described the concept in terms that continue to be used today, i.e., "Vital few" and "trivial many". "Main issue" and "critical few" are also used in various management applications.[6]

Insurance companies use their extensive computer capability to analyze accident data in terms of causes as well as costs. The "critical few" accident type causes, injuries, or body parts injured, (those resulting in the largest percentage of total costs) can become immediate targets for concentrated management effort. The objective, of course, is to maximize dollar results for the least expenditure of time and effort.

The critical few principle with its many efficiency benefits to management, can be aimed at losses from theft, fire, or any other loss control target.

A large segment of the chapter on property damage and waste control described an application of the critical few principle referred to as the loss improvement project team concept. Utilizing this system, the critical few items causing the majority of management's damage and waste loss can be accurately identified and costed within a relatively short period of time.

The Employer's Insurance of Wausau reveals a very dramatic illustration of the critical few principle in the booklet, "Keeping Score on Injury costs".[7] Their breakdown (in Fig. 7) reflects an interesting comparison of injury types:

INJURY	% OF TOTAL INJURIES	% OF TOTAL COSTS
Medical Only	67	7
Temporary Total	25	30
Permanent Partial	6	50
Permanent Total	1/8 of 1	4
Fatality	1/3 of 1	7

Figure 7.

Note: Since figures have been rounded off, they don't total 100 percent. It is interesting that while permanent partial injuries account for only six percent of all injuries, they result in 50 percent or more of the total injury costs.

While it would seem that common sense and logical deduction would be adequate to guide the average manager in selection of the critical few in any accident or loss costing situation, such is not always the case. The authors highly recommend the book, *The Nine Master Keys of Management* by L. R. Bittel, published by McGraw-Hill Book Company, as basic reading on management techniques.

Chapter two of this publication entitled, "Identify Vital Targets" is completely devoted to a fine discussion of this "critical" — "vital" subject.

The Principle of Reciprocated Interest

People tend to be motivated to accomplish results you want, to the extent you show interest in the results they want to achieve.

It would be extremely difficult to select an economic principle more important than the principle of reciprocated interest. Enormous motivational power can be channeled through application of this principle to bring the kind of executive management cooperation that thousands of safety and loss control people talk about, but seldom receive. Let's take another look at the top of the motivational scale suggested by the opinion survey of 2,352 managers. (See Fig. 8).

The principle of reciprocated interest tells us that if we want to sell management on any safety or loss control idea, we must present it in such a way that the receiver will recognize the value that he will gain toward those things he is interested in. From the motivational scale we have strong indication that "reducing operating costs," "increasing production" and "improving quality" rank extremely high in the typical executive's interests. The big question is: "Do we use this knowledge to best advantage in promoting or selling our ideas?"

Nearly every communication or motivational skill utilized in a safety or loss control program is identical to those required in the production and quality control area. The supervisor who is taught

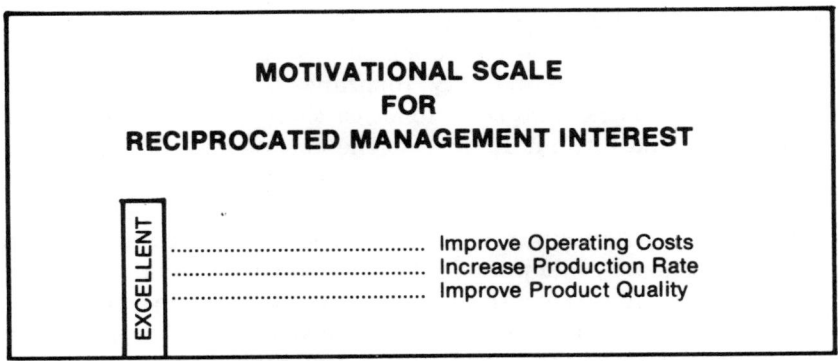

Figure 8.

149

group communications in order to conduct a safety meeting can apply the same skills in his production and quality control sessions. Likewise, the same personal communication skill necessary to effectively give job safety instructions to a worker being assigned a new or different job is of equal value in making sure he completes a quality product on time.

Perhaps no tool used in safety and loss control programming by leaders in industry possesses more potential for application of this principle than a job safety analysis/procedure program. The efficiency check, which is a critical part of developing the JSA, provides an opportunity to uncover inefficient practices and methods that, when properly changed, often result in substantial cost reduction. The principle of reciprocated interest can be employed even more significantly by proposing a standard job analysis/procedure program that includes one unified analysis/procedure, covering all three important elements of production, quality and safety. There are few, if any, elements of a safety or loss control program that cannot be sold on their value to all those items high on management's motivational scale, as well as their values to the attainment of humane objectives.

The Principle of Future Characteristics

The past performance of an organization or unit tends to foreshadow its future characteristics.

As indicated in the Chapter, "Management Control of Loss", the past performance of an organization or unit tends to foreshadow its future characteristics. We can study past economics involved with any area of a safety or loss control program and assume that, with no substantial change in those factors that would alter results, future results will continue the trend. The value here is to use past cost information as a guide to professional management action that will prevent history from repeating itself.

Perhaps one of the best applications of this principle can be found in the insurance industry. Insurance companies use past loss experience as a fine barometer to assist their accurate prediction of future loss, and ultimately provide the basis for establishing their underwriting rates of coverage. These companies also furnish periodic computerized loss analyses to insureds, to indicate the critical few loss problems in terms of costs. Costs are, of course, based on recent past history and provide the insured with a vivid warning of future losses unless something is done to alter their occurrence.

Industrial engineers and efficiency experts use this principle as the basis of much of their work in cost reduction. The past history of costs on machines or equipment, production maintenance, downtime and related damages are all used as a base line for many types of comparisons. Our chapter on Property Damage and Waste Control gives a detailed description of the "loss improvement project system" that makes extensive use of this principle. Of course, those using this valuable principle will seek adequate expertise to assure that their costing predications are statistically accurate and a valid representation of what they are meant to represent.

The Principle of Application

The more often a manager communicates a message, the more certain he can be that it is understood and will be retained.

Once the magnitude of loss problems is clearly defined, the safety or loss control manager has the challenge of communicating both the importance and the urgency of committing management resources to control of those problems. It's relatively easy for these staff experts to see, understand and accept the truth that accidents and other undesired events have an enormous effect upon an organization's profit, or the attainment of its established budget. It is considerably more difficult to gain timely acceptance of this truth by operating management of the enterprise. The principle of application can serve as a valuable tool in this regard.

The following items are examples of hundreds of ways that cost information can be utilized regularly to develop and reinforce management awareness of the importance of safety and other loss control activities:

... Circulate select articles on related costs from external publications to key members of management.

... Circulate regular reports utilizing local costs as indices, rates and/or averages by organization, department and division.

... Circulate select photographs of specific accidents involving property damage, with a brief explanation of estimated costs.

... Circulate general information on escalating compensation and liability trends.

... Circulate reports of individual large court awards or settlements involved with familiar circumstances.

This chapter contains dozens of other ways that costs of undesired events can be utilized, with the assumption that management understanding of the problem is directly related to the frequency of such application. The failure of safety and loss control managers to capitalize fully on the great benefits of this prinicple usually is not caused by lack of useful information, but by lack of its timely utilization.

The Principle of Economic Priorities

A manager will usually give priority response to items possessing the potential for the greatest proportion of results from the least investment of available resources.

This truth or principle is also expressed in a widely accepted economic corollary that a firm should choose from mutually exclusive cost control techniques the one which offers the highest rate of benefits to costs, when both are expressed as expected values.

It is difficult for many safety and loss control managers to understand why management will not financially support a specific program that promises a substantial potential return in cost reduction. There is little doubt that understanding of management's utilization of money would be greatly enhanced if concerned individuals were privileged to more abundant information on the amount of financial resources currently available. Of course, the application of skills and techniques that lead to cost reduction through effective loss control management also builds executive confidence in the individual and increases the probability that he will be more adequately informed on important aspects of the business.

It's ironic that an individual can understand why a person purchases the item for his family or house that gives the greatest return for the money invested, yet finds difficulty in understanding parallel management decisions. Like the average individual, management wants the greatest return for the limited resources available at any given time.

Hugh M. Douglas, co-author of the book, *Total Environmental Control,* and Loss Control Coordinator for the Imperial Oil Company of Canada, made a statement recently that highlights the wisdom of this management truth:

"We can take care of our people when we take better care of the business."

By utilizing available resources properly, management has more resources to utilize and, in effect, more money to spend on the needs of people, their safety and health. In the final analysis, this type of decision-making is absolutely necessary to assure the continuity and very existence of the business enterprise.

The Principle of Vested Interest

A manager is predominantly interested in those economic considerations affecting his own budget

One has only to ask any manager a few simple questions regarding economics, to quickly determine where his interests lie. He is surely interested in the operating costs of the company, his division, and to some degree, those of the employment or medical departments. But one thing will be made crystal clear in his responses—he is predominantly interested in the budget he is accountable to achieve, and by comparison, only slightly interested in anyone else's. This deep, vested, personal interest is very logical when we consider that promotions, merit increases and other things that really count, largely depend on his ability to contribute to the profitability of the organization. It is his budget, of course, that best reflects his management skills that count in this regard. Likewise, his longevity as a supervisor can spend its course rather quickly if he consistently fails to manage costs over which he has direct control.

Through the years, failure to consider this principle in the application of costs with management people has been a major shortcoming of many safety and loss control managers.

While terms like "insured - uninsured" costs can have significant motivational value when properly applied, they also can become sources of subtle resistance and underbreath jokes when used inappropriately. It is absolutely naive to believe there is any significant motivational value in using costs related to the medical, employment or safety departments with an operating manager. There are significant uninsured costs that have direct bearing on his personal budget, and in many cases he is not aware of the impact they have on the attainment of his budgetary objectives. Results of accidents and other undesired events that involve property damage, waste, overtime, rework and delays are buried in costing categories that camouflage their controlability, e.g., property damage costs are usually buried in a general maintenance category.

For similar reasons, terms like "direct" and "indirect" have become less and less part of the professional's vocabulary. The word "indirect" not only has been associated with many items external and remotely concerned with the individual manager, but also bears the inference of "remoteness" — something that would be difficult to do anything about.

For the reasons mentioned previously, there is an increased trend to use the terms, "ledger" and "non-ledger" costs. The term "ledger" carries the strong personal budget implications of vested interest for the manager.

It must be remembered that the budget of a plant manager is a composite of many smaller budgets of division managers, department heads and front-line supervisors. He may very well have a vested interest in a broader variety of uninsured costs than one of his subordinates. Regardless of the level of the manager involved, the principle remains the same. Cost items are relevant when they can be associated with the budget the particular individual is accountable to manage.

One of the greatest challenges facing the safety or loss control coordinator is to develop techniques and skills which enable him to clearly identify specific items of loss that relate in a meaningful way to the manager involved. Hopefully, some of the thoughts presented in this book will assist in that regard.

The Principle of Substantial Evidence

In the absence of adequate historical information, it can be assumed that a manager will require more substantial evidence of need.

Nearly every safety or loss control specialist has had an operating manager request evidence that losses in his immediate area of responsibility have occurred to justify the action suggested. Whether one is attempting to promote a 100% eye protection program or safety hat program, the ease with which a program can be sold to management is directly related to the loss experience to which managers can personally relate. This is the whole point of the principle of vested interest. Conversely, the more remote the loss experience, the more difficult it is to develop management interest, and the more substantial the evidence must be to motivate positive attitudes and actions.

Not only is it necessary to present evidence that can be accepted as relevant to the particular manager's operations, but it is also necessary to relate in a meaningful way to the likelihood of its occurrence. Experience has proven that, in the absence of historical information for the manager involved, it is usually helpful if he can visit other organizations engaged in similar operations to see the program in action, and to evaluate the needs and results.

A wide variety of communicative and motivational techniques must usually be employed to build the base of credibility that leads to acceptance.

The safety and loss control organizations of many large corporations circulate major loss announcements and reviews to their various companies, in order that those with similar conditions or practices can learn from the loss experience of others. Application of the truths involved with the majority of the other principles will also help to build the case. By recognizing that an even "harder sell" is necessary when historical local information is lacking, the safety or loss control coordinator is in a better position to plan, organize and develop more substantial evidence to support his position.

INSPECTION REPORT

AREA INSPECTED: Plant # 2	DEPARTMENT: Machine Shop	
TYPE OF REPORT: ~~Initial~~ Final	DATE: ~~8-3-73~~ 8-31	SUBMITTED BY: Charles Doner

ITEM NUMBER AND HAZARD CLASSIFICATION	ITEMS DETECTED AND ACTION TAKEN
* (1A)	6-10-73 guard missing. cutting head machine # 2046 - NE corner bay #2, bldg. B.
	wooden guard in use; W.O. issued scheduled to complete 8-12
* 2C	6-16-73 door at south end bldg 2 warped, hard to open
	WO issued to carpenter shop - work scheduled for 7-22.
* 3B (X)	7-8-73 heavy accumulation of oil and trash under main motor in bldg 2 pump house
	Cleaned out Aug 20. discussion with workers all turns, permanent signs posted.
4B (X)	two pallets of chemical 265 in yard, column B14. Directions warn against outside storage. Reported to R Jones - area supervisor
	Pallets moved to inside storage Aug 6.
5A	major building column P32 is receiving severe damage from bull dozer operation, cellar bldg 4
	WO issued for Concrete barricade base Aug 12, meeting held with all operators on major damage
6(A) (X)	Severe leaks in natural gas line between columns A1 and A2 in north bay, bldg 2. Strong odor of gas in entire area. D. Sutter, area foreman
	Roped area off. Gas dept. called. Leaks repaired Aug. 6.
7C (X)	Sharp edge on latch - tool shanty door bldg 2
	Maintenance worker Filed surfaces Aug 7
8(B)	compressed air pipe N side of truck door, E end press shop damaged guard or relocation could prevent major loss
	W.O. issued 8-9 for permanent striped metal guard - hazard striped board wired to pipe Aug 10
9	two empty whiskey bottles found in empty locker #15 - washroom bldg 1
	Plant protection notified 8-6 - request area sup. to permit observation before discussing

* indicates item detected previously ○ around a number indicates intermediate action X over a number indicates item corrected

"A" HAZARD:	Any condition or practice with potential for permanent disability, loss of life or body part and/or extensive loss of structure/equipment/material.
"B" HAZARD:	Any condition or practice with potential of serious injury/illness or property damage that is disruptive but less severe than Class "A".
"C" HAZARD:	Any condition or practice with probable potential for non-disabling injury or illness, or non-disruptive property damage.
INTERMEDIATE ACTION	Immediate temporary measures taken to reduce the potential loss occurrence until more suitable or permanent remedial action can be taken.

Note: These sample forms may appear somewhat different from those used in your company's safety program. They are included as a reference and are only intended to be representative of those generally used.

Figure 9.

The Principle of Adequate Information

The timeliness of a manager's decision-making is directly related to the adequacy of information he has upon which to act.

Through the years, one of the most common outcries of safety, environmental health, security and fire protection managers alike, has been directed toward failure of management to act promptly on decisions involved with hazard correction. While there may be some justification for the outcries of management inaction on matters involved with these important areas, on many occasions there is also a great deal to be said in defense of the delayed management action. It is not difficult to understand the frustration of a program coordinator who has organized a planned inspection program that provides detailed reports on all plant areas month after month, only to have them seemingly ignored. Neither is it difficult to understand that a busy manager has many important things to do, in addition to taking the time to read voluminous reports that require interpretation and organization of thoughts in order to give intelligent direction. These kinds of frustrations and problems for both staff and operating managers will become less the rule and more the exception as managers in safety and other loss control disciplines learn to manage more professionally. Providing management with an increased number of devices and techniques to aid decision-making will bring the application of loss control in general industry more in tune with the technological advances of our space age.

The inspection form shown in Figure 9 reveals how hazard classification and the use of symbols can save busy managers time in screening inspection reports, assist them in making decisions on spending money, and provide ready guidance on the priority of attention they should give to problems. (The A.B.C. classification establishes severity of the hazard.) Note that an asterisk (*) indicates an old item carried over from a previous report, and a circled number tells the busy manager that intermediate action has been taken on the particular item. An item crossed out (x) tells the reader that complete correction has been taken.

More and more organizations are including a small, yet very important, section on supervisory investigation forms to quickly guide the supervisor on how much time to spend on his remedial action, and give inference to costs that can be justified for control of future events. (See Fig. 10)

Figure 10.

One of the more important requests that come to safety and loss control coordinators from operating managers is their cry for help on the evaluation of potentially hazardous conditions. Figure 11 presents five questions that a staff professional or operating manager can

**MANAGEMENT GUIDE
ON
RISK AND REMEDIAL COSTS
IN
HAZARD EVALUATION**

CLASSIFICATION OF HAZARD

1. What is the potential severity of loss if the undesired event occurs?

   ☐ Minor - Class C - No. 2   ☐ Major - Class A - No. 4
   ☐ Serious-Class B-No. 3   ☐ Catastrophic-Class AA-No. 16

PROBABILITY OF OCCURRENCE

2. What is the probability that a loss producing event will occur from this problem or hazard?

   ☐ Negligible-Class D-No. 1   ☐ Moderate-Class B-No. 3
   ☐ Low-Class C-No. 2   ☐ High-Class A-No. 4

COST OF CONTROL

3. What is the cost of the recommended control?

   ☐ Minor-Class D-No. 1   ☐ Medium-Class B-No. 3
   ☐ Low-Class C-No. 2   ☐ Substantial-Class A-No. 4

DEGREE OF CONTROL

4. What degree of control will be achieved by this expenditure?

   ☐ Slight-Class D-No. 1   ☐ Moderate-Class B-No. 3
   ☐ Low-Class C-No. 2   ☐ Substantial-Class A-No. 4

EXTENT OF OF APPLICATION

5. What is the extent of application for the control recommended?

   ☐ Single-Class D-No. 1   ☐ General-Class B-No. 3
   ☐ Limited-Class C-No. 2   ☐ Broad-Class A-No. 4

Figure 11.

use to assist him in evaluating a potential or real hazard and arriving at a decision on the need for action, and justification for spending money on its remedy. Note that evaluation can be in terms of words, letters, numbers or a combination of evaluative terms. The important thing is that management can be given valuable information on all critical aspects of hazard evaluation, to help them give a timely response regarding cost justification and priority emphasis.

William T. Fine, Chief of the Safety Department of the Naval Ordinance Laboratory, Silver Spring, Maryland, has designed a similar system of evaluation that quantifies hazard-related data to provide management with a single number that can be translated in terms of specific action.[5]

It is the opinion of the authors that the Management Guide on Risk and Remedial Costs in Hazard Evaluation (Fig. 11) is the most practical tool for all general purposes.

An increasing number of companies are classifying accidents, as well has hazards, in their loss control programs. The classification is generally based on the degree of liability anticipated, and its related economic implications. The enormous court awards involved with product and general liability cases require that managers consider their involvement in accident investigation as a critical part of their loss control effort. Accident classification generally guides management in determining the types and numbers of experts necessary to conduct adequate scientific investigative procedures.

One utility company uses the classifications shown in Figure 12.

Loss control specialists in petro-chemical operations regularly employ sophisticated techniques to mathematically evaluate risks, clearly define the degree of risk involved, and determine the potential economic loss should the undesired event occur. These risk analysis techniques establish cost factors such as the maximum loss exposure; the maximum possible loss that might occur from the foreseeable event, with protection facilities not functioning; and the maximum probable loss to be expected when control facilities function as intended to limit the loss.

While fire protection engineers representing insurance underwriters apply these techniques in general industry, there is a great need for local program coordinators to expand their use, and provide management with additional decision-making tools.

One additional example of a tool to provide management with more adequate information upon which to act can be found in the suggested form shown in Figure 13. It is extremely difficult for managers to evaluate the hazard involved, as well as the economic value of a safety or health related suggestion, when adequate information is lacking. A tool like this gives more adequate information.

Critical	Personal Injuries	Property Damage
Class 1	More than one public (non-employee) death	$500,000 +
Class 2	One public (non-employee) death or hospitalization of five or more public.	$250,000 to $499,999
Class 3	More than one employee on-the-job death.	$100,000 - $249,999
Class 4	Hospitalization of 2 or more public or one employee Death or hospitalization 5 or more employees (OSHA report).	$50,000 - $99,999
Major	Disability (loss time) of 1-4 employees or of one public (non-employee).	$1,000 - $49,999
Recordable	Workman's Compensation Medical Care case.	$0 - $999

Figure 12.

An increasing number of organizations are designing into their suggestion forms the type of evaluative techniques described in this section. The need to provide management in general industry with these and other tools to aid them in sound and timely decision-making should be self-evident.

Understanding the principle of adequate information will hopefully motivate expanded use of such valuable techniques.

The Principle of Dimensional Value

The degree of management attention is directly related to the size of the problem.

A classic story is told of the corporate safety manager who went to the executive vice president of his corporation with plans to produce a motion picture on the prevention of finger and hand injuries. The executive asked what the cost of the film would be and what reduction in hand injury costs was anticipated by the proposed expenditure. When told that the film would cost $40,000, that the costs of finger and hand injuries exceeded $50,000 during the past year, and that any reduction in those costs was difficult to determine, the executive's response was most interesting. "Jack," he said, pulling his

LOSS CONTROL DEPARTMENT
SUGGESTION INVESTIGATION REPORT

Suggestion Number _____ Date _____

DETAILS OF INVESTIGATION ARE AS FOLLOWS:

Not original	1. Degree of hazard:
	Minor ......
Routine Maint.	Serious ......
	Major ......
Not suggestible	2. Frequency of Exposure:
Covered by former suggestion	Light ......
	Medium ......
	High ......
Violation of safety rule	3. Extent of Applications:
Present standard practice violation	Local ......
	Restricted ......
	General ......
Cost too great for expected results	4. Degree of Elimination
	Slight ......
No hazard with normal caution	Substantial ......
	Complete ......
Should become standard practice for same type items in future	5. Originality:
	Not new ......
	Partly new ......
	Entirely new ......

RECOMMENDATION

Acceptance _____ Rejection _____ Trial _____ Additional information required _____

Investigation made by _____ Discussed with _____ Approved _____

ALL SUGGESTIONS... must be answered in detail promptly following a thorough investigation at the location concerned. The investigation should always include thorough questioning and evaluation of the opinions of persons involved and should thoroughly answer the what, why, when, how, where and who of all the suggestion aspects.

SEND... original to the Suggestion Administrator — retain copy for file.

SEE REVERSE SIDE FOR ADDITIONAL DETAILS ☐

Figure 13.

operating budget from his middle desk drawer, "We paid $55,000 last year for toilet paper, and my manager of services tells me that by going to a 3½" roll rather than the standard 4½" roll we presently use, we can immediately cut our annual costs $10,000." Needless to say, as the story goes, this safety manager was made aware that this executive was not impressed with either the dimension of the total loss figure or the uncertain plan to reduce it.

Safety and loss control managers have recognized for many years that medical and compensation costs were not usually sufficient to provide the degree of motivation their programs deserved. To this day, many safety people use H. W. Heinrich's 4-1 cost ratio in an effort to show management a substantial enough problem to gain additional attention. Many of the colleges and universities conducting graduate programs in safety studies give considerable attention to the development of the insured-uninsured cost concept in order to equip future safety and loss control managers with the ability to demonstrate the magnitude of the accident loss problem.

The authors certainly concur with the generally accepted premise that the use of costs can add a significant motivational dimension to a program. They also believe deeply that medical and compensation costs alone have not generally provided a large enough base for significant comparison to the operating cost areas of a typical company or organization.

While it is entirely possible that the fast-rising insurance and compensation costs may change what has been the exception, to be the rule, it is not probable that this will occur for some time. This does not mean that management is uninterested in on-the-job injury costs. They are interested in all operating costs, especially during a period of inflation and reduced business activity that so adversely affects profitability. What the principle of dimensional value tells us is that we do not gain the management attention possible unless we present the costs of our losses in their true magnitude and economic perspective.

Check the lists of cost items related to typical safety programs to determine how many of them are accurately known and utilized properly in your own motivational efforts. (See Figures 14 and 15) A glance at the estimated annual accidental losses to industry gives insight to the magnitude of the total accident cost problem compared to that of the on-the-job injury area alone.

The authors believe, the reason that over 95 percent of the vice presidents of safety and loss control are in motor fleet and railroad operations is directly related to their application of this principle of dimensional value. They precisely identify accident costs involved with injury, property damage and liability, and direct well organized programs at these enormous targets of loss. The dimension of their cost targets is large enough to gain significant management interest and attention. Their personal recognition is further evidence of the value that management places on their effort.

While we use the safety discipline to illustrate their point, the coordinators of security, fire, environmental health and other individual loss-related programs should recognize that this principle has

ACCIDENT COSTS TO INDUSTRY	
On-the-job injuries	$ 6,800,000,000
Off-the-job injuries	4,500,000,000
Family injuries	2,750,000,000
General property damage	15,500,000,000
Cargo damage in transit	9,750,000,000
Material handling equipment damage	5,750,000,000
General Liability	11,750,000,000
Automotive Liability	2,275,000,000
Boiler and Special Machinery Damage	140,000,000
Product Liability	Unknown at this time
Total Costs	$62,715,000,000

The above costs include insurance and immediate direct costs paid by industry. They do not include money value of time lost by workers directly or indirectly involved. Neither is any investigative or clerical time included. These "other costs" could be conservatively estimated to be at least 25 percent of the total immediate costs given above. It is estimated that costs to Canadian industry would approximate 10 percent of the total costs above.

Figure 14.

identical application to their areas. As a matter of fact, it is also our opinion that the career development path for a specialist in any one of these fields clearly leads to loss control management. Precedence for this can be found in the petro-chemical industry in particular..

While the potential today for a catastrophic loss (multi-million) from a single loss-producing event, such as the death of a young outside contractor with a large family, is quite realistic, it will be some time before these individual risks provide the motivational force with management that their everyday common losses provide. When we combine the everyday costs involved with the inter-related disciplines of safety, environmental health, fire and security, we possess a package of cost reduction potential of unquestionable motivational dimension.

Robert Wright, Loss Control Advisor of the Gulf Oil Canada Limited, has received international recognition for a Loss Management Information System (referred to as LOMIS) he introduced in his company several years ago.

**LOSS MANAGEMENT RECORD
COST IN 000 DOLLARS
YEAR TO DATE DECEMBER 31, 1972**

DEPARTMENT	INJURIES	DAMAGE LOSS	VEHICLES	NO. OF INC.	TOTAL
Chemicals	$ 7.3	$ 254.5	$ 8.2	96	$ 270.0
E & P	$ 1.7	$ 144.6	$ 24.5	170	$ 170.8
Marketing	$ 23.2	$ 300.6	$ 237.2	1,270	$ 561.0
Pipeline	$ 1.2	$ –	$ 4.5	10	$ 5.7
Refining	$ 16.1	$ 5,040.9	$ 1.7	176	$ 5,058.7
Other	$ 11.0	$ 0.1	$ 0.7	64	$ 11.8
TOTAL	$ 60.5	$ 5,740.7	$ 276.8	1,786	$ 6,078.0

Loss per barrel thruput 5.5¢   Loss per manhour worked 39.6¢

Reprinted through the courtesy of Gulf Oil Canada, Ltd.

**LOSS MANAGEMENT RECORD
COST IN 000 DOLLARS
YEAR TO DATE DECEMBER 31, 1973**

	INJURIES				FIRE & EXPLOSIONS		MIXES & SPILLS		UNEXP. LOSSES		POLLUTION		LOSS & DAMAGE		MILES DRIVEN 000	VEHICLES AUTO		VEHICLES TRUCK		TOTAL	
	FREQUENCY	SEVERITY	M.A.	L.T.	OFF JOB	COST	NO.	COST	NO.	COST	NO.	COST	NO.	COST	NO.	COST		NO.	COST	NO.	COST	NO.	COST
CHEMICALS	10.50	291	44	13	5.2	4	178.4	3	4.9	1	–	–	–	5	8.5	–	–	–	–	–	70	197.0
EXPLORATION & PRODUCTION	4.06	66	6	5	1.0	10	326.8	208	90.4	2	1.3	–	–	29	69.9	8	1.4	9	2.4	277	493.2
MARKETING	5.99	84	148	81	23.7	26	38.3	221	388.2	180	68.9	15	7.1	360	160.7	130	48.3	320	465.1	1481	1200.3
REFINING	11.66	277	143	40	17.8	15	173.0	42	86.0	3	.6	1	5.5	39	1088.1	–	–	10	2.8	293	1373.8
RESEARCH & DEVELOPMENT	–	–	3	–	–	–	–	–	–	–	–	–	–	–	–	1	.2	–	–	4	.2
SUPPLY & TRANSPORTATION	4.79	396	4	–	5.1	–	–	4	62.0	–	–	–	–	–	–	2	.7	5	1.6	15	69.4
OTHERS	1.03	6	6	4	.4	1	.2	–	–	–	–	–	–	–	–	3	1.0	–	–	14	1.6
TOTAL	6.14	133	354	143	53.2	56	716.7	478	631.5	186	70.8	16	12.6	433	1327.2	144	15.6	344	471.9	2154	3335.5

M.A. MEDICAL AID
L.T. LOST TIME   Loss Per Barrel of Thruput .... 2.75 cents   Loss Per Manhour Worked .... 22.26 cents

Reprinted through the courtesy of Gulf Oil Canada, Ltd.

Figure 15.

Within the broad scope of his program fall all activities designed to preclude loss producing incidents or minimize their adverse effects.

The practical application of his effective loss management program involves:

1. The establishment of accepted standards of performance in every area of activity for every situation.

2. The assignment of resources and organization to accomplish the desired goals.
3. The assignment of responsibility with commensurate authority and accountability to ensure that the desired levels of performance are met and maintained.
4. Measurement and evaluation of Performance:
    (a) The systematic reporting of all loss-producing incidents by the employees at the level where the incidents occur.
    (b) The proper investigation of each incident to determine the immediate cause and the underlying causes.
    (c) The careful analysis of underlying causes to indicate specific breakdowns in the management system which lead to injury, damage and loss situations.
    (d) The costing of each incident by estimate based on the Supervisor's first-hand knowledge of the costs involved.
5. The commitment of management to take effective corrective action.

The results of Mr. Wright's program are evident in Fig. 15.

The economic value of an effective loss control program was made quite clear in a recent talk presented by Raymond H. Marks, President of Tenneco Chemicals, Inc. as part of A.S.S.E.'s program segment of the OSHA-USA Conference and Exhibit held in New York City. Mr. Marks made the following comments:

"For purposes of understanding, let me state that a sound loss control program is not limited to employee safety and health — although this is an important segment and our first consideration. A comprehensive loss control program is aimed at elimination of problems of occupational health, property protection, product safety, security, and any other area where an unintended incident can occur and detract from the company's profitability

... Let me assure you that by taking this broader view of the problem, we in no way diminish our emphasis on protecting the safety and health of our employees. To the contrary, emphasis on a broader range of subjects creates a continuing greater awareness and complements our efforts to protect individual employee safety and health.

... Safety and loss control are an important part of the executive suite. No longer second-class citizens, loss prevention and profit performance have become synonymous. There is no room for compromise. Total dedication and re-dedication are the key ingredients.

... How often do we get the chance to demonstrate our innate concern for the health and well-being of our fellow humans, while at the same time, we improve our profit performance. Think about it — total loss control programming is sound business planning, the best of both worlds under one roof."[8]

Perhaps there is no greater way to illustrate the cost reduction side of the double-edged sword referred to by Mr. Marks, than to consider the sales volume that must be developed to make up the profit lost by the undesired events in the absence of a good loss control program. (See Fig. 16.)

CHAPTER SUMMARY

These, then, are ten principles of economic application that can be utilized to motivate increased management interest and action in a loss control program. You can use them individually, or in combination; but most important of all, use them.

While there is no doubt that their effective application will require an enormous amount of innovative application of our best professional efforts and energy, the rewards that will come with success will be more than justified. We have everything to gain and only our losses to lose.

IN TIMES OF KEEN COMPETITION AND LOW PROFIT MARGINS, LOSS CONTROL MAY CONTRIBUTE MORE TO PROFITS THAN AN ORGANIZATION'S BEST SALESMEN.

It is necessary for the salesman of a business to sell an additional $1,667,000 in products to pay the costs of $50,000 in annual losses from injury, illness, damage or theft, assuming an average profit on sales of 3%. The amount of sales required to pay for losses will vary with the profit margin.

YEARLY INCIDENT COSTS	PROFIT MARGIN				
	1%	2%	3%	4%	5%
$ 1,000	100,000	50,000	33,000	25,000	20,000
5,000	500,000	250,000	167,000	125,000	100,000
10,000	1,000,000	500,000	333,000	250,000	200,000
25,000	2,500,000	1,250,000	833,000	625,000	500,000
50,000	5,000,000	2,500,000	1,667,000	1,250,000	1,000,000
100,000	10,000,000	5,000,000	3,333,000	2,500,000	2,000,000
150,000	15,000,000	7,500,000	5,000,000	3,750,000	3,000,000
200,000	20,000,000	10,000,000	6,666,000	5,000,000	4,000,000
SALES REQUIRED TO COVER LOSSES					

This table shows the dollars of sales required to pay for different amounts of costs for downgrading incidents; i.e., if an organization's profit margin is 5%, it would be required to make sales of $500,000 to pay for $25,000 worth of incidents; with a 1% margin, $10,000,000 of sales would be necessary to pay for $100,000 of the costs involved with downgrading incidents.

Figure 16.

BIBLIOGRAPHY

1. *Accident Facts*. Chicago: National Safety Council, 1975
2. Allen, Louis A. *The Management Profession*. New York: McGraw-Hill Book Company, 1964.
3. Bird, F. E. Jr. *Management Guide to Loss Control*. Atlanta: Institute Press, 1974.
4. Drucker, Peter F. *The Practice of Management*. New York: Harper & Row, 1954.
5. Fine, William T. "Mathematical Evaluation for Controlling Hazards." *Selected Readings in Safety*. Macon: Academy Press, 1973.
6. Juran, Joseph M. *Managerial Breakthrough*. New York; McGraw-Hill Book Company, 1964.
7. "Keeping Score on Injury Costs." Employer's Insurance of Wausau. *Insurance Facts*. New York: Insurance Information Institute, 1975.
8. Marks, Raymond H. "Total Loss Programming Becomes Sound Business Planning." *Professional Safety*, June, 1975.

VI.
Measurement Tools For Management

INTRODUCTION

One of the biggest roadblocks to progress in establishing measurement tools to help management has been the long search by loss control leaders for one universal measurement that would answer all needs. This quest for a measurement that would reflect the degree of safety or loss control for the occupational system has been one of the most perplexing problems throughout the entire history of the organized safety movement.

The difficulties involved can be more fully appreciated when one considers the multiplicity of events, circumstances and conditions involved in the interaction of a large group of elements; management, planning and design, work environment, machines, materials, supervision, job procedures, and worker relationships could enter into one single loss producing event. (12) The complexity of the measurement problem becomes even more evident when one breaks down the many subfactors that interact within each of these elements to produce influential effects upon the overall system (i.e., a seriously-ill family member, unpaid bills or a malfunctioning heater in his home could affect the worker on the job). It doesn't require intensive study to determine that the scientific measurement of total safety or loss control per se is virtually impossible. The professional safety or loss control manager can utilize a number of measurement tools that evaluate certain aspects or components of his program that are known to contribute to its overall effectiveness. (13) He can also use measurement tools that reflect the extent of certain loss results.

The important thing is that everyone recognize exactly what it is that's being measured and what significance the use of the measurement will have on producing the results desired.

It is hoped that this chapter will help to orient the thinking of those involved with the use and selection of measurement techniques toward a better understanding of the role various measurements *can* play in their loss control programs.

CHARACTERISTICS OF A GOOD MEASUREMENT TOOL

Regardless of the specific factor that is being measured, there are certain desirable characteristics that the user should keep in mind when selecting his measurement tools. Several of special importance are briefly discussed here:

1. *It should be administratively practical.*

   Many organizations fail to utilize various loss control program tools that could benefit them greatly because they are apprehensive about their ability to deal with the apparently sophisticated techniques involved. Most managers must come to grips with time, budget, and manpower at some point and, no matter how good a technique sounds, if it can't be administered within the practical limits, it can't be utilized.

2. *The measurement criterion should be quantifiable.*

   Dr. H. J. Kolodner, in his article, "Correlation of System Safety and System Reliability", said, "The history of science has adequately established that rapid progress is made when concepts being dealt with are reduced to quantitative terms which can be predicted, measured, evaluated, and finally communicated." In selecting a criterion that can be quantified, the loss control manager increases the probability that it will be accepted as a solid technique because of the objectivity implied (if not necessarily inherent) in quantification. Peter Drucker, author of *The Practice of Management* emphasizes that "The measurement used determines what one pays attention to." The more precise the quantification of the criteria involved in any measurement tool, the more probable it is that management can and will use the tool effectively.

3. *It should be a valid measurement of what it is supposed to represent.*

   For many years, safety managers considered a disabling injury rate as a measurement of safety performance rather than a measurement of the occurrence of a very limited number of reported accidents that terminated in certain loss. To this day, many organizations around the world base safety contest awards on disabling injury frequency records and present awards as recognition for outstanding safety performance. Creating a better understanding of what measurements used in a loss control program *actually measure* is one of the key objectives of this chapter and it is certainly a major need in any system that is to provide management with tools to manage effectively.

4. *It should be as objective and error-free as possible.*

Techniques must be employed that minimize the possibility of individual bias entering the measurement system. (i.e., random sampling of workers to determine the level of group safety meetings would be better than asking supervisors to report their own performance). One of the oldest arguments against the disabling injury rate system under the ANSI Z16.1 Code has been the bias that favors the large company with more available established light-duty jobs or the company that permits workers on the job with crutches, walking casts, canes, etc.

While an essential discipline of management should be an attempt to precisely quantify what one is trying to achieve, management will accept that there are important areas with limited related information that can only be measured in qualitative and subjective ways. This is not desirable and should be avoided as much as possible, but it is much better to include important areas that do not lend themselves to objective quantification than to rely solely on incomplete measurement data. (11) Specific ways and means to minimize bias and maximize objectivity will also be discussed later.

5. *A good measurement system should be understandable.*

Too often, we think management people and workers alike will continue to understand terms and relationships involving a vague figure that was explained on several (infrequent) occasions. The easier it is for a measurement to be understood on sight and to stand on its own, the more certain we can be that we have provided management with a tool it can relate to and can utilize with others.

6. *It should be sensitive to change.*

A good measurement tool must accurately reflect changes that take place within the system it is designed to measure (i.e., measuring the same representative factors in a program could, in the course of time, cause a concentration of management effort on those specific areas of concern, resulting in diminished effort in others). Allowance for such probabilities should be considered in planning (i.e., random periodic selections of representative program factors or elements could keep the system effective and the measurement technique more sensitive to overall change.) (16)

No conscious effort was made in this listing to arrange these characteristics in any order of importance; neither is this list intended to represent *all* of the characteristics important to a good measurement system. These six were selected because of their special importance and frequent reference in this chapter.

CLASSIFICATION OF MEASUREMENTS

The need for some simple point of reference that aids the manager in his understanding of the use and capability of recognized measurement tools becomes self-evident by the complete disparity of approaches suggested when you ask ten different specialists how to measure safety or loss control performance. No other facet of the safety or loss control discipline is more poorly understood or organized in the mind of the average individual responsible for this activity than is *measurement*.

The following discussion of the three classifications of measurement may serve to clarify this subject in general and aid in the proper application of currently-available measurement systems.

MEASUREMENTS OF CONSEQUENCES

For discussion purposes, this broad classification will be subdivided into two groups that we will refer to as *Actual Loss Measurements* and *Potential Loss Measurements*.

Actual Loss Measurements

These are, as the words imply, measurements of the *results* of contacts above the threshold limit of the body or structure. They could include personal injury or property damage and are usually expressed in terms of frequency and severity.

While the general classifications of loss for personal injury and property damage are quite similar, measurement techniques vary considerably. For this reason, we will discuss the categories of injury measurement separately from property damage.

Injury Rates

The Disabling Injury Frequency Rate could also be referred to as a major or lost-time frequency rate since the degree of injury *usually* requires lost time. It is probably the most widely-used measurement involved with loss control in the world today. Under the ANSI Z16.1 Code, this rate would be calculated in the following manner: (2)

Disabling injury frequency rate =

$$\frac{\text{Number of disabling injuries} \times 1{,}000{,}000}{\text{Employee hours of exposure}}$$

The definition of disabling injury will vary from country to country, the point of difference usually being the time off the job before the injury is considered disabling or major. The popularity of this measurement has been greatly influenced by its extensive use by safety and business associations, in their publications and contests through the years, as a base to compare performance of individual units and business groups. Because this is a measurement of a very small statistical base of rare events, and since the decision to count or not count the individual case can be influenced by many variables, the measurement has been the subject of considerable controversy in recent years. There is an increasing trend among leading organizations to diminish the emphasis on this measurement. This is not as much the result of the disparity in counting practices as it is a recognition of the need to increase the size of the base employed, for purposes of statistical validity.

The Disabling Injury Severity Rate has also been used extensively as a measurement of days lost or charged for death, permanent disability and temporary total disability. It expresses the number of days lost for this death and injury in terms of a million hour unit by use of the following formula: (2)

Disabling injury severity rate =

$$\frac{\text{Total days charged} \times 1{,}000{,}000}{\text{Employee hours of exposure}}$$

While the specific time charges involved with death and permanent disability would be constant, this rate would be affected by variables in the counting procedure followed by the individual company, as mentioned earlier. This rate is usually used in conjunction with the disabling injury frequency rate.

The Disabling Injury Index is one of numerous measurements used through the years to combine disabling injury frequency and severity rates into a single measure. The index is computed by multiplying the disabling injury rate by the disabling injury severity rate and dividing the product by 1,000 as shown below. (2)

Disabling injury index =

$$\frac{\text{Disabling injury frequency rate} \times \text{Disabling injury severity rate}}{1{,}000}$$

This measure reflects both frequency and severity, yielding a combined index of total disabling injury. It has been used primarily for ranking different establishments, organizations, and industries.

In its raw form, it can be correctly applied only for ranking total experience from "best" to "worst". When using the measure to determine percentage of improvement or to compare the degree of difference between two units, the square root of the index must be used.

The strong desire to utilize measurements of consequence that would have improved statistical validity and be more in tune with modern loss control thinking has led several large organizations to include disabling lost time off-the-job accidental injuries in the disabling frequency rates used on the job. More organizations are placing increasing emphasis on the frequency of "serious" injuries described as a non-standard measure in the ANSI Z16.1 Code and the rate of reportable injuries introduced by the Occupational Safety and Health Act of 1970.

The number of *serious* (or reportable) injuries occurring in the average occupational establishment is 5 to 20 times larger than the number of *disabling* injuries. While the number of serious injuries (in comparison to the total number of accidents) is still very small from a statistical standpoint, the broader base for analysis certainly gives a more valid point of reference than the extremely small population segment involved with disabling injuries alone. The criteria used to classify serious or reportable injury are presented below for readers who may be unfamiliar with them.

Definition of Serious Injury

Serious injuries include the following work injuries:

A. All disabling work injuries.

B. Nondisabling injuries in the following categories:

1. Eye injuries from work-produced objects, corrosive materials, radiation, burns, etc., requiring treatment by a physician.

2. Fractures.

3. Any work injury that requires hospitalization for observation.

4. Loss of consciousness (work related).

5. Any other work injury (such as abrasion, physical or chemical burn, contusion, laceration, or puncture wound) which requires: a) treatment by a medical doctor, or b) restriction of work or motion, or assignment to another regularly-established job. (2)

Serious injury rate

Serious injury frequency rate =

$$\frac{\text{Number of serious injuries} \times 1{,}000{,}000}{\text{Employee hours of exposure}}$$

(Note: No severity measure can be computed for the serious injury classification.)

Definition of Reportable Injury

Reportable non-lost-time cases are those involving loss of consciousness, restriction of work or motion, transfer to another job, or any medical treatment except for treatment variously defined as first-aid, even when the first-aid is performed by a physician or nurse. The first-aid cases not to be reported are those that involve a single treatment of minor scratches, cuts, burns, splinters, etc., of the type not ordinarily requiring medical care. (9)

The rates are calculated on a base of 100 full-time employees, or 200,000 man-hours. A full-time employee is considered to work 40 hours a week, 50 weeks a year, or 2,000 manhours. While this rate will be expressed in terms of reportable injuries per 100 full-time employees, safety and loss control managers will generally relate it to man-hours, as they do other injury rates.

The All-Injury/Accident Frequency Rates

A good number of organizations with more advanced loss control programs have maintained injury rates that largely reflect the inclusion of minor injuries. These rates vary from organization to organization, in that they *may* include all injuries reported and therefore could be referred to as an "all-injury frequency rate". Still others may exclude disabling or recordable injuries, and be referred to as a "minor injury rate". Loss control leaders seldom encourage the use of the whole body of these statistics as a valid barometer of injury occurrence, since many conditions are known to affect reporting (i.e., inclement weather, distance from the treatment area, time of day, etc.). In addition, many of these injuries may be the result of such agencies as nuisance dust, splinters, etc., inflicting injuries under such circumstances that cause of major loss is highly improbable. While the much greater number involved increases the statistical base, the probability of a reasonably error-free rate is not likely.

W. W. Allison of the Sandia Corporation was the first to encourage the establishment of a *high potential accident index* (HYPO), or a rate based on the occurrence of *all* accidents reported, whether or not they resulted in loss, that under slightly different circumstances *could* have resulted in serious personal injury or property damage. (1) Regardless of factors (similar to those that affect the reporting of minor injuries) that could introduce error or bias into this rate, many new thoughts and valuable components of present-day loss control knowledge were contributed by this progressive line of thinking.

Property Damage Rates

Major Property Damage Frequency Rate is based on the total number of property damage accidents that reach or exceed the locally-established value for a major property damage accident which occurs during the period covered by the rate. (While this figure varies, the most common figure used is $1,000 or more.)

It is significant to mention that non-operating employee hours are not usually included in this rate, since this group has little to do with the exposures that cause such loss and the inclusion of these hours could distort the accuracy of the measurement. The rate relates these accidents only to the operating or production exposure hours worked and expresses the number of such damage cases in terms of a million man-hour unit, by use of the following formula:

Major property damage frequency rate =

$$\frac{\text{Number of major property damage accidents} \times 1{,}000{,}000}{\text{Operation (non-clerical/office) man-hours worked}}$$

(Note: Refer to Chapter 8, "Control of Property Damage", for additional information.)

Major Property Damage Cost Severity Rate is based on the total costs of all major property damage accidents which occur during the period covered by the rate. Like the major property damage frequency rate, non-operating production man-hours (office workers, etc.) are not included. It is based on the use of the following formula: (6)

Major property damage cost severity rate =

$$\frac{\text{Total costs of major property damage accidents} \times 1{,}000{,}000}{\text{Operating (non-office) man-hours worked}}$$

Total Property Damage Cost Severity Rate is based on the total cost of all reported property damage accidents which occur during the period covered by the rate. Like other damage rates, it does not include non-operating employee man-hours. It is based on the following formula:

Total property damage cost severity rate =

$$\frac{\text{Total costs of property damage accidents reported} \times 1{,}000{,}000}{\text{Operating (non-office employee) man-hours worked}}$$

(Note: Certain organizations have established a minimum dollar value for reporting property damage accidents, such as those resulting in $20.00 or more of damage. A similar rate is used to measure reportable motor fleet property damage accidents. This latter rate is based on the total number of reportable accidents during the period covered, related to the miles driven.)

Potential Loss Measurements

These are techniques that can be utilized to measure the rate of undesired events that do not result in loss but, under slightly different circumstances, *could*. The whole premise in using these techniques is that many more incidents do *not* result in loss than *do*, and that many of these undesired events not only could have resulted in *loss* but could have resulted in *major loss*. By utilizing measurement techniques to report and analyze these no-loss incidents, the measurement system utilized not only has a much larger statistical base but has assured a *predictive* posture rather than the widely-used *reactive* ones dependent on loss-related data.

The Critical Incident Technique

The critical incident technique has been used to obtain near-loss or near-accident data as well as accident data. (Since more accidents are *not* reported than *are*, useful information not previously reported can be detected through this system. No attempt is made to differentiate between the two during the application of the technique.) Because the frequency of no-loss incidents is so gross compared to those that result in loss (refer to the ratio study in Chapter 1, "History and Philosophy"), the preponderance of data obtained will be related to incidents rather than accidents. This technique involves a trained interviewer, obtaining the recall of near-loss incidents, accidents, or causal-related factors through a special interviewing process by which the worker is encouraged to report any incident, accident, or error in which he was involved or of which he has knowledge. The large body of statistics provides data that can be analyzed to determine critical problems, practices, or conditions that could lead to accidents or may have been factors in previous ones. (8)

The primary purpose of the critical incident technique (CIT), as generally employed in the past, has been to analyze causal data that would identify individual problem areas, and *not* to measure the rate of occurrence of critical incidents or utilize individual segments of data to any great extent. The tool has essentially been employed by highly-trained interviewers or research personnel, and very little information has been made available to loss control specialists through the publications or journals they normally read.

Incident Recall

Incident recall is a practical modification of the critical incident technique. Like CIT, it involves an interview that encourages the reporting of near-loss incidents. Its major purpose is not to obtain information for analysis, but to gather individual pieces of data that can be used before-the-loss (in the same manner as individual accident cases) to prevent or control future occurrences. Since its application is generally directed through the front-line supervisor, the special employee relationship required by the technique creates an atmosphere conducive to efficient, continuous, voluntary reporting of accident data, as well as near-loss incidents. Chapter 6 of this book discusses this technique in further detail.

MEASUREMENTS OF CAUSE

This second major classification can be subdivided into *Measurements of Actual Cause* and *Measurements of Potential Cause*.

Measurements of Actual Cause

Analyzing the causes actually involved with those accidents reported, in order to establish critical targets for safety and loss control work, is a long-established practice. Raw figures are frequently expressed in terms of percentages of cause related to the total number of accidents reported. Two methods of causal measurement are briefly discussed below.

Standard Cause Measurement

Historically, the cause analysis reports of many organizations have included a measurement of related factors, in addition to the common factors analyzed by percentages of total cases, including nature of injury, part of body, source of injury, accident type, hazardous condition, agency of accident and unsafe act. An analysis of a single case taken from the ANSI Z16.2 illustrates these terms:

A forklift truck went out of control when one wheel struck a piece of stock lumber which projected into the aisle. It ran out of the aisle and struck a machine operator, breaking his leg between the ankle and knee. (3)

Analysis

Analysis Category	Answer
Nature of injury	Fracture
Part of body	Lower leg
Source of injury	Forklift truck
Accident type	Struck by
Hazardous condition	Improperly-placed lumber
Agency of accident	Lumber
Agency of accident part	None
Unsafe act	Unsafe placement of material

For the most part, cause analyses have historically related to the *immediate causes* involved with accidents and have been, in effect, analyses of *symptoms* rather than the *basic* or *underlying causes*.

Measurement of Organizational Error

A measurement in which W. C. Pope, formerly Chief, Division of Safety Management, U.S. Department of the Interior, has pioneered is the development of a computerized accident analysis system that identifies organizational errors and managerial weaknesses involved with accidents, by permitting the line supervisor to input his personal assessment of causal factors involved. A printout of a case or group analysis does not require a code dictionary to be understood. The example in Figure 1 is typical. While the bias of individuals would naturally affect the degree of error in the measurement, management officials do determine what key line managers think their problems are.

Recognizing that people's actions reflect what they believe, such data can serve a useful purpose in cutting across organization and managerial lines to the heart of many problems affecting results in accident losses.

```
BUREAU
DATE            06-29-70
NAME            JOHN SMITH        RESULT                       PI WITH PD
STATE           S. DAKOTA         SEVERITY                     DISABLED
HOUR            11 A.M.           NAT INJ                      CONTUSION
EMP STAT        TEMPORARY         PART INJ                     BACK
OCC CODE        LABORER           SOURCE                       FORK LIFT
WORK ENV        WAREHOUSE         HUM ER A                     INEXPERIENCE
AGE EMPL        25                HUM ER B                     FAIL TO CHECK DEFEC
PROP OWN        INTERIOR          DEFECT A                     MECH-MATERIAL DEFEC
AGE PROP        15 YEARS          DEFECT B                     WORN, CRACKED, FRAYED

MGMT PROB                         INSTRUCTIONS NEEDED          COMP COST $778.
SUP, TRNG, CONT                   NO WRITTEN INSTRUCT          PROP DMGE $350.
FIT DUTY                          NONE                         TORT PAID NONE
PER SERV                          LABOR-MGMT PROBLEM
MAINT INV CONT                    POOR REPAIR FACILITY
FIT FOR USE                       OVERAGE, OBSOLETE
PROP SERVICE                      EQPT. OPERATIONS             TOTAL $1,128.
```

Figure 1.

Since this unique system is one of the few approaches that goes substantially beyond a measurement of basic causes or symptoms, the accident investigation form which serves as the computer input document is shown in Figures 2 and 3. (15)

Measurements of Potential Cause

Measurements that utilize accident-related causal data are essentially after-the-fact and reactive because of their dependency on events that have *already resulted in loss*. (This does not imply that they do not serve a valuable purpose in the prevention or control of future loss events.) In the search for predictive and before-the-fact measurement systems, an increasing number of loss control leaders are measuring factors related to substandard practices and conditions (*potential* loss causes) that can be detected *before* loss occurs. Several of the more interesting approaches are briefly described below:

Behavior Sampling

Random sampling is an accepted management tool that has a solid background of use for measuring

FORM DI-134 *(face)*

UNITED STATES DEPARTMENT OF INTERIOR

(FOR SAFETY MANAGEMENT USE ONLY)

SUPERVISOR'S REPORT OF ACCIDENT

Refer to DI-134A for Instructions, Definitions and Standard Coding Details

Field Report No.

DATE OF THIS REPORT

SECTION A. IDENTITY *(Supervisor to Complete)*

(1) ORGANIZATIONAL UNIT *(Area, Region, etc.)*

Last No. Here

(2) REPORTING STATION *(Name & Address)*

LAST NO. HERE

(3) STATE IN WHICH ACCIDENT OCCURRED

(4) Mo. ______ Day ______ Year ______

(5) NEAREST HOUR OF ACCIDENT ______ a.m. ______ p.m.

(6) NAME *(Employee, visitor or other involved)*

Last ______ First ______

Soc. Sec. No.

Use separate form for each employee involved

(7) EMPLOYMENT STATUS *(Circle one)*

1 Permanent*
2 Y.O.C.
3 Temporary
4 Emergency
5 Contractor
6 Vista
7 Job Corpsman
8 Public *(visitor)*
9 Public *(other)*
0 Other *(specify)* ______

*Include Job Corps Staff

(8A) CSC OCCUPATIONAL CODE

Last No. Here

(8B) WORK ENVIRONMENT

(9) RESULT OF ACCIDENT *(Circle one)*

01 PI *(personal injury)*
02 PI with PF *(fire)*
03 PI with PM *(motor veh.)*
04 PI with PB *(boat)*
05 PI with PA *(aircraft)*
06 PI with PO *(all other)*
07 PF *(fire)*
08 PM *(motor vehicle)*
09 PB *(boat)*
10 PA *(aircraft)*
11 PO *(all other property damage)*

(10A) IDENTIFICATION OF PROPERTY DAMAGE *(if any)*. *(Give name, model, number, size, make, type, etc.)*

Employee operated:

"Other" operated:

(10B) Make of Govt. Vehicle ______

(11) PROPERTY OWNERSHIP *(Circle one)*

0 No prop. involved
1 Interior owned
2 Other Federal
3 Contractor
4 Concession
5 Employee-owned on O.B.
6 Inter-agency (GSA) motor pool
7 Interior Leased
8 Privately owned
9 Other *(Explain in No. 18)*

(12A) AGE:

a. Of Employee *(Year of birth)* ______

(12B) b. Year of Mfg. ______

(13) IS TORT CLAIM EXPECTED? Circle one: Yes or No
If "Yes", has Tort Claim Officer been contacted? Circle one: Yes or No

SECTION B. MEDICAL *(Supervisor to Complete)*

(14) NATURE OF INJURY

(15) PART OF BODY INJURED

(16A) SEVERITY OF INJURY *(Circle one)*

0 No injury involved
1 First aid attention only
2 Medical attention only
3 Disabling injury *(fatal)*
4 Disabling injury *(temporary)*
5 Disabling injury *(permanent)*

(16B) Z16 COMPLIANCE

___Chargeable

___Not Chargeable

(16C) BEC FORMS PREPARED

Circle: Yes No

(17)

a. Leave date: Mo. ______ Day ______ Year ______
b. Return date: Mo. ______ Day ______ Year ______
c. Death date: Mo. ______ Day ______ Year ______

d. No. days lost

SECTION C. STORY OF ACCIDENT *(By Supervisor or Employee)*

(18) NARRATIVE: INCLUDE: WHO, WHAT, WHEN, WHERE, AND HOW
Facts are important—fault finding not.

NOTE: Condense Story Here

SECTION D. SUPERVISORY OPINION *(Supervisor to Complete)*

(19) I THINK ACCIDENT MIGHT HAVE BEEN PREVENTED IF:

Employee Fault Finding Adds Nothing to Management Improvement.

(20) I SUGGEST THE FOLLOWING POLICY OR PROCEDURE CHANGE BY MGMT. TO HELP PREVENT SIMILAR ACCIDENTS:

SIGNATURE AND TITLE OF REPORTING SUPERVISOR

AN. 1970

(Over)

Figure 2.

ANALYSIS OF MANAGEMENT PROBLEM

FORM DI-134 *(back)* If the title you need is not in DI-134A, tell your safety officer so that it may be added. Information on this side represents a "team" effort by safety and other functional managers to assist the Line Officer to discover underlying causes of accidents and to plan for their ultimate correction. All items require an entry. Follow bureau instructions for analyzing the problems on this side.

SECTION E. LINE MANAGEMENT PROBLEM *(Operations, Design, Construction, Maintenance, Plant Mgmt., etc.)*

(21) TYPE OF ACCIDENT *(Event)*   (22) WHAT WAS USED, DONE, CONTACTED *(Source)*

(23A) HUMAN ERROR *(First selection)*   (23B) HUMAN ERROR *(Second selection)*

(24A) CONDITION DEFECT *(First selection)*   (24B) CONDITION DEFECT *(Second selection)*

(25) REVIEW OF THE MGMT. PROBLEM CITED IN SECTIONS D AND E

Sig. _____ REVIEWING MGMT. OFFICIAL

SECTIONS F, G, H, I, and J to be completed by responsible mgmt. analysis identifying problems in the system to be resolved that will reduce accident loss. Keep remarks brief. Use coded information. Leave no blanks. In each section, consultation with supervisor and appropriate mgmt. official is desired.

SECTION F. PERSONNEL PROBLEM *(Consultation with Supervisor and Personnel Official is Desired)*

(26) SUPERVISORY CONTROL AND TRAINING   (27) FITNESS-FOR-DUTY EVALUATION

(28) OPINION OF THE MGMT. PROBLEM RELATED TO PERSONNEL SERVICES

Sig. _____ REVIEWING PERSONNEL OFFICIAL

SECTION G. PROPERTY/EQUIPMENT/ENVIRONMENTAL PROBLEM *(Consultation with Engineering and Property Official Desired)*

(29) MAINTENANCE AND ENVIRONMENTAL CONTROL   (30) FITNESS-FOR-USE EVALUATION

(31) OPINION OF THE MGMT. PROBLEM RELATED TO PROPERTY SERVICES

To be repaired: Yes or No Est. Cost $ _____
To be replaced: Circle: Yes or No Tentative date: _____

Sig. _____ REVIEWING ENGINEERING OR PROPERTY OFFICIAL

SECTION H. FINANCE PROBLEM *(Consultation with Administrative and Finance Official Desired)*

(32) AMOUNT OF PROPERTY LOSS   To Govt. Prop. $ _____   (33) AMOUNT OF TORT CLAIM AWARD   To Government $ _____
To "Other" Prop. $ _____   To "Other" Party $ _____

(34) OPINION OF MGMT. PROBLEM RELATED TO FINANCIAL SERVICES

Sig. _____ REVIEWING FINANCE OFFICIAL

SECTION I. LEGAL PROBLEM *(Consultation with Tort Claim or Legal Official Desired)*

(35) OPINION OF THE PUBLIC SAFETY PROBLEM   (36) POSSIBILITY OF RECOVERY FROM A 3RD PARTY? *(Check One)*
___YES ___NO If "No," explain why in (37)

(37) OPINION OF THE CAUSE, NOT RELATED TO A GOVERNMENT EMPLOYEE OR OPERATION

Sig. _____ REVIEWING TORT CLAIM OR LEGAL OFFICIAL

SECTION J. CORRECTIVE ACTION *(Taken to Make Less Probable the Recurrence of this Accident)*

(38) LOCAL CORRECTIVE ACTION TAKEN OR PLANNED: When: Now _____ Fiscal Year _____

Sig. _____ MGMT. OFFICIAL TAKING ACTION   TITLE

(39) RECOMMENDED BUREAU OR DEPARTMENT ACTION TO ASSIST IN SOLVING IDENTIFIED PROBLEMS: *

*A Bureau response is expected if request for action is made here.

SIGNATURE OF REVIEWING SAFETY OFFICER	DATE	SIGNATURE OF REVIEWING AUTHORITY	DATE	Initial of Bureau Safety Officer	DATE

Figure 3.

product quality and work effort. It has also been employed in recent years to measure the degree of safe behavior of groups of workers. Utilizing a list of unsafe behaviors, a trained specialist can observe the workers (using a theatre counter to indicate safety violations) and accurately estimate the safe and unsafe behavior (as related to the list) of the group being studied. Care must be taken to make sure that the principles of random selection are followed and that observations are made *before* the observer is observed. Bias is minimized by using a trained specialist, and error-free results become a function of training the observers to *see* the violations included in the system. One of the important "musts" is that management must recognize and accept the tool as a reputable measuring device.

The number of observations required of a selected population or group is determined by the results of a preliminary survey and the degree of accuracy desired. Let's assume, for example, that a specialist observes 100 workers and decides that 25 are engaged in unsafe practices listed within the program; the percentage of unsafe observations could be calculated as being 25%. By utilizing the percentage (P) and the desired accuracy (let's assume we have selected plus-or-minus 10%), we can calculate the total number of observations that will be required (N) by utilizing the formula shown below:

$$N = \frac{4(1-P)}{Y^2(P)}$$

N = the total number of observations required
P = the percentage of unsafe violations
Y = the desired accuracy

Using the results of the preliminary study, a 10% desired accuracy would mean that 95% of the time the correct answer would fall between 22.5% and 27.5% of the total.

$$N = \frac{4(1-P)}{Y^2(P)}$$

$$N = \frac{4(1-.25)}{(.10)^2(.25)}$$

$$N = \frac{3}{.0025}$$

N = 1200 observations required

While the study must have 1200 observations as a minimum, they could be conducted during different times, turns, and days, all selected at random.

The use of this technique has great potential to provide management with a measurement of cause that is *predictive* and *before-the-fact*. One must keep in mind that the degree of safety measured is limited to the scope of the behaviors included in the system. (14)

Environmental Sampling

Random sampling can also be applied to physical conditions in a loss control program in a similar fashion to its application in quality control. For example, a housekeeping rating could be established through the application of this technique. Good housekeeping is one of the most sought-after goals in any management efficiency campaign. By applying the safe and unsafe determination to a listing of items that would include a further breakdown of conditions related to machinery and equipment, stock and materials, aisles, floors and buildings, etc., a very similar system could be developed. Care would have to be exercised to maximize objectivity.

Since physical conditions do not usually change constantly, a special random sampling chart could be obtained from a quality control or industrial engineer that would probably not require as many observations as would be required for a population with constantly changing conditions.

Mathematical Evaluation for Controlling Hazards

William T. Fine, Chief of the Safety Department, Naval Ordinance Laboratory, Silver Spring, Maryland, has developed a most interesting method of mathematical evaluation, aimed at meeting two important needs: (1) to determine the relative seriousness of hazards, in order to guide management in assigned priorities to preventive work, and (2) to measure the justification for the estimated cost of contemplated remedial action.

To accomplish these needs, Mr. Fine suggests the utilization of a formula that "calculates the risk" of a hazardous situation and provides a numerical evaluation for the urgency of corrective attention to the item. The calculated risk scores can then be used to establish corrective action priority. Another formula weighs the estimated cost and effectiveness of the contemplated corrective action, and gives a quantitative estimate of the justification for the cost. (10)

MEASUREMENTS OF CONTROL

"Control" in this context does not mean "to hold back", "to keep within limits", or "to restrain", but refers to the "control" function of the professional manager described in Chapter 2. These measurements are based on the accepted professional premise that the work a manager must do to produce desired results in a safety or loss control program can be clearly *identified* and *measured*, and that these results can be *evaluated* in an accepted quantifiable language that enables management at all levels to *correct* performance deficiencies of subordinates to the level of their established *standards*.

Since they are a direct application of the proven tenets of the profession of management, as described earlier, emphasis of the brief comments below is noteworthy:

Advantages of Use

1. They are easily understood. There is nothing vague about the fact that 50% of accidents reported are being investigated, or that 75% of required general or critical parts inspections are being conducted.

2. They give inference to control. An extremely high frequency rate may have great motivational value but all it can tell management is that "something" or a "great deal" must be done. On the other hand, a 50% coverage of critical jobs by Standard Job Procedures gives inference to *why* people may be making mistakes. Likewise, all other evaluations of management's compliance to its work standards in the safety or loss control program contribute to a much broader

capability to draw inferences about the probability of loss for a supervisor, a department, a division and the organization.

3. They are not dependent on *losses*. They completely involve measuring compliance to program standards that produce results that *control* loss. In this context, they are pre-active and before-the-fact.

4. They are suitable for inter-and intra-plant comparisons. Within a multiple-location organization (with common standards such as the investigation of all reported accidents, the conducting of a proper job indoctrination with all new employees, the presentation of a group communication to all employees weekly, etc.), these measurements can be utilized more equitably than measurements greatly influenced by variable exposures.

5. They minimize bias and can be statistically valid. Since they are based on management's compliance with its own standards, there is no undue pressure or influence to inject bias. A high degree of quantification can be built into a system that can be administered by trained safety or loss control personnel. Measurement tools already accepted by management as valid barometers of system effectiveness can be utilized.

A Dynamic System at Work

There is substantial evidence from the aerospace program to indicate that (with unlimited resources) 99.9% of all downgrading incidents could be prevented. The average businessman knows this is neither practical nor economically feasible. The role of the safety or loss control manager should be to optimize the effectiveness of his existing system by managing control of the established standards for work performance. His realistic objective should be to upgrade his program and its standards, over a planned period of time, to equal the leaders in general business, in order to achieve the probable level of efficiency that studies have indicated is attainable. While he measures management performance to existing standards, his program should reflect the *dynamics of change* to achieve longer-range goals. As new work and new standards are added to the program, the base by which he measures management performance must reflect these continuing changes. (4) The following Management Work and Related

Standards charts contain the work activities and general standards believed to be common to leaders, as a guide to those organizations whose program level has not evolved to that point. The organization which presently has a program considered to be equal to the leaders will find a challenge to optimize its effectiveness through measurements of control.

Practical Utilization of Control Measurements

Once the safety or loss control manager decides to establish control through professional management techniques, he must decide how to best utilize time and resources to accomplish the job in the most efficient manner. It would be utterly impossible for the average program manager to accomplish the measurement of performance levels for each management group in his organization each month in all work activities. It is far more practical to devise a system that measures management performance in *representative* work areas to detect substandard performance on a timely basis in order to maintain continuous "control". (See Figure 4) Selecting six to ten work activity areas for measurement and evaluation on a bi-monthly or quarterly basis would be a manageable practical goal. A *comprehensive* measurement of the entire program (all major work areas) could be made on a less frequent basis. A minimum goal should be every six months; quarterly would be considered ideal. (See Figure 5)

As experience is gained in administering the management process involved, measurements become increasingly objective as well as efficient. A manager could elect to measure five or six work activities that are considered the "critical few" from a basic need standpoint (i.e., Incident Investigation, Planned Inspections, Protective Equipment, Group Communications, and Personal Communications), continuously and randomly select four or five different areas each measurement period, in order to maintain a desired level of performance at all times in all areas. This minimizes the possibility of concerted effort on only a few areas receiving maximum attention continuously. Further reduction of overall effort can be achieved by keeping score of "year to date" performance levels as well as the performance at the time of the measurement. These measurements can become the basis of recognition for management personnel at the department, division, and organization levels. A similar measurement system can be utilized to measure the performance level of the individual front-line supervisor and can be utilized as an important part of his

LOSS CONTROL RATING

"It is seldom that a tool provides a double edge sword of opportunity that we have in our loss control effort."
T.R. Pledger
President

DEPARTMENT	1 Weekly Meetings	2 Supervisor Investigations	3 Standard Job Procedures	4 Jobsite Inspections	5 Comprehensive Inspections	6 Protective Equipment	7 Skill Training	8 Employment Procedures	9 Loss Improvement Project	10 Pre-use Check	11 General Promotion	12 Rules	Performance this Quarter	Performance Year to Date	Honor Status
TELEPHONE GROUP															
Burnup & Sims, Fla.	90	96	88	92	80	85	90	85	86	89	90	93	88.6	86	··
Deviney	85	89	84	80	80	78	80	76	90	88	92	90	84.5	85	·
Dysard	87	90	93	89	92	90	96	89	97	92	93	92	91.6	92	··
Fitton & Pittman	96	89	90	87	92	96	93	87	89	87	91	89	90.5	91	··
Burnup & Sims-Texas	92	97	89	92	93	89	97	94	96	90	93	96	93.1	92	··
ELECTRICAL GROUP															
Burnup & Sims, CATU	86	90	83	76	91	86	90	86	89	86	90	92	87	88	··
Burnup & Sims, Electrical	75	69	72	80	63	76	84	87	79	82	86	79	77.6	80	·
Georgia Electric	85	90	87	89	86	73	88	79	84	90	83	82	84.6	83	·
Line Dismantling	96	92	89	97	88	99	97	86	97	91	98	96	93.8	94	··
Southeastern Printing															
West Coast Line	69	73	84	79	82	67	84	76	83	72	69	81	76.8	79	
UTILITY GROUP															
BSI, Inc.	93	92	98	97	94	99	96	92	99	98	96	91	96	93	··
Ford-Wehmeyer, Inc.	79	80	78	81	83	76	82	81	90	79	76	82	82	85	·
Greenbank	86	89	91	87	93	88	92	94	90	87	91	96	90.3	91	··
Smith - Sweger	97	94	91	96	93	89	97	94	87	96	91	99	93.6	90	··
TOTAL - CORPORATION															

PEOPLE ..... PROFITS ..... PROPERTY

Figure 4.

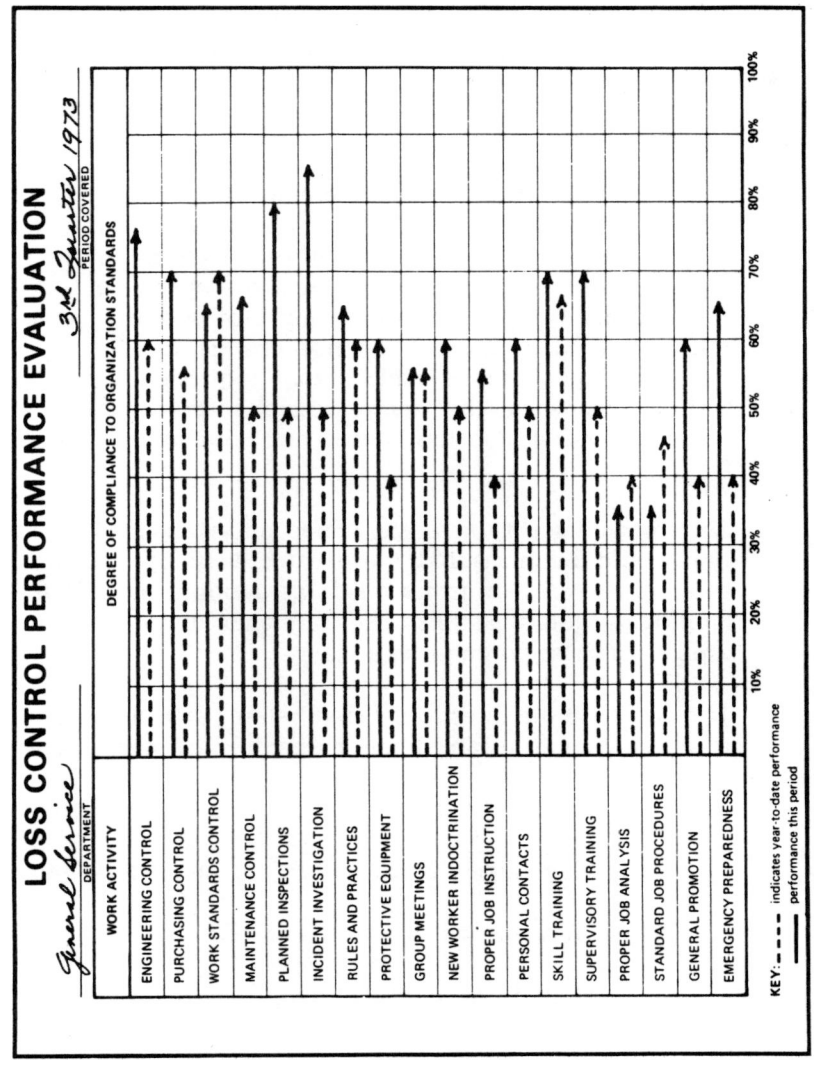

Figure 5.

annual job performance review. This will be discussed in Chapter 7.

Common Methods of "Control" Measurement

While a variety of methods can be employed to measure management performance levels for the various work activities in the program, the three briefly discussed below will probably be utilized more than others.

1. *Random Sampling*

    This accepted tool of management (suggested earlier to determine the composition of the whole by examining a large enough representative part) can be used as an effective measurement and time saver. Since the particular statistical populations being measured in most of the areas involving measurements of control do not change rapidly, a special sampling chart can be utilized that requires fewer observations than needed when sampling such items as safe behavior, which has a fast, constantly-changing population. (See Figure 6)

    Let's consider a typical work area and its related standards that could be measured quite effectively by utilizing this tool. In the area of group meetings, random interviews with employees could establish the frequency and certain quality aspects of these communications.

    The three simple questions listed below (with their tabulated responses) could establish the percentage of meetings held in accordance with the standard, whether or not visual aids were being used, and an estimated degree of meeting effectiveness.

    1. When did you last attend a group safety meeting?
    2. What was the subject discussed?
    3. What do you recall seeing that was shown by the group leader?

    Random sampling is especially helpful in measuring personal communications of various types, as well as compliance with protective equipment requirements.

NUMBER OF EMPLOYEES IN DEPARTMENT	SAFETY PERFORMANCE SAMPLING TABLE USE WITH NON—REPLACEABLE POPULATIONS						
	95% confidence ± 10%						
	EXPECTED PROPORTION						
	95%	90%	85%	80%	70%	60%	50%
	NUMBER OF CONTACTS OR OBSERVATIONS REQUIRED						
20	10	13	14	15	16	17	17
30	12	16	19	20	22	23	23
40	13	19	22	25	27	28	29
50	14	21	25	28	31	33	33
60	14	22	28	31	35	37	38
70	15	24	29	33	38	40	41
80	15	25	31	36	41	44	44
90	16	26	33	37	43	46	47
100	16	26	34	39	46	49	50
120	16	28	36	42	49	53	55
140	17	29	37	44	52	57	60
160	17	29	39	46	55	60	62
180	17	30	40	47	57	63	64
200	17	31	41	48	59	65	67
220	17	31	41	50	61	67	69
240	18	31	42	51	62	69	71
260	18	32	43	51	63	70	72
280	18	32	43	52	65	71	74
300	18	32	44	53	66	73	75
320	18	32	44	53	67	74	76
340	18	33	44	54	67	75	77
360	18	33	45	54	68	76	78
380	18	33	45	55	69	77	79
400	18	33	45	55	70	78	80
420	18	33	45	55	70	78	81
440	18	33	46	56	70	79	81
460	18	34	46	56	71	80	82
480	18	34	46	57	72	80	83
500	18	34	46	57	72	80	84
520	18	34	46	57	72	81	84
540	18	34	46	57	73	82	84

See Reverse Side For Directions On Use.

DIRECTIONS FOR USE OF TABLE

If a preliminary survey of 110 employees in a department of 300 revealed that 77 had received safety instruction for the task they were doing, and 33 did not, the estimated proportion of jobs for which safety instruction was given would be 70% (77 divided by 110). Since there are 300 employees in this department, we would go down the "number of employees" column to the nearest figure (300 in this case) and across the right to the 70% column. You will note that 66 employees should be sampled in this department in order for the sampler to be 95% confident that the sample statistic (in this case the proportion of employees having received safety instruction) will lie within ± 10% of the proportion believed to exist (i.e. 60% to 80%).

Let us suppose that there had been no preliminary survey conducted and that the proportion of employees having received the desired safety instruction had been merely estimated to be 70%. In the event the sample of 66 employees indicates that less than 60% had received the desired safety instruction, the sample size should be raised accordingly. For example, assume the sample of 66 employees indicates that only 50% have received proper safety instruction. Go down the left hand column again to 300, and across the 50% performance column. An additional 9 employees or a total of 75 must be sampled in order for the sampler to be 95% confident that the sample statistic (in this case the proportion of employees having received safety instruction) will lie within ± 10% of that proportion believed to exist (i.e. 40% to 60%).

After the first study has been made, the resulting measurement can be used to determine the size of the next scheduled sampling.

RANDOM SAMPLING GUIDEPOSTS FOR INTERVIEWER OR OBSERVER

* Select program areas for sampling that will provide useful data.
* Determine what information will be of most value.
* Develop questions for interviews that will not lead to bias by giving cues or by being emotionally loaded. Have alternate questions to avoid misunderstanding and enable respondent to consider all alternatives. Establish good rapport quickly before questioning. Make sure questions never degrade supervisory image.
* A theater-counter facilitates quick recording of observations.
* Determine whether population to be sampled is replaceable or non-replaceable, and use appropriate chart to select sample size.
* Sampling should be random by turn, day, plant area, etc.
* Observation should be undetected, and sampling should stop when presence is known.

Figure 6.

2. *Actual Count*

The simple, objective approach of actually counting the items involved and comparing results to the number prescribed by the standard can be used in such work areas as inspections, job procedures, job observations, investigations, rules, and skill training. For example, an objective may have been established which asks each supervisor to complete four Standard Job Procedures on critical jobs for the year. Let's assume that there are 20 front-line supervisors in a hypothetical department. The total job procedures to be completed for the year would number 80 with 100% compliance. The loss control manager counts the number completed at the end of the first quarter, and discovers that only ten have been completed. The management performance level for Standard Job Procedures would be 50%, since 20 should have been completed. Measurement by actual count is probably the most widely-used technique; it may also be used in conjunction with professional judgment when quality factors are included as presented in the example under professional judgment.

3. *Professional Judgment*

When a measurement system is initially introduced to determine the degree of management control, the stress is usually placed on *quantity* rather than *quality* (i.e., holding group meetings 100% of the time is usually considered more important than having great concern about the degree of their planning and content). As time and experience create confidence in the system, the manager will desire to encourage *quality* in communications, job procedures, inspections, observations, etc. He can then employ a value factor method of scoring (described next) and utilize a judgmental system to measure quality.

Experience has proven that whenever judgment is employed, accuracy can be increased by applying these steps (illustrated by the example in Figure 7) for measuring the quality of an investigation report form:

    A.   Break the item down into its components.

    B.   Place a value factor on the parts.

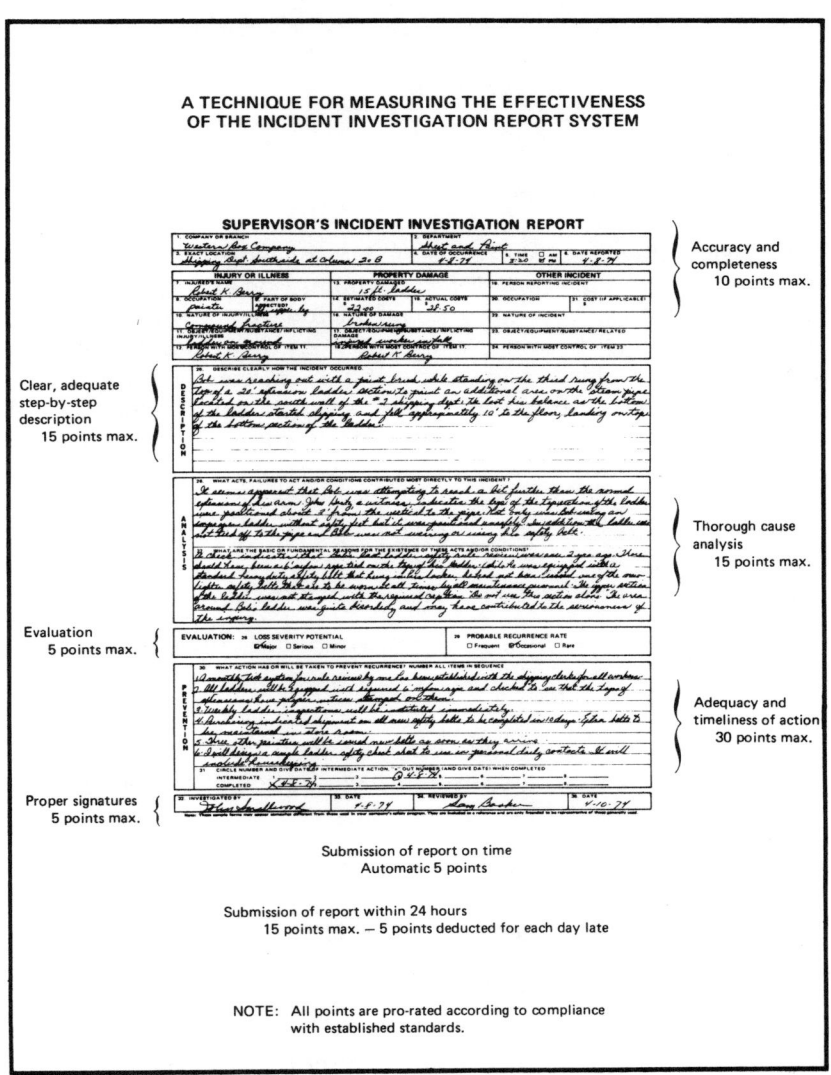

Figure 7.

C. Ask enough questions about each part to increase certainty of judgment.

D. Record findings accurately.

In other words, accuracy is increased when a method is utilized to facilitate a systematic evaluation of whatever is being judged.

The Use of Value Factors

A value factor could be defined as the worth of a component as compared to the worth of the whole. For example, standards for Standard Job Procedures could include a specific amount of prescribed formal training, a specific objective number of procedures to complete, and a specific requirement for updating procedures. These three component areas of work involved, with related standards, could be broken down and assigned value factors:

WHOLE	PARTS	V.F.
Standard Job Procedure	Supervisory training	30
	New procedures completed	50
	Degree of updating	20

Likewise, in order to determine a total loss control performance rating, each work area would have to be given a value factor in order to combine scores into one figure:

WHOLE	PARTS	V.F.
Loss Control Performance Rating	Investigation	10
	Group Meetings	15
	Inspections	20
	Personal contacts	20
	Protective Equipment	20
	Rule Review	15

While the establishment of value factors is a judgmental thing, based on the opinion of the professional staff, any degree of difference in opinion is offset by the overall average of the system. The biggest benefit of value factors is that this system prevents an insignificant component factor from disproportionately influencing the evaluation of the whole. (5)

Systems Safety Techniques in "Control" Measurements

Systems safety techniques were used extensively to control the safe performance of very expensive and potentially very dangerous products of the aerospace industries, such as rockets, rocket fuel, atomic power units, etc. The greatest value served by these techniques was the predictive capability of probable failure modes at pre-design stages, which enabled management to engineer safer design by elimination of the risk or by controlling it at a tolerable level. Practical use of systems safety has been generally limited to the aviation industry and applications involved with atomic energy utilization. Other large industries plagued with consumer pressure to produce safer products and services — such as rapid transit authorities and automotive manufacturers — have begun to apply these techniques at an ever-increasing pace. Since they are very expensive and time-consuming, they have not gained significant utilization in general industry. A number of factors, including the new Consumer Products Safety Act of 1972, will probably motivate management in general industry to make greater application of these techniques in order to prevent catastrophic losses that could end their business missions overnight.

Since the potential for utilization of these advanced predictive measurements of engineering control in general industry is so significant, a subsequent chapter of this book is devoted entirely to this subject.

GENERAL SUMMARY

Before selecting any measurement that will provide management with useful tools for the control of losses, it is necessary to determine what it is one wants to measure and what purpose the measurement will serve in the achievement of desired goals.

The nature of the interactions of multiple causal factors that lead to various types of loss is of such complexity that no single measurement can possibly fulfill all needs of management.

It has been suggested in this chapter that management's needs can best be served by providing measurements of *consequence*, measurements of *cause*, and measurements of *control*, each providing a different type of information and serving a separate useful purpose. While all three measurements are important, a major portion of this chapter has concentrated on the measurement of management compliance with established standards in the local safety or loss control program. This was referred to as measurement of *control*. The timely measurement and evaluation of work performed to standards provides management (at all levels) with information needed to correct performance and produce desired results (standard compliance). This measurement, more than any other, applies the principles and skills of professional management. With diligent application, we could effect the sure results in loss control which have been achieved in quality control, production and other important areas of management concern.

BIBLIOGRAPHY

1. Allison, W. W., "High Potential Accident Analysis." *Journal of the American Society of Safety Engineers*, Vol. X, No. 7, July, 1965.

2. ANSI Z.16.1 (1967). New York, N.Y.: American National Standards Institute.

3. ANSI Z.16.2 (Rev., 1969). New York, N.Y.: American National Standards Institute.

4. Bird, Frank E., Jr., "Measurement of Management Action and Employee Response." *Insurance Buyers Guide*, October, 1968.

5. Bird, Frank E., Jr., "Measurement Tools for Safety Management." *Canadian Mining Journal*, September, 1971.

6. Bird, Frank E. Jr., and Germain, George L., *Damage Control*. New York, N.Y.: American Management Association, 1966.

7. Bird, Frank E., Jr., and O'Shell, Harold E., "Incident Recall." *National Safety News*, October, 1967.

8. Chapanis, Alphonse, *Research Techniques in Human Engineering*. Baltimore Maryland: The Johns Hopkins Press, 1959.

9. *Federal Register*, April 1, 1971.

10. Fine, William T., "Mathematical Evaluation for Controlling Hazards," *Selected Readings in Safety*. Macon, Georgia: Academy Press, 1973.

11. Humble, John W., *Management by Objectives*. New York, N.Y.: American Management Association, 1973.

12. Johnson, W. G., *MORT, The Management Oversight and Risk Tree*. Prepared for the U.S. Atomic Energy Commission. Washington, D.C.: U.S. Superintendent of Documents.

13. Miller, C. O., "Systems Approach to Accident Investigation," presented at the Flynt Safety Foundation Seminar, Montreux, Switzerland, 1969.

14. Polina, V., "Safety Sampling", *Journal of the American Society of Safety Engineers*, August, 1962.

15. Pope, W. C., and Nicolai, E. R., "In Case of Accident, Call the Computer." U.S. Dept. of the Interior, Personnel Management Publication No. 23.

16. Tarrants, William E., "Applying Measurement Concept to the Appraisal of Safety Performance," *Journal of the American Society of Safety Engineers*, Vol. X, No. 5, May, 1965.

VII. Evaluating Individual Performance

INTRODUCTION

Most progressive organizations recognize supervisory personnel as one of their most valuable assets. Proper supervisory motivation and developmental needs are usually considered a prime concern of upper management. Likewise, management has long recognized the need to compensate this valuable asset according to individual contribution to the success of the enterprise and on a basis consistent with individual work performance.

In recent years, the critical need to determine the quality of manpower resources and the vital importance of a program of planning to improve performance level have achieved organizational recognition.

The annual (or periodic) "performance review program" has proven one of the best tools for stimulating "total performance of the total *individual*" in the interest of "total performance of the total *system*".

For the individual supervisor, this tool is recognized as the measuring stick of his own contribution to the organization and the catalyst to improved personal income. It is very important to the individual that he be provided with the information he needs to know specifically "where he stands" and what his future management growth prospects really are.

Historically, the individual's safety and loss control performance has not been a considered factor in the performance reviews of most companies. Without doubt, this has not resulted from any feeling that loss control performance was not important. As difficult as it might be for the reader to accept, in scanning a book with much of its content related to measurement, managers have long raised the question, "How do you get a handle on safety or loss control performance?" Performance is a continuing pattern of significant job-related activities or actions that can be readily observed and objectively assessed, and ways and means to measure performance related to these subjects have baffled management through the years.

Regardless of how we feel about management attitude in this regard, the individual supervisor's natural interpretation has been that this attitude simply emphasizes that we *preach* safety and loss control but, "we sure don't put our money and action where our mouth is".

There are many reasons, then, for suggesting that the area of safety or loss control become an integral part of every supervisor's performance review. None, however, is more important than the identification of this subject in the mind of the supervisor as one deserving proper perspective in relation to other important subjects like quality and production.

AN OPPORTUNITY

Many of us tend to think that every organization is as progressive as the one we are (or have been) a part of. In effect, the "average" usually proves to be far behind the leaders in *all* aspects of management activity, including the performance review program area. It is, therefore, not inconceivable that many readers or their organizations will not have established performance review programs. Here again is an opportunity for the safety or loss control manager to demonstrate "the principle of reciprocated interest" and motivate the adoption of this program that can make a major contribution to management's goals in production and quality, while simultaneously advancing the cause of safety or loss control.

CURRENT AND PAST PRACTICES

Historical records of work injury scores for individual supervisors date back to 1913. Anyone who has had close association with occupational safety programs is fully aware that keeping records of injuries by organization, division, department and work group units is as common a practice today as it was 50 to 75 years ago.

Large charts with the names of supervisors and the injury record of their work groups will frequently hang in management conference rooms. A modification of this type of chart is sometimes displayed openly in large work areas, so that everyone can "associate" with a supervisor and work group injury score. These charts will sometimes have color codes to distinguish minor injuries from disabling injuries.

Since safety performance has generally been measured in terms of injury records for so many years by so many important organizations, it was probably only logical to think of individual safety responsibility and accountability in terms of injury record results. The wide use of the injury record association with names of supervisors would also seem to indicate that management has been aware of the need to measure and evaluate performance of the individual as well as the group.

The inequities involved in using injury records in work performance evaluations of supervisors must have been recognized early. It is very difficult to find organizations that have actually applied this measurement criterion and accompanying performance review programs in a meaningful way. In conversation, the authors have heard of companies and organizations that include safety or loss control in their performance review programs. Questioning has consistently revealed that performance is based on disabling injury records. It is very significant that our records, files, and library are devoid of one single previous article or sample form where this important subject has been considered. This does not mean that the practice is non-existent; it does imply that pride and confidence in this facet of the overall safety or loss control program is apparently not what it ought to be, nor even a very important consideration.

THE HYPOTHESIS AND ITS GOALS

The application of measurement and evaluation in the performance review system is obviously only an extension of the management concepts that have already been discussed in earlier chapters. We are now merely extending that thinking by saying:

> Safety/loss control performance can be measured and evaluated for the organization, division, department, unit, and for the individual as well.

Our goals in adding this important area of management concern to the performance review system could be summarized to include, but certainly not be limited to, the following:

- To upgrade the importance of safety/loss control by making it an important part of the overall performance review.

- To strengthen the safety/loss control management system by upgrading each component at this point of control.

- To objectively measure and evaluate the individual's efforts, in order to correct substandard performance.

- To establish a system of management development in safety/loss control that includes organized guidance of upper management.

- To provide a systematic way for management to manage the achievement of the safety/loss control work.

THE BASIS FOR BELIEF

The hypothesis that safety or loss control performance can be measured and evaluated for the individual supervisor is based on five major theorems that have been discussed in one fashion or another in several portions of this book. Without further detailed explanation, let's mention them.

Theorem 1. Accidents and other losses are largely the result of inadequate management performance in safety or loss control work.

Theorem 2. The level of management's safety or loss control performance depends on the composite effort of the individuals on the management team.

Theorem 3. When practical standards for safety or loss control are clearly defined and understood, the supervisor should be expected to comply with them.

Theorem 4. A manager can only correct the safety or loss control performance of a subordinate supervisor when he has measured and evaluated that supervisor's conformance to the work standards required of him.

Theorem 5. The safety/loss control work required of a supervisor can be clearly identified and measured.

With this background and basis for the strong belief that the performance review program can achieve the values and results outlined, let's proceed to the actual doing of the job.

IMPLEMENTING A PERFORMANCE REVIEW PROGRAM

Orientation

The first important step in any program of this type should be to make the supervisor or manager who is to be reviewed aware of what the program is and what is expected of him. In this regard, he should be called in by the reviewing supervisor or manager for a thorough explanation of how his performance will be measured and evaluated.

Since the addition of safety or loss control to a performance review program would probably only be considered after the overall program had been well organized, it is assumed that the programs, organization, procedures and standards have been accomplished and are well defined in written form. Of course, it would be good to request that the supervisor bring to the review session his copy of the company manual or any other previously-furnished materials on management standards of the program. This will assure the reviewing manager that the reference guideposts for measurement have been given to the supervisor and are available for his periodic reference. In addition to making sure that the organization's standards for each work area are well understood, the supervisor must also understand how *his work* in each area will be measured. (Since the subject of measurement is also covered in another chapter, details will not be presented here.) Before the discussion ends, the supervisor can be given a self-appraisal form, to measure and evaluate his own performance. This self-appraisal form (See Figure 1.) would include the identical work areas that would be evaluated by the reviewing manager at performance review time. A request should be made to have the form returned by a specific mutually-agreeable date about 90 days away. This provides ample time for the supervisor to take whatever actions are necessary to improve his own performance so that his self-appraisal can be a good one. It must be kept in mind that the purpose of the performance review program is to give recognition for proper performance and to develop management needs and skills that will help the supervisor improve areas of deficiency, and that management development in safety/loss control is the primary purpose of the whole program. Additional details of the self-appraisal part of the program will follow. Finally, the supervisor should be informed of the importance of keeping all records required by the loss control program standards in good order. He should be told by his manager that all required forms and records will be reviewed as part of the actual measurement exercise. The record review itself can take place at the supervisor's desk in his own work locale, or certain forms can be requested by the manager prior to the actual performance review. While this exercise may seem childish to a few, the majority will view it as the proper way to do the required job. After all, if the performance review helps determine important future events for the supervisor involved, he should want to have the facts known and properly evaluated. This record review will be mentioned again several times. It need not be personally completed by the reviewing manager, but could be accomplished by a management trainee or a safety/loss control specialist under the reviewing manager's direction. During the pre-review explanation, it is important for the reviewing manager to determine whether the supervisor has any needs that require help from him (or from other sources, through him) at this time. Just as with his quality of production effort, we can't expect top performance if the supervisor doesn't have the knowledge, motivation, and tools to get the job done. The orientation discussion may establish that some additional training is necessary in communications, or that there is insufficient lighting in the area where meetings are held.

The big message that comes across to the supervisor in this session will hopefully be that loss control . . . and everything related to his loss control performance . . . will serve to assist him in his needs, help him do a better job, and make his review as good as possible.

"BEAVERS BUILD HOUSES; BUT THEY BUILD THEM IN NO WISE DIFFERENTLY OR BETTER NOW THAN THEY DID FIVE THOUSAND YEARS AGO. ANTS AND HONEYBEES PROVIDE FOOD FOR WINTER; BUT JUST IN THE SAME WAY THEY DID WHEN SOLOMON REFERRED TO THEM AS PATTERNS FOR PRUDENCE. MAN IS NOT THE ONLY ANIMAL WHO LABORS, BUT HE IS THE ONLY ONE WHO IMPROVES HIS WORKMANSHIP."

— Abraham Lincoln

The Self-Appraisal

The self-appraisal system itself has a number of benefits that are briefly outlined below. The form itself can be relatively simple, to provide a powerful motivating influence for the supervisor (See Figure 1.). It should clearly and concisely present the standards by which he will be measured and evaluated in his performance review, and it should seek his honest self-evaluation. The benefits include:

SELF APPRAISAL OF LOSS CONTROL PERFORMANCE

NAME _____ DATE _____

DEPT. _____ REVIEWER _____

Calculate your level of performance for each of the work areas and related standards listed below. Utilize personal records available and base calculations of the measurement system that will be utilized in your annual Job Performance Review.

1. **Rules:** Thorough initial review for each employee; annual review for everyone, with testing to assure knowledge retention. 100% compliance motivated.
   My performance is _____

2. **Personal Communications:** (Job indoctrination program properly presented and records maintained on each new or transferred employee.) Proper job instruction given with record maintained on every new or different job.
   My performance is _____

3. **Inspection Program:** Thorough inspection of areas recorded bimonthly, with effective follow-up program. Immediate temporary action on "Class A" (potentially serious) conditions. All individually reported conditions recorded on condition report form.
   My performance is _____

4. **Proper Job Analyses:** Critical jobs analyzed as assigned. All existing PJA's reviewed annually. Related procedures reviewed annually with employees.
   My performance is _____

5. **Proper Job Observation:** Total job observation recorded annually on all critical jobs. Routine job observations recorded monthly. Follow up activities recorded.
   My performance is _____

6. **Group Meetings:** Evidence maintained of planned weekly group meetings, utilizing effective visual aids.
   My performance is _____

7. **Personal Equipment:** All required personal equipment issued, its use properly explained, and 100% proper use maintained. Effective program maintained to control wasteful use.
   My performance is _____

8. **Investigation:** All undesired events that could (or do) result in loss properly investigated, with effective follow up of recommended action.
   My performance is _____

9. **Behavior Reinforcement:** Reinforcement objectives are reviewed bi-annually for each employee with results of positive recognition recorded.
   My performance is _____

10. **Personal Attitude:** Personal example of proper behavior and enthusiasm toward proper job performance of all employees exhibited at all times.
    My performance is _____

Figure 1.

A. *Reinforcement of the organization's safety/loss control standards.*

It is surprising how often a well-established, well-documented program will atrophy and fall into disuse because books are not opened, reviews are not held, reinforcement is not given. This system provides a unique way to provide strong reinforcement of the standards that serve as the "heart" of any program.

B. *The reminder that performance efforts matter.*

There is no factor involved in inter-managerial relationships that encourages *substandard* work more than the feeling that *performance doesn't matter*. Management authorities have consistently discovered that in order to keep good people from growing frustrated, it is necessary to recognize their developing talents and capabilities, in order to put them to the best use. When people get the impression that their work performance in any area doesn't really matter, the quality of their efforts tends to slump visibly.

C. *Creation of self-awareness of substandard performance.*

It is wrong to assume that a supervisor is fully and accurately aware of the degree of his compliance with program standards. It has been said that substandard performance in the absence of a management review program is, to a large extent, a subconscious incompetency in the individual; the first objective of a good manager is to properly create a consciousness of that incompetency. This is what happens with the self-appraisal. Psychologists have long emphasized the validity of the biblical quotation, Proverbs 23:7 . . . "for as he thinketh in his heart, so is he . . ." Psychologist David Guy Powers, a well-known safety conference lecturer and speaker, has said, "What the mind attends to, it considers; what the mind does not attend to, it dismisses; what the mind attends to regularly, it believes; and what the mind believes, it eventually does."

The great dynamic of self-appraisal is that it creates a conscious awareness of competency or incompetency, so that the reviewer will know what the present "thinking" of a supervisor is regarding his performance. If, for example, the reviewer discovers that the supervisor regards his investigation as nearly 100%, he can logically assume that the supervisor is doing very little to improve it. On the other hand, if the reviewer knows the supervisor is aware of his sub-standard performance and its degree, he can at least be more confident that he is equipped mentally and psychologically to cope with it; in effect, the supervisor will now guide himself *consciously* to a state of *competency* in safety/loss control work. This is not to imply that the reviewer can assume all of this will be done most efficiently by self-motivation. He must give his support and resources, to whatever extent they are needed. Self-awareness of one's degree of performance is, therefore, one of the most important benefits of the self-appraisal.

D. *Provision of before-the-loss control information.*

If we are not communicating the loss control message effectively, not educating employees on rules and proper procedures, not investigating losses or inspecting to see that substandard conditions are detected, etc., we should assume that barriers to potential losses have not been established. Conversely, we can have assurance that loss control has been optimized to the degree that compliance with established standards has been optimized. The self-appraisal gives us insight (although biased) into the existing level of compliance. In effect, we have, therefore, before-the-loss information that can assist us to take additional action or give additional attention before loss occurs.

E. *Timely self-improvement.*

As mentioned earlier, the self-appraisal permits the supervisor to improve his performance immediately and to maintain the improvement during the interim period between the initial orientation to the program and the performance review. This is a major reason and goal for the self-appraisal system.

F. *Easier understanding of the system.*

   When the steps suggested for conducting this segment of the total job performance review are carried out, the supervisor certainly cannot say, "No one ever told me," or "I didn't know what was expected of me."

G. *Teaching of supervisory management skills.*

   All the activities (standards, measurement, evaluation, correcting performance) involved with the "control" function of professional management (planning, organizing, leading, *controlling*) are reviewed and reinforced in this program. Their application will also tend to reinforce in the supervisor's mind the fact that all these principles are applicable to all *other* important aspects of his job (production, quality, cost control, etc.).

H. *Increased management participation.*

   The self-appraisal system involves several levels of management and various groups of staff personnel. This involvement has mutual benefits for everyone, concerning education and motivation related to loss control.

Self-Appraisal Follow-Up

It would seem to be a fair assumption that the reviewer should have a reasonably good idea of the objectiveness and validity of the supervisor's personal performance evaluation. This assumption is based on the fact that, unless program standards are established with a system to measure supervisory compliance, any entry into a performance review program for safety/loss control would seem unprofessional and premature. It is imperative that the reviewer act promptly and respond to the supervisor by reinforcement of those areas believed to accurately reflect good performance and by reviewing the need for assistance where it is deemed necessary to upgrade substandard performance. Prompt, effective follow-up will serve as a powerful motivational tool in gaining desired performance during the period that bridges the orientation and the actual performance review.

Records Review

Most programs designed to make significant contributions to management's goals and objectives require documentation. Major work areas in the safety/loss control program that should have well-documented records to provide evidence of new employee indoctrination and standard compliance would include Job Orientation, Group Meetings, Proper Job Instruction, Engineering Designs, Job Analyses and Procedures, Job Observations, Investigations, Inspections, Skill Training, Rule Reviews and Personal Contacts.

It is deemed significant to mention again that supervisors will usually pay attention to those things they feel serve a useful purpose in the management system. Records that are deemed significant by the supervisor will reflect, in their quantity and quality, the value he places on them. Once the performance review program is established and the orientation of all supervisors has been carried out properly, significant positive changes should occur in all related forms.

From these forms, the reviewer can determine with a high degree of accuracy the quantity and timeliness of work performed, as well as its quality. The chapter on Measurement Tools for Management describes the use of this information in a measuring system.

Dependent on the degree of sophistication of the organization's loss control program, an ongoing score for each supervisor could be maintained as a safety function of the loss control department. One of the major goals of the loss control specialist should be to switch his emphasis from being a "doer" of loss control to being a manager of the system, so that supervisors and management people gain the effective tools to be good "doers". Certain hospital management systems have already established a pattern of sophistication in individual performance measurement of a physician's patient care, pointing the way for measurement of individual performance in other industries. Information from the patient's record is computerized, by the illness or condition and by individual physician. Performance is based on the degree of compliance of the *actual* care to the *standards* of care established by the local medical group.

While computerization of data may not be practical for the average loss control or safety specialist, many practical systems can be devised to channel information for measurement and evaluation on an ongoing basis.

It may be more practical to "sample" each area of work included in the program by a staff visit to the local area where the supervisor's records are maintained as organized for the program. Statistical experts within the organization can offer advice on the number of items and "samples" in each area of measurement.

It is this type of work organization that the average loss control specialist rebels against, since it "takes too much time" and "there aren't enough hands" now to stretch far enough into such an approach. However, exactly this type of work *must* be devised to effect positive control and optimize assurance of safety . . . and control of losses that could spell profit-or-loss, or perhaps the actual demise, of a business.

Of course, even a system dependent upon the subjective opinion of the reviewer in assessing a score to each area being measured would be much better than nothing. In fairness to the latter suggestion, it should be pointed out that most performance appraisal programs are based on subjective determinations, predicated on professional knowledge gained from day-to-day observation and review of the supervisor's work.

"EVERY PLAYER IS GIVEN A PERFORMANCE GRADE ON EVERY PLAY IN EVERY GAME. THAT GIVES US THE TRUE STORY OF A MAN'S PERFORMANCE."

> Don Shula
> Coach,
> Miami Dolphins

THE PERFORMANCE REVIEW

Evaluation of Past Performance

This is the evaluation of the supervisor's performance in each work area of the program. The supervisor has already evaluated himself and now the reviewer adds his own performance rating in each of these areas. As previously discussed, this evaluation can be accomplished by staff personnel or by the reviewer; it can be almost completely objective, depending upon the level of program growth; it can be subjective, yet definitive in subject content. It can include an evaluation for each activity of the program involving the supervisor's work input.

Scoring Alternatives

The use of percentage figures can actually compare the level of compliance by the supervisor to the standard desired (100%). There is a big difference between 70 and 79, or 80 and 89 when using objective information; therefore, this type of scoring is possible with a high confidence level. The subject of safety and loss control management is becoming important enough to justify a level of program sophistication in many companies that would make this type of scoring highly desirable. On the other hand, if a subjective evaluation system is utilized, a scoring method such as that indicated below could be used, and would prove of great value, compared to no performance review program.

- A. Performance is consistently exceptional (and at a level rarely achieved by others).
- B. Performance is consistently superior.
- C. Performance is very satisfactory.
- D. Performance is for the most part satisfactory.
- E. Performance is predominantly unsatisfactory.

Evaluation of the supervisor's past performance should be completed before the reviewer meets with him. Most of the time involved in the first meeting of the review should be allocated to a discussion of the performance level of each work area, giving reinforcement for acceptable levels of performance, and seeking information or reasons for substandard performance where indicated. This entire discussion should be positive, in order to gain the greatest amount of cooperation and lead to meaningful results.

Evaluation of Personal Development Accomplished

Assuming this is not the first performance review, the next phase of the discussion centers on an evaluation of the degree of completeness of objectives established together at the last review. The reviewer should keep in mind that he should have had a major role in the attainment of such objectives, and failure to attain

SUPERVISOR'S LOSS CONTROL PERFORMANCE REVIEW

Paul T. Jackwood NAME	Conditioning DEPT.	12-10-73 REVIEW DATE	4th quar. REVIEW PERIOD
Conditioning Supervisor POSITION TITLE	18 mos. HOW LONG	8-7-65 EMPLOYMENT DATE	12-6-41 BIRTH DATE
Frank K. Roberts REVIEWER	Frank K. Roberts DEPARTMENT HEAD	William M. Wood DIVISION HEAD	

1. **FOR ORIENTATION ONLY:** Entire program, including all related performance standards, measurements to be utilized, records review, self appraisal forms, and final performance, explained on 12-8-72 by Frank K. Roberts

2. **SELF APPRAISAL REVIEW:** Date 4-5-73 Score 72%.

3. **EVALUATION OF PAST PERFORMANCE:** The number shown at right for each classification of loss control work represents the actual percent efficiency achieved by the individual, measured against the applicable standards established or expected.

WORK ACTIVITIES AND PERFORMANCE RATINGS:

1.	Rules	70%	6.	Group Meetings	65%
2.	Personal Communications	60%	7.	Personal Equipment	66%
3.	Inspection Program	95%	8.	Investigation	93%
4.	Proper Job Analyses	86%	9.	Behavior Reinforcement	64%
5.	Proper Job Observation	92%	10.	Personal Attitude	75%

4. **EVALUATION OF PERSONAL DEVELOPMENT ACCOMPLISHED:** Summarize how well the development objectives you and this employee agreed upon at last review have been carried out. Express rating in terms of total percentage of achievement.

Three of five objectives were completed during the past year. There has been partial completion of objective #2 and #4 since Paul did complete the home study course in communications and has visited four group meetings and three P.J.I. sessions of fellow supervisors during the year. Two important objectives 2a and 4b were not achieved and probably could be considered the most important related to Paul's basic problems. In the light of the value on these two incomplete objectives it would be fair to state that the objectives established were not more than 67% completed.

5. **NEED ANALYSIS:** Summarize the principal developmental needs of this supervisor that would improve his loss control performance. Give priority in your consideration to the most significant areas that you feel can be corrected.

There is no doubt that the complexity of Paul's communication problem requires counselling from Charles Bowen in Personnel for both Paul and myself. I believe that failure to complete 1973 objectives were in part my own lack of in depth knowledge of the real problem. Unless determined otherwise participation by Paul in a live communication training program will be an important objective in 1974. I plan to establish a personal objective of one additional review follow-up session with each supervisor in order to tighten our departments effort to control performance in 1974. Objectives may be added or modified following our counselling sessions with Chas. Bowen.

Common problems involved with sub-standard performance: Inadequate knowledge of specific technical subjects . . . Inadequate knowledge of motivation . . . Inadequate knowledge of problem-solving techniques . . . Inadequate knowledge of company/department/unit loss control work standards . . . Not properly motivated . . . Help needed in personal qualities (attitude, behavior, etc.)

6. **DEVELOPMENTAL OBJECTIVES:** List the specific objectives (in order of priority which you and this employee have discussed and agreed upon) in order to fulfill his developmental needs.

1. arrangements will be made with Charles Bowen - Personnel - for counselling with Paul Jackwood on or before Feb 28, 1974. Follow-up sessions determined necessary by Chas. will be part of the objective.
2. Frank K Roberts, supt. - will meet with C. Bowen on or before 3-7-74 to receive his evaluation and discuss best ways and means to follow up program suggested.
3. Paul will complete at least 1 participative communication course on or before 9-1-74 per advice of C. Bowen.
4. Frank K. Roberts will conduct review follow-up meeting with each supervisor whose overall performance is below 90%. with 1 less review for those meeting or achieving the level, on a quarterly basis.

Areas to consider in establishing objectives: Self-development through reading, self study, etc. . . . Formal training courses within the company/organization . . . Outside education courses . . . Counseling or coaching by peers . . . Counseling or coaching by upper management . . . Counseling or coaching by safety/personnel/training or other staff specialists . . . Team-up with strong supervisory peer.

Date of most recent review and objective-setting discussion 1-17-74 By whom Frank K. Roberts

Tentative schedule for future discussions 2-28-74; 3-7-74; 3-15-74; 6-13-74; 9-1-74; 9-16-74; 12-10-74

Note: These sample forms may appear somewhat different from those used in your company's program. They are included as a reference and are only intended to be representative of those generally used.

Figure 2.

them could reflect on the reviewer's own lack of follow-up. The importance placed on the accomplishments related here will set the direction for the next year's program of objectives. If this is the beginning of the performance review program, this phase of the review is not applicable.

Analysis of Needs

Following the discussion of past performance, and based on the interactions the reviewer has had with the supervisor during the year, he should be in a fair position to evaluate what needs may exist that prevent or impair the supervisor's ability to perform to standards. Common problems involved with substandard performance could include inadequate knowledge of specific technical subjects, such as "proper job analysis" or "job observation"; inadequate motivation; inadequate knowledge of problem solving techniques; inadequate knowledge of organization/division/department/unit loss control work standards; or a need for help with personal qualities (attitude, behavior, etc.). Once the reviewer feels that major needs have been identified, he is in a position to assist the supervisor in developing objectives they both recognize as necessary to achieve the desired level of performance.

Developmental Objectives

Specific objectives may be agreed upon during one performance review meeting, but it could be necessary for both the reviewer and the supervisor to do some thinking or investigating before finalizing objectives.

Each objective should be written as a specific statement of what should be accomplished, expressed in such a way as to clearly indicate primary emphasis; identify meaningful, quantifiable indices for measurement; and quantify a valid, precise target or goal.

Areas to consider in establishing loss control performance objectives could include self improvement through reading, self study, etc.; formal training courses within the organization or from outside educational sources; counseling or coaching from peers; counseling or coaching by upper management; counseling or coaching by loss control specialists, personnel/training, or other staff people; or a team experience, with a strong supervisory peer. Once completed, the objectives provide the avenue to improvement, but the level of development growth for the supervisor will greatly depend on the importance placed on the achievement of these important performance targets.

SUMMARY

The values of the performance review for a supervisor are almost identical with those of the self-appraisal that served as a preparatory step. The major difference is that there is now no speculation about the meaning and worth of loss control. He is not only well aware of his deficiencies in performance, but has been motivated to improve them. His pride in those areas for which he received recognition will quickly spread as he is helped to improve in other areas. With his increased knowledge of the loss control program and his relationship to it, his performance growth will also reflect itself in results for his department, division, and (ultimately) the organization.

Review of the Choices

Of those choices listed below, one should now stand out as the most logical route to take:

- Evaluate performance in terms of accident records.
- Evaluate performance by measuring personality.
- Evaluate performance by subjective judgment criteria.
- Evaluate performance by measuring objective compliance with program standards.
- Let the program manage itself.

It is hoped that this chapter will have motivated the reader to take a giant step forward in loss control, by having its precepts included in your supervisor's regular performance review program, and that the reader will have been given some direction in the selection of criteria and methodology to use in the actual performance measurement and evaluation system.

BIBLIOGRAPHY

Allen, Louis A., *The Management Profession.* New York: McGraw-Hill Book Company, 1974.

Bird, Frank E., Jr., "Measurement of Management Action and Employee Response", *Insurance Buyers Guide*, August, 1968.

Bird, Frank E., Jr., "Measurement Tools for Safety Management," *Canadian Mining Journal*, September, 1971.

Drucker, Peter F., *The Practice of Management.* New York: Harper & Row, 1954.

Tarrants, William E., "A Definition of the Safety Measurement Problem," *Journal of Safety Research*, September, 1970, Vol. 2, No. 3.

Tarrants, William E., "Applying Measurement Concept to the Appraisal of Safety Performance," *Journal of the American Society of Safety Engineers*, Vol. X, No. 5, May, 1965.

Incident Recall Techniques

VIII.

INTRODUCTION

Eight people were killed when a rescue helicopter's twin rotor blades separated from the plane's body, causing the craft to hit the ground in a shower of fire and smoke. There were no survivors. Testimony following the accident indicated that oil had sprayed all over the roof of this plane's cabin the day before, during the pre-flight cockpit check. Unusual vibration was also evidenced during the first few seconds of start-up. Neither of these incidents had been reported.

A combustible materials fire was detected within a building containing several tons of high explosives. Fortunately, all employees were evacuated before the explosion occurred that completely demolished the entire structure. Investigation revealed that three employees had observed small flash fires on different occasions in the same general area; they did not report them because these fires immediately extinguished themselves and did not seem to be important or dangerous at the time.

The driver of a truck was killed in a crash when tons of fish entrails dropped from another trailer truck, slicking the highway for almost a mile; a massive pile-up blocked the entire area, causing thousands of dollars worth of property damage and backing up traffic for miles. Investigation revealed that the same piece of equipment had dropped an identical load in the same general location a few weeks before, but cleanup was instituted before any major loss occurred.

An industrial worker was struck and fatally injured when a cable broke, causing a 2,500-pound block-and-hook assembly to drop from an overhead crane. Investigation following the accident revealed that several persons had witnessed crane operators deliberately misusing the safety limit switches on this crane's hoist as a brake when raising the block-and-hook assembly, rather than manually activating the regular hoist brake.

These four accidents occurred hundreds of miles apart, but they had one factor in common: one (or more) less serious or "near-loss" incidents had occurred prior to each serious loss-producing accident. Unfortunately, the majority of people who witness such incidents as these do not report them to someone who could utilize the information in time to save lives or avert other serious loss.

This chapter will discuss practical methods for motivating people to recall and to report near-miss accidents to a trained interviewer, who will on most occasions be the recaller's immediate supervisor. Techniques will also be described for encouraging people to voluntarily report near-miss accidents when they occur. These techniques will assist the supervisor in handling these reports in a manner most effectively utilizing the information gathered as well as reinforcing the reporting system itself.

INCIDENT AND ACCIDENT DEFINITIONS

In keeping with the overall loss control objectives of this book, the term "incident", as defined in an earlier chapter entitled "The Causes and Effects of Downgrading Incidents", was broad enough to include all undesired events that could result in downgrading the efficiency of the business operation. The term "incident", as associated with incident recall techniques, will refer to a particular category of this larger family of undesired events and, for the purpose of this chapter's only, will be defined as "an undesired event that under slightly different circumstances could have resulted in physical harm or property damage."

The term "accident", as defined earlier in this book, refers to "an undesired event that results in physical harm to a person or damage to property. It is usually the result of a contact with a source of energy above the threshold limit of the body or structure."

With these two functional definitions in mind, it is not difficult to understand why the "incident" has been appropriately referred to as the "near-miss accident", "near-accident" or "near-loss incident." It is significant to point out that many supervisors inappropriately refer to accidents that result in property damage with no physical harm as "near-miss accidents". This improper reference is, of course, a holdover from earlier days when the term "accident" and "injury" were generally considered to be synonymous.

WHY INCIDENT RECALL SHOULD BE UTILIZED

While the techniques for obtaining incident information will vary because of circumstances, the values to be obtained for the professional are essentially the same.

Important reasons for the use of incident recall include:

1. It is a predictive or "before-the-fact" tool.

2. It has been tested and proved successful.

3. It can be an efficient method for reporting undesired happenings.

4. Its application encourages the development of pride of performance.

5. It can strengthen important supervisory skills.

Since an understanding of these reasons would be quite helpful to the supervisor as he applies the techniques described later, each will be discussed in some detail.

1. *It is a predictive and before-the-fact tool.*
The examples of actual incidents mentioned in the introduction to this chapter give some indication of the predictive or before-the-fact value of incident information. Had the early indications been properly heeded, the incidents would have been reported. Their predictive values could have been utilized to correct the problems that recurred and resulted in serious losses. The study which pointed up the 1-10-30-600 accident ratio discussed earlier established that 17 times more incidents were occurring than all accidents reported involving personal injury and property damage. As we look at the number (repeated in Figure 1), we can assume that the 600 is silently saying what has been said thousands of times in different ways by respected loss control leaders, "Get at the greater number of near-misses, the no-injury/damage accidents, and you will stand a much better chance of reducing the relatively few rare events that terminate in serious losses."

Other studies have indicated that 75% of all accidents resulting in personal injury or property damage are preceded by similar near-miss accidents that could serve as advance warnings of loss to come.

One of the authors of this text was associated with industrial engineering studies of materials handling accidents that involved continuous observations of four judgment level employees on a round-the-clock basis for two 30-day periods. During the first observation period, 3,240 individual handling operations were observed with the occurrence of three

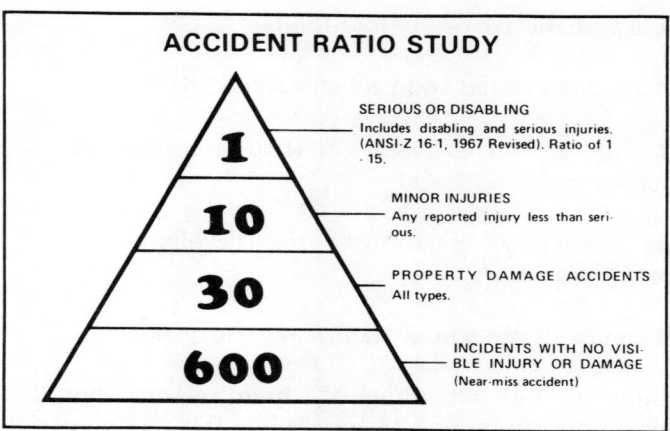

Figure 1.

property damage accidents and 31 near-miss damage events. The second observation period included 4,856 operations during which seven cases of damage were observed to occur while 143 incidents occurred that these specialists believed could have resulted in property damage or injury under slightly different circumstances. The knowledge gained from the total of 184 events enabled the study group to make recommendation that eventually reduced losses appreciably. Previous analysis of accidents in damage above did not reveal adequate information upon which the engineering group felt they would act.

The fact that a vast amount of reliable information is available for a supervisor on loss-producing problems, before-the-fact, is perhaps the most exciting and interesting aspect of incident recall.[1]

2. *It has been tested and proved successful.*
One of the incident recall techniques that occupies a major part of the following discussion is a modification of the critical incident technique, which grew out of studies in the aviation psychology program of the United States Air Force. One of the early studies using the critical incident technique surveyed psychological and man-machine systems problems involving the use and operation of aircraft equipment. Investigators asked a large number of pilots whether they had ever made, or seen anyone else make an error in reading or interpreting an aircraft instrument, detecting a signal, or understanding instructions. A total of 270 "pilot error"

incidents were collected during the study, and many reports of similarity were found. The summary classification of these errors suggested some logical remedies. For example, the fact that so many pilots reported serious errors in reading multi-revolution indicators suggested that perhaps the instrument was too hard to read. This study led to a number of important design and procedural changes that are believed to have prevented many aircraft accidents that could otherwise have taken place.[4]

A 1954 study by McFarland and Moseley reported how this technique had been used to determine the types, frequency, and conditions pertaining to near accidents in long-haul trucking operations. This application involved the collection of data by a trained observer and former truck driver. Observations of 17 drivers on 20 trips totalling approximately 5,000 miles recorded 48 incidents, approximately one for each 100 miles. Under slightly different circumstances, such incidents could have resulted in an accident. The most significant aspect was that this study revealed much more data related to potential accidents than could have been obtained by waiting for serious or major accidents to occur.[5]

The critical incident technique, which provided valuable reference information for the techniques described in this chapter, was also tested several times in industry by accident researchers. The Division of Accident Research of the Bureau of Labor Statistics in Washington, D.C. conducted one of the more recent studies at an industrial site, to evaluate the usefulness of this technique as a method for identifying potential accident causes and developing procedures for its practical application.

The results of the study reported by this reliable government research group are:

A. The critical incident technique dependably reveals causal factors in terms of errors and unsafe conditions which lead to industrial accidents.

B. The technique is able to identify causal factors associated with both injurious and non-injurious accidents.

C. The technique reveals a greater amount of information about accident causes than presently available methods of accident study, and provides a more sensitive measure of total accident performance.

D. The causes of non-injurious accidents, as identified by the critical incident technique, can be used to identify the sources of potentially-injurious accidents.

E. Use of the critical incident technique to identify accident causes is feasible.[6]

In addition to these major studies involving the critical incident technique, an ever-increasing number of organizations in the United States and Canada are incorporating techniques in their loss control programs for the utilization of incident information. The United States Department of the Interior now has an ongoing system that includes incident and accident information input by supervisors into its computer analysis program.

The tremendous risks inherent in the ever-increasing application of the technological advances of the past two decades are placing greater demands than ever before on the safety and loss control professional for predictive rather than reactive actions. Catastrophic risks that were restricted to a few select industries, chiefly those associated with aviation and atomic energy development, are now commonplace in general industry.

There seems to be little doubt that there will be an ever-increasing utilization of incident information in loss control programs, as the logic and results of its application become known.

3. *It can be an efficient method for reporting undesired happenings.*

Psychological studies have shown that people are much more willing to talk about "close calls" than about personal injury or property damage accidents in which they were involved. The psychologist interprets this as an implication that the individual feels his special skill or ability allowed him to avoid injury or property damage and, since no loss resulted, no blame for the incident should be forthcoming.

Numerous confidential attitude samplings of industrial employees by consulting organizations have revealed that it is common for employees to hide accidents that caused injury or property damage. The reasons most commonly given in these confidential studies and interviews for not reporting accidents are:

- Fear of discipline.
- Concern about the record.
- Concern about reputation.
- Fear of medical treatment.
- Dislike of medical personnel.
- Desire to prevent work interruption.
- Desire to keep personal record clear.
- Avoidance of red tape.
- Concern about attitude of others.
- Poor understanding of importance.

    Most of these barriers to efficient reporting of events terminating in loss are eliminated by the application of incident recall. Since employees need have few (if any) of these fears or anxieties when reporting incidents, it takes very little extra effort on the supervisor's part to gain considerably more accident causal information through incident recall techniques than would normally be obtained on all types of accidents resulting in loss.

    Another important aspect involved with the efficiency of incident reporting relates to the smaller organization. Since accidents are rare events, relatively speaking, small organizations are frequently misled about the seriousness of their problems. They also have great difficulty collecting a sufficient amount of causal data to make any meaningful analysis of their problems. Since "incidents" (near-miss accidents) occur at a much more frequent rate than injury/damage type accidents, a small organization can collect a sizable amount of reliable causal information in a relatively short period of time. For these reasons, management personnel in the smaller organization can also be made more aware of the true potential for loss through incident reporting.

4. *Its application encourages the development of pride of performance.*

   To obtain effective incident recall through any of the techniques to be described later in this chapter, the interviewer must create within the individual employee the feeling that he is making a significant contribution to the loss control program. Each "recall" by the worker of an incident that could have resulted in serious physical harm or property damage is a most significant opportunity to apply the behavior reinforcement techniques that encourage future corporation. Not only will this reinforcement of his reporting result in the maintenance of a high level of recall efficiency, but it will also stimulate his desire to contribute to the prevention and control of the incidents themselves. The wise supervisor recognizes how much the individual can contribute to loss control in his area and will capitalize on every opportunity to utilize this valuable knowledge. The worker's knowledge regarding important problem solutions related to incidents reported by him becomes more readily available as he comes to realize the importance his supervisor or other management leaders place on it. There are few areas in the entire loss control program where greater application of behavior reinforcement can be made than in the application of incident recall techniques.

5. *It can strengthen important supervisory skills.*

   With all the technological developments that have produced the marvelous products of our space age, management continues to rely on the actions of people as the hub around which the major spokes of an organization's operations revolve.

   Since each person is a unique personality, different things motivate and cause him to strive for attainment of his goals in an individual way. Some of the major goals described earlier in this book that motivate people are mental security, belongingness, affection, acceptance, accomplishment, self respect, recognition, status, and self-expression.

   The fulfillment of these goals, through personal motivation by the supervisor, is important because it is frequently the conversion point between what the employee is *capable* of doing and what he *actually does*.

   Whether the objectives involve incident recall, product quality or production, the success or failure of the program is frequently determined by the degree of motivation the supervisor instills in his people. The presence of motivation within the individual makes all his tasks and assignments more pleasant, meaningful and satisfying.

To get people to perform their total jobs at a maximum level of efficiency, the supervisor must continually strive to cultivate and develop those avenues of motivation that affect their behavior.

There is conclusive evidence that good incident recall can be obtained consistently if proper interviewer/employee rapport is established and maintained. The exercise of recall requires that the interviewer follow a pattern of relations that strengthens the employee's feeling of mental security, makes him feel he is contributing, demonstrates personal interest in him, recognizes his contribution, and provides an opportunity for his self-expression.

Each successful application of incident recall reinforces the interviewer's knowledge and ability to gain the cooperation of people he is working with or supervises. He quickly recognizes that improved employee response to many other areas of the business operation can be gained by application of the same good human relations he has applied with incident recall.

In effect, the application of incident recall techniques helps substantially to develop and maintain those qualities and characteristics that make an effective supervisory leader.

DIFFERENCES IN TECHNIQUES

It was stated earlier that "incident recall", as described in this chapter, is a modification of the critical incident technique that has been used so successfully and described in detail in several major research reports (See references, end of chapter). While the application of "incident recall" is essentially quite similar to the application of the "critical incident technique", there are several important differences. Most of the successful application of C.I.T. involved the collection of data by outside persons who could guarantee anonymity, so important to reporting success, to the recaller.

Incident recall is designed as a tool for the general supervisor, as well as the staff professional who by virtue of special training, proper motivation, and knowledge of upper management policy, recognizes that anonymity is frequently the difference between program success or failure.

Those using the critical incident technique seem to place great importance on a repetitive pattern of similar incident occurrences. Incident recall users, on the other hand, lean towards the utilization of decision-making tools concerned with the potential for major loss and the probability of occurrence of the individual incident.

This is not to infer that values are not obtained from critical incident analysis for common causes that provide special targets for a concentration of management attention.

Since the "critical incident technique" has generally been applied and reported on by researchers, scientific persons, or specialists with significant in-depth training, its practical application has not been communicated in a manner that has gained acceptance by general supervisory personnel. This chapter is an attempt to accomplish this desirable objective through a modification of C.I.T., referred to as "incident recall".

TWO TYPES OF INCIDENT RECALL

Experience has proven that two types of incident recall can be utilized and have significant value in a loss control program, depending on the results one desires to achieve. There is no question about the fact that the planned or formal incident recall interview will produce optimum results in the number of incidents reported. It logically follows that any system that would reveal the greatest number of potential downgrading problems would also possess the greatest potential for loss control.

The planned incident recall interview requires the supervisor or user to take the time to utilize a special interviewing technique. The success of recall in terms of the number of incidents reported is directly related to the proficiency with which the interviewer conducts the recall interview. Properly administered, a system utilizing the planned interview technique with a group will consistently reveal a substantially greater number of incidents with potential for loss than the number of loss-producing accidents voluntarily reported by the same group.

For best results, the informal type of incident recall should be employed *in addition* to the planned incident recall interviews to encourage the on-going voluntary reporting of incident information and to reinforce the need for cooperation in future interviews. Should an organization or supervisor so elect, any one of the informal incident recall techniques to be described later could be used in the absence of the planned interview type of incident recall program. As a matter of fact, whether or not a supervisor uses the planned incident recall interview, he would greatly advance his loss control effort by employing one or more of the informal techniques.

THE PLANNED INCIDENT RECALL INTERVIEW

The basic objective of the planned incident recall inverview is to gain the willing cooperation of the employee, so that he freely relates all incidents that he can recall. Since the success or failure of the program depends so heavily upon the results of the interview, it is very important that the interviewer understand the importance of the major related factors.

Front-line Supervisor as Interviewer

Generally speaking, the front-line supervisor is preferred as the interviewer for his own people when any broad application is being made of planned incident recall interviews. While he may not possess the ability to interview as well as certain staff specialists or even other selected supervisors with in-depth psychological training, he does have important qualities and characteristics not possessed by an "outsider". In addition, the exercise can prove to be an outstanding tool to strengthen certain weaknesses while maintaining or reinforcing those characteristics of management leadership recall.

The major reasons why the front-line supervisor is the most recommended interviewer for general incident recall applications are identical to those that make him the best-equipped investigator of accidents in his area.

1. He has a personal interest to protect.

2. He knows the most about his people and conditions.

3. He knows how to get the information.

4. He will be initiating the action anyway.

The safety or loss control professional who is part of an organization or industry permeated by habits, customs, even laws and accustomed to handling reports of substandard practices with a view to discipline or punishment — such a professional must recognize that interviews by general supervisors would be impractical or impossible. In such situations an outside or independent agency could conduct all interviews. An alternative would be the use of selected staff personnel who, because of their academic and professional backgrounds, can guarantee anonymity and ensure that the negativity of the past is overcome.

One large company found that the benefits of a limited trial were significant enough to justify a full-time professional interviewer utilizing incident recall on a continuous basis. His scope includes as well as special problem application, regular incident recall applications.

For most organizations, these latter choices may not be practical; and in climates deemed unfavorable to the application of incident recall through general supervision, a concentration should be placed on adjusting attitudes and preparing people for program application at a future time. With complete upper management support, a sound policy, adequate training and follow-up, an incident recall program can effectively be administered through regular supervisory channels. The many indirect benefits to be derived from the application of a positive management response to volunteered information on problems will in themselves justify the time and effort spent on the entire program.

Privacy Preferred

Since the interview requires a good degree of concentration by both the interviewer and the employee being asked to recall incidents, interruptions can have a damaging effect upon the results. Best results are achieved when a location for the interview has been selected where there will be as few interruptions as possible. It should be pointed out that excessive noise can be just as disturbing as the telephone or other interruptions. These can be minimized by planning. Failure to achieve anticipated results in recall has sometimes been traced to the interview location. Attempting to obtain recall is difficult while standing in a busy shop area next to an operating machine or while sitting on an uncomfortable locker-room bench; these are examples of interview locations that would not be conducive to attaining best results. An office, where both the supervisor and the employee can sit comfortably in relative freedom from unnecessary outside noise, is generally best. People also tend to be more self-conscious about such conversations with an individual, or their supervisors, conducted in the view of their fellow-workers. It should also be kept in mind that the employee's complete cooperation sometimes relates to the importance that is placed on his ability to recall. The interview location can have a very positive effect on him by demonstrating the importance that management places on this special exercise.

Selection of People to be Interviewed

While the selection method could vary with the specific technique being employed, there is a general practice that can be recommended. Whenever the supervisor is using a recall technique that will involve all his men, it is best to use a system of selection that enables them to understand why specific people are being interviewed. (For example, they could be selected by alphabetical order of their last names or by numerical order of their organization's personnel number identification. This type of selection places everyone on notice that he will get his chance in fair order, and it tends to minimize unnecessary guessing by employees as to why certain people get such attention from the supervisor.) An orderly system of selection is also best because interviews are probably not everyday occurrences. This fact alone would tend to focus special attention on the person being interviewed. Exceptions to this general rule would be where the planned recall interview is being used as a problem-solving tool with people on a specific job or occupation. Interviews would then be restricted to the specific group, or individuals, as the case may be. The others quickly understand exceptions when there is some logical explanation — as there obviously would be in these cases.

Adjusting Interview Time

The time required to conduct a planned recall interview will also vary with the particular method of recall application. Just as with worker selection, there are several recommended general practices relative to the required time when the supervisor is utilizing a recall technique that will eventually involve all his people. The immediate objective of the recall exercise is to obtain clear, accurate descriptions of incidents. Once this important objective has been fulfilled, the supervisor can determine the degree of criticality in obtaining any additional help on cause analysis or remedy. It may very well be that many of the incidents recalled will have little or no potential for serious or major loss consequences. Likewise, the probability of recurrence may be negligible for a considerable number. In most cases, it is best to use whatever time is allotted by management to that aspect of the interview involving getting the incidents reported. The supervisor can always make one or more personal follow-up contacts at a later mutually-convenient time. (This is not meant to demean the important role the employee can play in assisting the supervisor to determine the causes and remedies of incidents with serious potential. It is merely suggested as the most practical way to use the time available to the best advantage.) Experience has proven that a well-planned recall interview can be conducted in a period that could vary from 10 to 20 minutes, depending on the interviewing experience of the supervisor. The first few recall interviews would naturally take the supervisor longer, in order to establish the special rapport which is necessary for effective recall. Subsequent interviews with the same man would naturally require less time, since experience and knowledge reinforced by informal incident recall techniques would facilitate an easier exchange of information.

When incident recall is being applied to problem-solving, select interviewers could be used and the time involved may be dependent on the dimension of loss involved. Organizations accepting the potential values to be received by this technique have, on occasions, permitted extended periods of planned time to be utilized with certain employees whose special knowledge could help provide answers to costly problems.

No-Fault Assurance a Must

As has been pointed out so often in training on accident investigation, many employees have been conditioned through experience to associate blame-fixing and fault-finding with the reporting of accidents. This well-founded association is one of the biggest reasons that many accidents are never reported by the people who could furnish the best information on the prevention of similar occurrences in the future. The decision to use the planned incident recall interview in an organization's safety or loss control program should only be made if no-fault, no-discipline assurance can be given to every man participating. This type of decision is, of course, made by top management before the program is ever introduced. The reason for its need must be thoroughly understood by every member of management or the program will be doomed to failure. Loss control leaders feel so strongly about the importance of this point to the success of the program that most organizations utilizing this valuable tool make the listing of a name on the incident report form an optional choice of the person reporting the incident. Actual experience in diverse types of business organizations has proven over and over again that any indication of fault-finding or discipline for information revealed through the incident recall interview system will close off the flow of vital information faster than any other single factor. The big question that must be answered by the organization and supervisor is simply, "Am I willing to give positive no-fault assurance to everyone involved in a planned recall interview, in exchange for an abundance of information that could be of value in preventing future serious losses?"

Most organizations that adopt the planned interviewing technique with its no-fault assurance requirement also encourage and endorse confidentiality whenever it is deemed to be best by mutual agreement of the supervisor and the employee being interviewed. This important aspect of the planned recall interview demands the highest degree of leadership maturity possessed by the supervisor or interviewer. It is this type of experience that will reinforce and strengthen the individual.

Preparation for Interviews

One of the first things the supervisor or other interviewer should do in preparation for the interview is to review personal facts and information that could acquaint him with anything that might be helpful in putting the employee at ease and stimulating his participation in the recall exercise. This background will enable the supervisor to personalize his comments and demonstrate the type of interest and relationship that encourages a spirit of cooperation.

It is easy to understand that employee transfers, replacements for illness and vacations, seasonal layoffs and other such factors would make it difficult for every supervisor to know this information in the depth desired without preparation.

Another very important step is to prepare a check list of potential sources of accidents involving the employee's job. The same check list can be utilized with other people on the same job who will be interviewed at a later date. The check list will assist the supervisor to stimulate the recall of incidents if it doesn't come easily. Experience has also proven that suggestions of specific areas of thought are great stimulators of recall.

Like priming a pump, one word or thought is frequently all that is needed to stimulate the recall of incidents that may be quite removed from the suggestion, yet can be brought back to mind by association with a familiar word or thought. The job description, job analysis or procedures and safety rules are excellent sources to procure items for the recall aid check list.

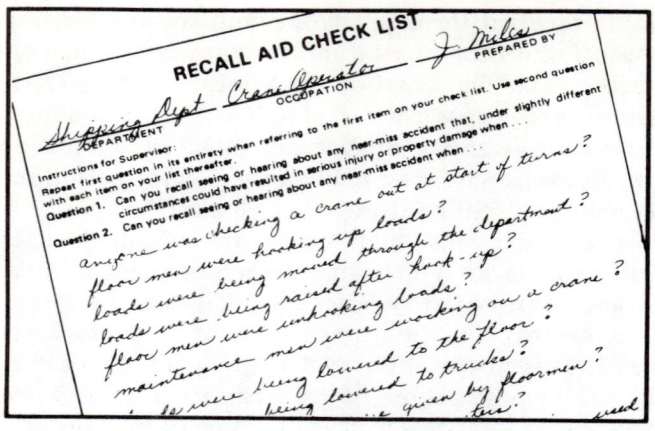

Figure 2.

Figure 2 shows a typical check list that might be prepared to aid recall for crane operators.

The supervisor or interviewer should recognize that the only purpose of the recall aid check list is to prime the pump (the memory) and not to be a complete guide for recall. The check list may not even be necessary with certain people, but the good supervisor will always have it ready if recall does not come freely.

Conducting the Interview

In the final analysis, the success of the entire program rests on the incident recall interview itself. The rewards for effective application are too great to jeopardize success by not following the necessary guidelines established by experience. Generally speaking, the step-by-step procedure described in this section must be followed routinely. After the first interview has been conducted with everyone in his group, the interviewer will certainly modify the details he uses, but the general format must remain the same to produce the most effective results. The quotation below was taken from an actual recording of an interview with a general plant manager who had introduced a planned incident recall program throughout his organization. The interview was conducted several months after a broad program had been introduced, and the questioner had inquired as to whether or not he would have done anything differently if he had the introduction of the program to do over again. His answer is as follows:

"Yes, there are little things we've learned. For example, the importance of insisting that the initial training instructions be accepted on faith that recall will only work effectively when a proper rapport with the employee is established. We have had instances when supervisors have deviated from the original instructions and, just as we warned in advance during the training, recall was not as effective as it had proved to be with those supervisors following the recommended interview technique. The importance of the supervisor's check list cannot be overemphasized."

The important step-by-step procedure that the supervisor must follow in his interview is described below:

1. *Put the employee at ease.* One of the best ways to create the favorable rapport necessary for good recall is to talk briefly about the employee's family or a subject of interest to him. The interview's advance planning will assist in this regard. Make sure you use his first name rather than his last. Be friendly, warm, cheerful, sympathetic and sincere. This step appeals to the basic psychological needs of the average person and provides motivation that encourages cooperation.

2. *Explain the purpose and state the importance of recall.* Don't keep the employee waiting. Tell him what incident recall *is*. (While most supervisors or interviewers will have explained the program to groups involved in advance, there are usually several in every group that didn't get the word or didn't hear it properly.) Remind him of the great aerospace successes because of the capability of people to predict potential system failures in advance. This is what incident recall will do. Try to create within him a desire to be part of a program that can play an extremely important role in the correction of many problems that could cause serious injury to him or others, or inflict extensive damage to equipment or property. While the interviewer's comments can vary on subsequent occasions and may not require as much detail, he must keep in mind that motivation varies in its intensity and an appropriate reminder of the important contribution the employee can make would always be in order. At this point, the interviewer is appealing to his needs for acceptance and accomplishment, and his desire for self-respect and status. Creating recognition of the real value and importance of the employee's role in accomplishing the objectives of the

program can make a big difference in his willingness to cooperate.

3. *Give assurance that recall is confidential.* While every step in the recall routine is important, this is probably more important than others. The degree of recall the supervisor achieves on his first attempt is usually directly related to his ability to convince the employee that the conversation is privileged. The supervisor should point out that the only purpose of recall is to aim at the facts related to near-injury/damage accidents. Depending on the organization's policy, the interviewer or supervisor can emphasize his point by letting the man choose whether or not his name should be placed on any form being utilized. The employee should be reminded that only by his willingness to share information with others will his name ever be associated with the incidents discussed. The no-fault assurance position of the organization, mentioned earlier, should be utilized throughout this part of the recall interview.

4. *Point out that recall benefits everyone — the organization, the department, the employee, and his family.* Explain that less injury and damage means less lost time from work and less downtime of equipment. Fewer accidents mean increased operating efficiency, which results in an improved business climate, conducive to increased profit and job security that benefits everyone.

5. *Explain the use of the recall aid check list.* Show the employee the check list you have developed of items related to his job or jobs like his. Point out that you have selected aspects of his work activity that you believe could have had some involvement in near-miss accidents. Explain that the list was designed as an aid to assist him in remembering or recalling incidents he has seen or heard about, and that it need not be used if he can recall without his tool. The interviewer could explain that the check list words provide associations with experiences stored in the memory that become unlocked when some key word or phrase is heard. If he has good recall without this help, don't spoil a good thing. Lay the list aside and pursue his own line of recall. Always have the list available for the next person who may need its assistance, and add to it as recall or experience gives you additional items.

6. *Apply recall, using the check list.* Simply ask the employee to recall each near-miss accident he can remember seeing or hearing about that, under slightly different circumstances, could have resulted in serious injury or property damage. Use the recall aid check list to help. With each incident recalled, be sure to determine how many times he has seen or heard of its occurrence. This information will help determine the probable recurrence rate of this incident and serve as a guide to the urgency and extent of action necessary. Explain that the primary purpose of the interview is not to determine why it happened or what to do about it. Politely advise him, if necessary, that you plan to come back to these aspects another time. It is extremely important to get all of the incidents reported in whatever time is allotted for recall. Experience has proved that people will sometimes report accidents during incident recall that were not previously reported. Whenever this happens, the supervisor should treat it in the same manner as the near-injury/damage incidents being reported. As a matter of fact, the same principles for recalling incidents can be used for the recall of accidents that resulted in unreported minor injuries or costly property damage. "Accident Recall" will be mentioned later in this chapter.

7. *Ask questions to fill in the gaps.* Try to avoid interrupting his chain of thought. Whenever possible, let him complete his description of what happened, but do ask questions if his recall is not clearly stated. Don't make it a bull session; just clarify important points necessary to clearly identify the incident.

8. *Review your understanding of the incident with him.* Quickly repeat your understanding of each incident, just to make sure your information is accurate. Again, make sure your approach is tactful. Be warm and friendly during the review. Correct your detailed notes on any changes made in your understanding of the incident.

9. *Discuss causes and remedies if time permits.* Time permitting, and depending on the completion of the recall of incidents, the supervisor may want to invite the employee's valued opinion on the possible causes, as well as any suggested remedy or control. If the supervisor plans to contact him later for this information, he should say when the contact

will be made. Prompt follow-up contacts, as much or more than anything else, will give him a genuine feeling that he is contributing and that he is needed. Remedial action may frequently involve the procurement of help from several other sources, just as it does in accident investigation. The most important thing of all is to get the record of events that will identify serious problems needing attention.

10. *Express your sincere thanks for the employee's cooperation.* Never fail to feed one of the deepest desires every person has — let him know his contribution has been helpful. Be sincere and honest in whatever reaction is given, but always be sure to underline your personal appreciation with a thank-you. The supervisor who believes the employee has been helpful is best able to convey a message of appreciation.

Figure 3 shows the actual incidents reported to supervisors during planned incident recall interviews. The value of obtaining information of this type to control future similar occurrences should be obvious.

INCIDENTS RECALLED THROUGH PLANNED INTERVIEWS

- While sweeping railroad switches during a snowstorm, a trackman caught his foot between a flat rod and a railroad tie. His helper had gone ahead to sweep other switches. The trackman could not get his foot released and he noticed the engine approaching. He screamed to attract the brakeman's attention. The brakeman saw him in time and stopped his train 10 feet away.

- Employee reported narrowly being struck broadside by another auto when he drove through a red light on his way to work. Wind caused branches from the tree to block out a hanging stop light in the center of the intersection.

- Employee reported catching the heel of her shoe in a crack on one of the stair treads at the top of the front stairs in the main office. She caught the guard rail and narrowly missed falling down the stairs.

- An employee was working on top of a vertical mill. A crane passed over, knocking off his hard hat. Fortunately, it was only the hat and not the individual under it.

Figure 3.

- Instructions were issued to change from gas to oil fuel because there was a shortage of gas due to the cold weather. In pulling the gas burner out and in the process of installing the oil burner, the employee had his face close to the fitting while trying to line up the burner. There was a backfire causing smoke and fire to come out the end of the tube and almost burning the employee.

- Lift truck operator attempted to start a lift truck that was parked near a local floor heating unit. He had trouble starting it, so he opened up gas valve on propane tank to release condensation and in doing so, the gas hit the heating unit causing flames to strike the man. Luckily he was not burned.

- The mail delivery boy tripped over the packing boxes at the top of the stairway on the landing platform outside the shipping room and nearly fell headlong down the stairs.

- A trailer was parked in a building. When the tractor backed into the trailer to move it, the air leaked out of the trailer brakes causing it to move backward about 6 feet. A few seconds before a man had walked from behind the trailer. This could have resulted in a very serious injury.

- The electrician approached a buffing machine in the shop with the intent of changing an indicator lamp. He opened circuit breaker marked "Buffer". But this switch was for lighting only — not for drive motor on same. As he turned to work on the machine, he barely missed walking into the buffer wheel while at full speed because operator last using this machine failed to turn buffer off. This was a surprise to the electrician, and certainly a near miss.

- A heavy piece of lumber came down off the high roof on Building "H" and narrowly missed one of the office girls walking across the cafeteria.

- A repairman was changing the light bulb in top of the building on an elevated platform and he dropped bulb to the ground but did not see the man nearby who was fortunately blocked by a column.

- A scooter was approaching 4th Avenue from the south area; when the operator applied the brakes, the brakes failed to operate. The scooter continued out on and across the avenue, narrowly missing a car travelling north on the road.

Figure 3, continued.

- A bridge wheel packing box fell off crane going between JBT office and Charlie Bowan's office and nearly ran into the awning window that was opened into walkway.

- Safety lights were turned on and switch tagged out. Two men were replacing parts on the scrap transfer car on the floor. Overhead crane operator picked material up from the car with men working under it; they could have been mashed and easily killed.

- The bottom roll in a pile of carpets moved out of position causing the whole pile to come down rolling across the aisleway about 11:50 a.m. yesterday. Some of those rolls weighed about a ton and could have caught several people if the incident had occurred a few minutes later.

- Two men pulling on 18" pipe wrench to loosen nuts on a machine while standing on work platform. Handle of the pipe wrench broke throwing the men down on platform. They could have been struck by the pipe and thrown off the platform to the ground below which might have caused broken bones or something even more serious.

- A lift truck operator had a near-miss during last rain storm due to unloading platform being slippery when wet. He almost damage crate containing valuable instruments.

- Electrician was running conduit in tool work room "B". In process of drilling holes through wall, when drill broke through far side of the wall and pushed large tool box from top of cabinet. It fell 8 feet to floor. Employee in this room had just walked by where box fell.

- A maintenance man was working under a machine which had not been locked out. Another employee — unable to see the maintenance man — closed the switch and started the machine. Fortunately, at that instant, the maintenance man's hands were away from the moving parts of the machine.

- A tow motor was backing up in a comparatively dark area of a warehouse. The tow motor brushed the hop of an employee who had been taking a short-cut through the building to get to the canteen.

- An employee slipped on an oil slick which had dripped from a machine into the aisle. He recovered his balance without injury.

Figure 3, continued.

Making Out the Report

Most organizations using planned incident recall interviews through their supervisors require the submission of an individual report form on each incident with the potential for serious injury or damage, within a 48 to 72 hour period following the recall. This gives the supervisor an opportunity to follow up and complete his cause analyses and remedial recommendations with the person interviewed (or others if indicated). Detailed notes are usually made during the incident recall interview, with the final incident description written on the report form immediately following the interview, while all facts remain clearly in the supervisor's mind. The step-by-step procedure for conducting the interview has proven so important to effective recall that many organizations print it on top of a special supervisor's incident recall report form to serve as a guide and reminder during the interview. (See Figure 4.) This special form, including the discussion procedure, is most frequently utilized during the early introductory stages of the program. While "Incident Recall" requires an additional form, experience has proven its value. It minimizes the tendency to blame fixing, fault finding, and punishment so frequently associated with the usual accident investigation procedure.

Since there is usually some permissible lag time between the conducting of the interview and the completion of the supervisor's report, another special form (such as shown in Figure 6) is sometimes required to be submitted within 24 hours of the interview. This latter form provides loss control personnel with a record that enables them to measure whether or not program goals are being achieved, and gives them a tool to follow up completion of individual incident report forms.

Practical Application of the Planned IR Interview

There are several ways that the incident recall interview can be used to significant advantage in a loss control program. Several specific applications that have proven of special value are briefly discussed below:

- *Continuous Plant-Wide Application*

    This application requires that each supervisor be given a frequency objective for conducting interviews with all people in his group. Because of the planning and concentration of effort required to do an effective job, organizations usually establish this rate at 2-4 incident recall interviews per month. This enables the supervisor to have recall with an average-size work group of 15 several times each year. Management groups have come to recognize that the benefits to be derived from this program are too great to risk losing them through the improperly-conducted interviews that can result when objectives are not realistic. The number of incidents that will be reported through a program utilizing this

SUPERVISOR'S INCIDENT RECALL REPORT

COMPANY OR BRANCH	DEPARTMENT-SHIFT	RECALL REPORT NO.
Baldwin Works	Manufacturing 3-11	3
PERSON INTERVIEWED (NAME USE OPTIONAL)	OCCUPATION	DATE OF INTERVIEW
George P. Stall	Operator A	Feb. 15, 1974

REQUIRED RECALL ROUTINE

CHECK AS COVERED

1. PUT THE EMPLOYEE AT EASE ✓
2. EXPLAINED PURPOSE AND IMPORTANCE ✓
3. GAVE ASSURANCE RECALL IS CONFIDENTIAL ✓
4. POINTED OUT BENEFITS FOR EVERYONE ✓
5. EXPLAINED RECALL AID CHECK LIST ✓
6. APPLIED RECALL, USING CHECK LIST ✓
7. ASKED QUESTIONS TO FILL GAPS ✓
8. REVIEWED UNDERSTANDING WITH HIM/HER ✓
9. DISCUSSED HIS/HER PREVENTIVE IDEAS ✓
10. EXPRESSED SINCERE THANKS FOR HIS/HER COOPERATION ✓

(Item No. 9 should only be done if recall is completed and time permits)

DESCRIPTION

25. DESCRIBE CLEARLY HOW THE INCIDENT OCCURRED.

Recaller was riding maintenance scooter down hill from the # 2 power station when the brakes failed to operate causing the vehicle to run wild while going down entire hill between buildings and across main street, just missing a collision with an automobile traveling north. Maintenance personnel determined that the brake bands were worn substantially beyond normal wear.

ANALYSIS

26. WHAT ACTS, FAILURES TO ACT AND/OR CONDITIONS CONTRIBUTED MOST DIRECTLY TO THIS INCIDENT?

Maintenance records have verified that a substandard brake condition caused the mechanical failures involved. The recaller volunteered information indicating that a preuse check of the brakes was not made at the start of the work turn. A personal check indicates that about 20 % of operators are completing their preuse equipment check forms.

27. WHAT ARE THE BASIC OR FUNDAMENTAL REASONS FOR THE EXISTENCE OF THESE ACTS AND/OR CONDITIONS?

A systems to determine that operators of this equipment are checking critical parts prior to use is not working. In addition, these vehicles are not getting to the garage on schedule for the required quarterly inspection and routine maintenance check.

EVALUATION: 28 LOSS SEVERITY POTENTIAL ☐ Major ☑ Serious ☐ Minor — 29. PROBABLE RECURRENCE RATE ☐ Frequent ☑ Occasional ☐ Rare

PREVENTION

30. WHAT ACTION HAS OR WILL BE TAKEN TO PREVENT RECURRENCE? NUMBER ALL ITEMS IN SEQUENCE

1. I have requested our shop clerk to notify me each day before 9 A.M. of any scheduled operator that has not turned in his pre-use equipment check form.
2. The clerk will also compile each operator's conformance to this pre-use check on a monthly report.
3. Scooter rules have been reissued to all operators with special reinstruction on rule #10.
4. Charles Wright, garage supervisor will recommend a system of notification for inspection of all vehicles including a colored inspection sticker identification

31. CIRCLE NUMBER AND GIVE DATE OF INTERMEDIATE ACTION; "x" OUT NUMBER (AND GIVE DATE) WHEN COMPLETED.

	1	2	3	4	5	6	7	8
INTERMEDIATE	(1)	(2) 2-15-74		(4) 2-15-74				
COMPLETED	X 2-15-74		X 2-15-74					

32. INVESTIGATED BY	33. DATE	34. REVIEWED BY	36. DATE
[signature]	Feb. 16, 1974	John L. Lucas	Feb. 18, 1974

Note: These sample forms may appear somewhat different from those used in your company's program. They are included as a reference and are only intended to be representative of those generally used.

Figure 4.

reasonable rate of interviews will far exceed the number of accidents voluntarily reported through normal program channels. As mentioned earlier, continuous plant-wide application could also be conducted by a special staff professional. This may be in conjunction with a program directed through general supervision or in lieu of such a program. This approach generally requires

SUPERVISOR'S INCIDENT INVESTIGATION REPORT

1. COMPANY OR BRANCH	2. DEPARTMENT		
South Bend Works	Electric		
3. EXACT LOCATION	4. DATE OF OCCURRENCE	5. TIME ☐ AM ☑ PM	6. DATE REPORTED
extreme south end bay #2 bldg. 4	2-10-74	2:30 PM	2-10-74

INJURY OR ILLNESS		PROPERTY DAMAGE		OTHER INCIDENT	
7. INJURED'S NAME		13. PROPERTY DAMAGED		19. PERSON REPORTING INCIDENT: Charles Edgar	
8. OCCUPATION	9. PART OF BODY AFFECTED?	14. ESTIMATED COSTS $	15. ACTUAL COSTS $	20. OCCUPATION: motor repairman	21. COST (IF APPLICABLE) $
10. NATURE OF INJURY/ILLNESS		16. NATURE OF DAMAGE		22. NATURE OF INCIDENT: near fall from overhead	
11. OBJECT/EQUIPMENT/SUBSTANCE/INFLICTING INJURY/ILLNESS		17. OBJECT/EQUIPMENT/SUBSTANCE/INFLICTING DAMAGE		23. OBJECT/EQUIPMENT/SUBSTANCE/RELATED: crane bridge parking brake	
12. PERSON WITH MOST CONTROL OF ITEM 11.		18. PERSON WITH MOST CONTROL OF ITEM 17.		24. PERSON WITH MOST CONTROL OF ITEM 23: unknown at present	

DESCRIPTION

25. DESCRIBE CLEARLY HOW THE INCIDENT OCCURRED

Recaller was sent to adjust hoist brake on crane. After pulling main switch in cab, he proceeded to top of crane to adjust brake. As he started to perform job, crane started to drift down track. Crane operator depressed bridge brake pedal to stop crane causing recaller to lose his balance and almost fall off crane; only ladder that extends about 3 feet above crane prevented him from falling to ground.

ANALYSIS

26. WHAT ACTS, FAILURES TO ACT AND/OR CONDITIONS CONTRIBUTED MOST DIRECTLY TO THIS INCIDENT?

It happened because the bridge parking brake did not set when main switch was pulled open. The recaller should have checked to determine that the brake was set before going on the bridge. In addition he should have been wearing or using the required safety belt to avoid falling.

27. WHAT ARE THE BASIC OR FUNDAMENTAL REASONS FOR THE EXISTENCE OF THESE ACTS AND/OR CONDITIONS?

The improper working conditions of the brake was discovered to be a failure of adjustment by maintenance personnel. Records indicate that this critical item was not inspected as required on the turn following the incident. The obvious need for adjustment would probably have been discovered if the critical inspection had been made

EVALUATION: 28 LOSS SEVERITY POTENTIAL ☑ Major ☐ Serious ☐ Minor — 29 PROBABLE RECURRENCE RATE ☐ Frequent ☑ Occasional ☐ Rare

PREVENTION

30. WHAT ACTION HAS OR WILL BE TAKEN TO PREVENT RECURRENCE? NUMBER ALL ITEMS IN SEQUENCE

1. I reinstructed my crew at our weekly meeting to make sure parking brake is set and working before going on top of crane.
2. Also have improved my job instruction when assigning work.
3. Have recommended a change in the method of parking brake inspection for electrical maintenance
4. My entire crew has been equipped with the new type safety belts and lanyards for 100% wear with regular work pants.
5. The related rules have been completely reviewed with all employees

31. CIRCLE NUMBER AND GIVE DATE OF INTERMEDIATE ACTION. "X" OUT NUMBER (AND GIVE DATE) WHEN COMPLETED

	1	2	3	4	5	6	7	8
INTERMEDIATE			(3) 2-10-74					
COMPLETED	X 2-10-74	X 2-10-74		X 2-10-74	X 2-10-74			

32. INVESTIGATED BY	33. DATE	34. REVIEWED BY	35. DATE
Paul Gruendman	2-10-74	Frank Jackwood	2-12-74

Note: These sample forms may appear somewhat different from those used in your company's program. They are included as a reference and are only intended to be representative of those generally used.

Figure 5.

excellent cooperation between the special interviewer, the supervisor of the area, and the employees involved.

- *Periodic Plant-Wide Application*
Experience has proven that a good alternative to continuous planned recall application is the scheduling of interviews at a reasonable rate until all employees have been covered, then interrupting the program for a pre-determined period of time before starting the routine again. During this interim period, an organization-wide awareness of values received can be created or reinforced, by sharing the latest information on the incident recall program through individual case histories and causal analyses.

- *Problem-Solving on an Area or Job Basis*
Planned recall interviews can be directed at the solution of specific accident problems, where sufficient data on causal information may be lacking. Based on the logical assumption that many unreported accidents and incidents of a similar type are probably occurring, the supervisor can conduct incident recall interviews to obtain additional information that could be helpful in approaching a solution to his specific problem. One

SUPERVISOR'S INCIDENT RECALL INTERVIEW REPORT
TO BE COMPLETED FOR EACH WORKER INTERVIEWED

RECALLER (NAME OPTIONAL)	OCCUPATION machinist	REPORT NO. 4
DEPARTMENT machine shop	DATE OF INTERVIEW 3-10-74	LOCATION OF INTERVIEW shop conf. room

REQUIRED RECALL ROUTINE

	CHECK AS COVERED			CHECK AS COVERED
1. PUT THE EMPLOYEE AT EASE	✓	6.	APPLIED RECALL, USING CHECK LIST	✓
2. EXPLAINED PURPOSE AND IMPORTANCE	✓	7.	ASKED QUESTIONS TO FILL GAPS	✓
3. GAVE ASSURANCE RECALL IS CONFIDENTIAL	✓	8.	REVIEWED UNDERSTANDING WITH HIM/HER	✓
4. POINTED OUT BENEFITS FOR EVERYONE	✓	9.	DISCUSSED HIS/HER PREVENTIVE IDEAS	✓
5. EXPLAINED RECALL AID CHECK LIST	✓	10.	EXPRESSED SINCERE THANKS FOR HIS/HER COOPERATION	✓

(Item No. 9 should only be done if recall is completed and time permits)

TOTAL NUMBER OF INCIDENTS RECALLED _7_
NUMBER OF INCIDENTS RECALLED WITH SERIOUS OR MAJOR LOSS POTENTIAL _3_

INTERVIEWER W. R. Bunting	DATE 3-10-74	REVIEWED BY George B. Kahler

Figure 6.

particular such problem solved through the recall technique involved the damaging of end gates in railroad cars used by a manufacturer to transport bulk materials. Accident reports were grossly inadequate in revealing causal information of any significant value. Recall by equipment operators identified many occurrences of near damage and actual damage to the end gates by operators who were using methods they knew could be improved by the establishment of a simple procedure that had never been recognized by management. The common practice of the operator was to lower unloading devices into the cars by yard crane and move them back and forth to scoop up materials. This practice was causing the heavy devices to nearly strike the end gates with every movement. The simple procedure that evolved was to lower unloading devices at the end of the car and always lift them at the car's mid-point. A standard job procedure for this job, with a reasonable frequency of planned observations, reduced the costs of end gate replacement 82% for this company.[2]

The majority of supervisors will recognize the general problem-solving value of the incident recall technique, but special problems of unusual loss dimension deserve the utilization of the best interviewers possible. Staff professionals or selected supervisors, who have proven their skill at recall, are generally considered to be the best interviewers for this important assignment.

- *Use in the Termination Interview*

An employer of several thousand short-term employees per year has found that one of the best applications of the planned Incident Recall is their use with this special group upon termination. The normal practice had been to simply ask the departing employee whether or not he had any suggestions for the improvement of safety, quality or production. Historical records indicated that an average of 75 to 100 suggestions were offered each year through the previous system. The first season's application of the incident recall interview technique brought the reporting of over 2,000 incidents that were believed to possess serious potential for injury or property damage and that had not been previously reported. This employer feels that application of the recall interview in this manner has been a major contribution in continued improvement of safety and health conditions for his organization.

While the average employee being given an close-off interview prior to lay-off or termination is certainly not overjoyed, he does generally want to insure the possibility of his return, or at least a positive recommendation for a future job. By utilizing the planned interview, a supervisor or person assigned the recall exercise can be more certain that an abundance of helpful accident prevention information can be gained.

INFORMAL INCIDENT RECALL TECHNIQUES

Whether or not a planned Incident Recall interview program is utilized, the voluntary reporting of near-miss accidents can be greatly increased by the application of informal incident reporting techniques. When added to a program that has incorporated the recall interview, they reinforce the overall effort and encourage voluntary reporting of incidents on an on-going basis.

Many supervisors have discovered that asking for brief reports of any near-miss accidents following each regular group meeting can become an excellent stimulator of incident reporting. The good supervisor will keep incident reports brief and make sure that any group comments are always directed in a positive way, so that reporting is what was done to avert the accident is one of the best techniques to accomplish this. The regularity with which this practice is followed becomes a key to success in obtaining reports.

Other supervisors prefer to close off each personal contact with an invitation to report incidents. By making this practice a routine habit, many incidents that would otherwise not be reported become known. This personal approach appeals to the individual who has any concern about being embarrassed in front of the group.

A number of organizations appoint one man as a safety or loss control observer for a stipulated period of time. His duties include discussing with the supervisor any substandard working conditions or practices encountered by the group. Organizations using this approach believe that the presence of someone else in the group with special added interest in preventing losses greatly aids the supervisor. When the safety or loss control observer system is utilized, the individual assigned can be an excellent source of incident information.

Employee publications, bulletin boards and incident report boxes are other avenues of communication and motivation that can be utilized to encourage reporting of incidents. (See Figure 6 for one example of a poster which can reinforce your program.)

Regardless of the techniques employed, the three most important practices by the supervisor to stimulate the employee's desire to voluntarily report incidents on a continuous basis are: the frequency of his communications that encourage reporting, his immediate behavior recognition of those who report incidents, and prompt effective action to control their recurrence.

Shortcomings of Incident/Accident Statistics

While there is little doubt that the use of incident recall has benefits that far outweigh its shortcomings, it is only proper to indicate the potential limitations of the technique. Several have been briefly mentioned previously:

1. While "incidents" occur much more frequently than loss producing accidents, they, too, are relatively infrequent events. It is a false assumption to believe that an incident recall system will replace those program activities, such as observation and inspection, that detect substandard practices at an even earlier stage in the cause and effect sequence.

2. Since incidents are "near accidents" that occurred in the past, the investigator cannot see or examine the circumstances and evidence that led to the occurrence. Most of the related information must be obtained from the recaller who may have limited or biased knowledge of what really happened or brought about the happening. While there are sometimes ways to check what really happened, we must largely rely on the memory of the recaller for our data.

3. It must be borne in mind that neither incident nor accident recall information will necessarily tell us the causes that led to their occurrence or possible control of the problems. The interviewer will frequently place too much value on the limited knowledge revealed by one witness or one knowledgeable person when, in reality, this should be the beginning of a search for as many facts as possible to determine the real causes.[5]

4. Incident recall is no panacea for the accident problem. It is simply another useful tool to add to our arsenal of program activities. The value received from its application will be a

direct function of the planning and effectiveness with which the program is introduced.

Critical Stages in the Program

The potential value to an organization from the incident recall technique justifies a review of those stages most critical to its success.

1. *Complete upper management commitment and appropriate statement of policy:* It has been found that the dramatic changes in management attitudes at all levels in dealing with information of this type require complete endorsement of the program and a statement of company policy respecting it. An incident recall program should never be started without this complete endorsement by the highest operating management official.

2. *Thorough indoctrination of all upper management people:* A complete review of the entire program should be presented to department head level personnel by division executives following their own review and indoctrination. This is not intended to replace their participation in the formal training program for all management but it does serve to assure their complete cooperation. Experience has proven that this important step helps to prevent resistance to the changes that these key people can express in subtle ways.

3. *Explanation and indoctrination of lower levels of management prior to formal training:* This mind preparation and involvement of lower management levels through the positive expressions and endorsements of their supervisors is especially important. It is this level of supervisor that will provide the "point of control" in the program. No other aspect of the program deserves more attention than the building of positive attitudes within this important group.

4. *Development of forms, standard practice for application and training materials:* Adequate time spent in preparation of these items prior to training will avert many hours of retraining and unnecessary problem solving later. A representative group of management personnel should be involved in the development of all written communication items necessary for the success of the program. Incident recall

techniques will require additional time to be spent on safety/loss control by the supervisor, and every care should be taken to make sure there are no last-minute unpleasant surprises because people were not included in or unaware of developments in this critical planning stage.

5. *Adequate formal management training:* This critical stage should be carefully planned, adequate training materials prepared and an opportunity given future recallers to have all their questions answered. Also one simulated interview in recall should be presented. Training is most accepted when presented by a key member of management such as the department head. Extra time spent in preparations and presentations of effective training materials will save many hours of re-training. It is best to have small group discussions and critiques following the first actual recall application. This follow-up should provide answers to any remaining questions.

SUMMARY

By the application of the planned incident recall interview and other informal recall techniques management can obtain reports on many near-miss accidents with the potential for serious injury or property damage that would not otherwise be reported. The management group of any organization using this tool will have new opportunities to identify *potential* loss situations *at the no-loss stage*. Incident recall is a practical tool for the supervisor or staff professional to use, regardless of the type or size of organization he represents. It can be tried with as few as five employees to prove that it works. "The proof of the pudding is in the tasting" — the only thing management has to lose is the accident losses they might have had.

BIBLIOGRAPHY

1. Bird, Frank E., Jr., and O'Shell, Harold E., "Incident Recall", *National Safety News*, October, 1969.

2. Bird, Frank E., Jr., *Management Guide to Loss Control*. Atlanta, Georgia: Institute Press, 1974.

3. Chapanis, Alphonse, *Research Techniques in Human Engineering*. Baltimore, Maryland: The Johns Hopkins Press, 1959.

4. Fitts, P. M., and Jones, R. E., "Analysis of Factors Contributing to 460 'Pilot Error' Experiences in Operating Aircraft Controls," Army Air Force Air Materiel Command, Engineering Division, Aero Medical Laboratory (Wright-Patterson Air Force Base, Ohio), Report No. TSEAA — 694012, July 1, 1947.

5. McFarland, R. A., and Moseley, A. L., *Human Factors in Highway Transport Safety*. Boston, Mass.: Harvard School of Public Health, 1954.

6. Tarrants, William E., "Applying Measurement Concepts to the Appraisal of Safety Performance," *Journal of the American Society of Safety Engineers*, Vol X, No. 5., May, 1965.

ADDITIONAL REFERENCES

Dunlap, J. W., *Manual for the Application of Statistical Techniques for Use In Accident Control*. Stamford, Connecticut: Dunlap and Associates, 1958.

Flanagan, J. C., "The Critical Incident Technique," *Psychological Bulletin*, 51 (July, 1954), pp. 327-58.

Flanagan, J. C., "Principles and Procedures in Evaluating Performance," *Personnel*, 28 (1952), pp. 373-86.

Kahn, R. L., and Cannell, C. F., *The Dynamics of Interviewing*. New York: John Wiley and Sons, 1975.

Kerlinger, Fred N., *Foundations of Behavioral Research*. New York: Holt, Rinehart, and Winston, Inc., 1964.

Selltiz, C., Jahoda, M., Deutsch, M., and Cook, S. W., *Research Methods in Social Relations*. New York: Henry Holt and Co., 1959.

Suchman, E. A., "A Conceptual Analysis of the Accident Phenomenon," *Behavioral Approaches to Accident Research*. New York: Association for the Aid to Crippled Children, 1961.

IX.
Behavior Motivation

MOTIVATIONAL ANALYSIS

Safety specialists often suspect there is not sufficient "energy" or motivational commitment to safe behavior. There is a simple and fairly logical procedure that will help to pinpoint motivational problems and solve them. A shotgun or even a dynamite approach to safety motivation may be effective, but the procedure outlined below will help to identify the problem more precisely and solve it effectively. Let's take a look at the steps you can take if you think you have a motivation problem.

A Human Factors Formula

One formula for the human factors in safety is as follows: Safe Performance = Knowledge x Skills x Motivation x Job Procedures. In this formula the human consists of three components: KNOW-HOW, CAN-DO, and WILL-DO. The first two, knowledge and skill, can be influenced by selection and training. The last, job procedures, can be improved by job safety analysis. Improving the relationship between the first two and the last, human knowledge, abilities and the job, is the focus of human engineering.

Methods for improving motivation are not as definitive. Motivation is an indicator of man's relation to the entire system. There seems to be no one place where you can get leverage on the amount of energy and commitment a man puts into proper performance. Motivation is related to the worker's knowledge, beliefs, skills, job procedures, his relationships with other workers, supervision, the work environment, company policies, etc. There is no one "handle" for improving motivation. However, there are a number of clear sources of motivational input and a number of ways this motivation can be influenced.

Do You Have A Motivational Problem?

First, how can you tell when you have a motivation problem? There is one straightforward and reasonably simple way to identify this kind of problem. There is a *difference* between what the employee *knows how to do and can do* in the way of *proper* performance and the *improper* method he actually uses.

You do not have to be an amateur psychologist to recognize a motivational problem. You have to make sure that two conditions exist:

1. The employee knows how to perform properly — he has the skill and knowledge and knows when to use it.

2. He does not follow the proper work procedure, but performs in such a way as to increase the probability of injury by such acts as:
   - using unsafe procedures
   - omitting safety precautions
   - failing to use safety equipment
   - using unsafe tools or equipment
   - any other improper work performance

This discrepancy between employee KNOW-HOW, CAN-DO and what he WILL DO is the signal that you may have a problem of employee motivation to deal with. The purpose of motivational analysis is not to blame the employee for choosing poor alternatives or to understand "what makes him tick." The safety specialist has one job to do — prevent accidents. He can do that by:

- Making sure that each employee KNOWS HOW and CAN DO the job the safe, correct way.
- Making sure that each employee WILL DO the job the safe way.

That is the process in brief. But how do you examine these three human factors and what do you look for? Let us turn now to a detailed examination of each of the three: KNOW-HOW, CAN-DO and WILL-DO.

KNOW-HOW AND BELIEFS

Is there a knowledge deficiency?

Does the worker believe proper performance is in his interest?

Knowledge gaps are of two kinds: Knowing the *content* of the job and knowing *outcomes* of doing it in prescribed ways. To find out whether the procedures are known is fairly straightforward. The worker who knows the content of the job should be able to:

- Describe the sequence of basic job steps that he goes through.
- Identify potential hazards or accidents associated with each step.
- Relate the recommended safe job procedures associated with each potential accident or hazard.

In sum, the employee, especially if he is new or doing the job on a temporary basis, may not know the job procedure, the connection between his behavior and accident possibilities, and safe procedures.

The possession of this knowledge is no guarantee that the employee will act accordingly. The knowledge may not square with what he actually believes to be true. Knowledge is strictly information. Belief or attitude introduces the motivational factor — what the worker feels is in *his* best interests.

Job Attitudes

It is more difficult to find out what the person believes can happen as an outcome of doing the job in different ways. There are many attitudes that lead to "unsafe acts."

The *KNOW-HOW* analysis has two facets: (1) knowing how to do the job as prescribed by management and (2) knowledge and motivated attitudes about the *outcomes* of doing the job in different ways. If the problem is the connection between performance and outcomes, it is a motivational, not a knowledge, problem ... as we shall see in the discussion of that problem below.

Views From Opposite Ends of the Telescope

One important implication does stem from this analysis. *Worker* knowledge and attitudes about the consequences of unsafe acts are often very different from management attitudes, for unsafe acts have entirely different outcomes for management than they do for the worker. As management looks at unsafe acts and adds up the costs and benefits, it may well decide that it cannot tolerate unsafe behavior. The cost-benefit ratio is too great. The employee's perspective may be just the opposite. As he balances the risks involved against the gains, he may be willing to take that chance. He may be willing to take a chance on an unsafe performance for the satisfactions gained. Management, which deals with the "laws of large numbers", recognizes that chance-taking will lead to predictable injury. In contrast with the individual worker, who rarely experiences injury, management knows that the company or organization will suffer the consequences of a certain number of injuries if unsafe acts and conditions persist. Management and the worker have quite different points of view about accident probability because they are looking at it from different angles — a single person vs. the entire plant. For management, accidents are much more certain than they are for the individual worker. Therefore, efforts to get workers to accept management viewpoints are likely to be quite limited in their effectiveness.

Unless this fundamental discrepancy in points of view is admitted as a possibility, efforts to increase safe behavior by communications techniques may be misdirected.

1. It is unlikely that the employees will accept management's view of the seriousness of an accident, in total.

2. Attempts to change employee behavior may neglect the real gains the employee sees from the unsafe act.

3. Fundamental differences in perspective cannot be solved by improved communications, or from insistence on management views and suppression of employee assertiveness.

Employee Safety Attitudes

Workers may not be concerned with safety from the management point of view. They may not be interested, for example, in the total number of accidents, the overall cost of accidents, etc., but they are interested in the impact accidents have on them. Health and safety hazards are severe and frequent in the eyes of the American worker, according to a recent survey by the University of Michigan's Survey Research Center. "38 percent of all workers report at least one health or safety hazard. Of all workers, 64 percent consider it 'very important' that they get some kind of protection against this problem, and of these, 51 percent regard the problem as 'sizeable or great.'"

"The magnitude of the concern over health and safety is also reflected in several other of the findings: one-third of all workers report physical working conditions which are, while not hazardous, at least unpleasant, and about one-half want protection against the problem. A full 13 percent of the workers report that they have actually experienced a work-related injury or illness in the last three years. Of these, 42 percent say that the injury had kept them off work for two weeks or more. Two-thirds of the latter group have had trouble meeting medical or living expenses during this period."

Summing up, worker knowledge comes first. Does he know the job steps, dangers and safe procedures? Secondly, does he believe that safe performance is in his interest? Attempts to influence worker knowledge and attitudes toward safety should begin from a view of the favorable and unfavorable outcomes of performance, *as they are viewed by the worker.*

CAN-DO

Is there a skill deficiency in the man?

Is there a skill deficiency in the job?

The CAN-DO explanation refers to the fact that many workers may not have sufficient skill and capability to perform the job or, for one reason or another, their skills may be impaired. Lack of skill is characteristic of the new man on the job, the younger worker, the man who is changing jobs, or the man who lacks the aptitude or intelligence to learn the job. Skill impairments may be chronic or temporary, mental or physical. Chronic impairments may be sensory — hearing and visual losses; decision-making — judgment impaired by mental illness or mental deficiency; and physical ability — lack of strength, precision or speed.

Skill degradation results from alcohol, medicines, drugs, emotional upsets, the effect of recent injuries and illnesses, the onset of an acute illness, extreme fatigue, etc. The list is long. Lack of skill and impairment to performance are well known, and We remind you of them for one reason. . . . to suggest that their contribution may not be the result of the vagaries of humanity, the mysteries of human behavior, but tangible characteristics of the job.

Human factors specialists have provided us with additional insight about skills. They suggest that many skill failures may more appropriately be assigned to the lack of fit between the operator and the job situation. They suggest, for example, that the displays may not be designed to match human sensory capacity and the controls not designed to match the operator's requirements.

Summing up then, we can ask the following questions to see if there is a skill deficiency:

• Does the man have the basic *ability* to perform the tasks? Does he have some impairment which may prevent him from using his ability?

• Has he demonstrated his ability to do it in the past?

• Is the skill *practiced* often enough for retention?

• From the human factors point of view, are the requirements of the job compatible with the abilities of the employee? (This, of course, requires a separate and longer treatment.)

Leftover Skills?

If these questions can be answered positively, we move on to a more complex question which relates *skill* to motivation. Let us then carry the human factors insight of match between man and job situation one step further. The human side of the man-job situation interface is not simply man as an information-processing, decision-making, control-manipulating machine. The human has a number of personal and social characteristics that also have to be fitted to the job. Failure to utilize a reasonable proportion of the resources of a person on the job results in human failure.

The job should occupy and fully engage most of the capabilities and powers of the worker; otherwise, unused capabilities tend to assert themselves anyway and make trouble. Specialized jobs may temporarily inhibit impulses, wishes, daydreaming, randomness. In the long run, they also jeopardize initiative, intrinsic motivation, imagination, self-reliance, freedom from inhibition and finally even health, safety, and common sense.

In the early thirties, culture critic Lewis Mumford said that human and systems engineering have to regard the job created for the human as the major component of the system. He urged the positive fostering of life-fulfilling occupations and the discouragement of those kinds of jobs that neglect to use a major part of human capacities without providing compensatory intensification.

Systems analysis and human engineering concepts can provide us with new leverage on the problem of building systems and occupations that promote efficient utilization of the whole man. A narrow focus on system and human reliability will succeed only in producing jobs that fail to stimulate attention and are accident or loss promoting.

1. *Does the Job Grab You?*

   A person at work is supposed to be engaged in a focused activity. He has a single visual and cognitive focus on the job before him. A job is defined by what the man pays attention to, and what actions he is to take when certain conditions occur.

   What gives a job its order are the rules specifying what shall be attended to and what shall be irrelevant. Safety specialists, for example, are paid to pay attention to safety and the costs associated with unsafe performance. Most people can exclude from attention nearly all events in the world but the ones they are immediately concerned with. They can focus on some few things while eliminating everything else. The man at work will, for example, define as temporarily irrelevant such momentously important facts (to him) as the fortunes or misfortunes of his family. An invisible framework is placed around the work activity which determines the current reality. Other factors, such as home problems, worries and concerns, are usually held at bay. They are outside the framework of the job.

   The prevailing mood at work is close to neutrality, and departures from that mood would be viewed as a source of disorder. Disruptions such as clowning and fighting are discouraged. The rules of the job require paying attention to the task itself. The man who is preoccupied with other things, who is anxious to leave for lunch, or restless for quitting time to come, is showing disrespect for the job and is suspected of having undesirable personality characteristics.

2. *The Prisoner's Dilemma*

   An alternate interpretation is that restlessness and clock-watching are symptoms of jobs that fail to be engaging. It is a paradoxical fact that prisoners attempt to escape as they come closer to their time of release. As the thought of another life becomes more realistic, the conflict between life on the outside and inside the prison becomes unbearable. The closer to release, the greater the tension, and the prisoner finally breaks through the boundaries of the prison role and attempts to escape. Workers approaching lunch time, quitting time, or vacation are in a similar, if less conflicting, bind. The boundaries are not effectively sealing them in, as alternatives beckon in reality or imagination.

   A job is, in effect, a small world. Workers who are fully engaged in these worlds with the boundaries closed around them are not likely to have accidents. It is only on the job that the sheet metal roller can display the kinds of mastery and smoothness of performance that define his identity and import. In like manner, the expert long-haul truck driver makes a coast-to-coast trip seem smooth and effortless as he becomes fully involved in predicting the scene ahead of him and adjusting to it.

3. *Degree of Involvement*

   The power of a job to maintain worker interest depends on its complexities and on worker capabilities. We tend to like both games and jobs that have some uncertainty, require us to perform skillfully, and whose outcome depends on our efforts.

   Involvement increases as the richness of the behavioral repertoire of the task increases. One study has shown that worker job satisfaction in assembly line work and repair increases with the number of different operations performed. In straight-line assembly work, the worker has no mastery over the task. In an automated plant, the worker's mastery increases.

4. *What Keeps Him on the Job?*

   The job situation can be described as a system. The boundaries around the worker are set by more than his interest in the job. A job is part of a homeostatic system with controls which keep the behavior of the worker within certain levels of performance. Some of these controls are within the immediate setting — the rules and regulations, time schedule, physical arrangements, supervision.

Some controls are extrinsic to the immediate setting. They include training, indoctrination, pay schedules, retirement policies, and so forth.

What happens when worker behavior departs from the level required? Additional controls are exerted. Discipline is one kind of immediate control that comes into play when worker behavior goes outside of the limits of the job. In fact, supervisors are advised to behave like thermostats when disciplining employees, to adjust the amount of discipline to both the frequency and severity of the infraction. As the worker departs from the boundaries of the job, discipline is a kind of aversive control to keep him within the frame by making it unpleasant for him to be outside.

5. *Who Takes the Job Seriously?*

The boundaries of the job, the focus of attention are defined by interest and controls and, in addition, by other people. The job is defined not only by the worker, but by his peers, his supervisor and the company. The reality of the job as a focus of attention is reinforced by other people who also take it seriously. When peers and supervisors help the individual on the job, when they form a cohesive work group, and support one another, the boundaries of the job are easier to maintain.

Research on worker morale indicates that these conditions lead to high productivity. The absence of these social conditions of peer, supervisory and company support lead to low productivity.

6. *The Drama of Work*

A job world does not exist by itself, but in relation to a hierarchy of other job worlds. Each job has a place on the ladder of heirarchical importance. The allocation of jobs to persons is strictly limited by general rules of seniority, union and guild rules, company policy, and informal rules that develop in the history of an organization. Each job has a name and a symbolic significance relating it to the other jobs in the plant.

The hierarchical significance of the job is part of the drama of work. When a worker is asked to play a part lower on the scale than he expects, he finds it difficult to stay within the confines of that role. It demeans him. A skilled worker who is asked to do a menial task in the absence of available unskilled help will certainly take umbrage. A skilled worker who has to learn a new machine has lost his place on the ladder.

The drama of hierarchy grasps us as we see ourselves significantly involved with our small worlds linked to the larger. We will refuse to be attentive to a job that does not give us the role that we deserve.

7. *The Drama of Goals*

Staying within the framework of the job is easier when personal goals, company product and national goals are consistently related. As we saw with the space effort, every worker (even those who lost their jobs shortly after the landing) was proud to be associated with the effort. Pride in performance, as in a zero-defects program, is possible when the worker sees some consistency between his own efforts, the company's product and national goals. These consistent motivations are more apparent in times of war, when organizations producing war products and materials manage high productivity and an enviable safety record despite working conditions that would otherwise lead to shoddy performance and a poor safety record. (Morale and productivity both dropped after the war in one plant when the company shifted from airplane doors to automobile doors.)

The dramatic interest of a job is heightened when worker motivations, company goals and national interest are aligned. The man whose job results in a product or service that is socially esteemed is more likely to take pride in his work. Contrast, for example, the role of the shipyard worker with that of the man who pumps jelly into the cavities of doughnuts.

8. *The Space of Free Movement*

   A job, we have said, is a single focus of activity, enclosed by boundaries. The ease with which the worker stays within the boundaries depends, in addition, on the narrowness or width of the boundaries, or what one psychologist has called the space of free movement. This space-time framework may be very narrow, as illustrated by assembly line work, or very wide, as exemplified by the job of the executive.

   As we narrow the space of free movement in order to channel the worker's actions, we increase the reliability of his performance at the cost of occupying his attention. The man who can modify his behavior in response to changes in the job situation is behaving as a human. The man whose performance is rigidly prescribed is behaving like a machine.

9. *The Overlapping Situation*

   What happens to the human potentialities that are left out of the world of work? The world carved out by the job may not coincide with the world in which we are spontaneously engaged. What the worker is obliged to attend to may not be congruent with what he *wants* to attend to. With spontaneous feelings bringing one world alive, and the job dictating another, the worker will be in a state of conflict. He will feel uneasy, bored, restless. He is strongly drawn to matters officially excluded by what is supposed to be the focus of attention. Caught between these two overlapping situations, he has to manage the tension involved in being physically present in one situation and being mentally or emotionally involved in another.

   On the other hand, when the person is spontaneously involved in the work that is officially defined as the focus of his attention, he is at ease with the world, and responsive to the hazards and the difficulties. He is alert, alive, and fully attentive. Many occupations are absorbing in their own right. They are as much fun as the games and sports we enter into, because they are completely engaging worlds.

10. *Incidents and Outbursts*

    Incidents occur when the person is not able to stay within the officially defined boundaries. These incidents break the boundaries around the job. Some incidents may be deliberate, such as horseplay or whistling at girls — incidents deliberately committed to indicate that the men on the job are not only what the job defines them as, but are holders of other and more interesting identities, male and sexual.

Frequently the shift in attention does not remove the person from the job entirely, but shifts his attention sufficiently so that he makes mistakes, has accidents. An examination of clerical work performed in the disbursing division of a large government agency indicates that the number of mistakes increases drastically on the day before a vacation. Like the prisoner, the man about to be released finds his attention drawn away from the job to more attractive matters.

The high level of tension that results from being in an overlapping situation may lead to emotional outbursts. The worker erupts into open anger, impatience or boredom which he no longer bothers to conceal. Bored workers will often tease one another until they discover a vulnerability which they then attack until their target rises to the bait and bursts out of the framework.

11. *Handling Tensions*

Each man tries to find some way to manage tension, either by himself or in cooperation with fellow workers. Some men may look as though they are working at their tasks but be preoccupied with home problems. Sometimes a coalition of workers reasserts some control over the job situation by their agreement about its frustrations.

While this is a threat to the official position, it does offer a safety valve. Another technique allows the workers to redefine themselves by participating in another set of encounters of more interest to them. They may be listening to a sporting event or telling stories, perhaps. It has often been reported that the isolated worker who cannot alleviate the monotony of the job by communicating with his fellow workers because of noise or physical barriers will simply leave the job entirely.

12. *Job Enlargement*

Accidents occur when the demands of the job conflict with matters that spontaneously inspire attention. This lack of congruence leads to lapses of attention, or to the direction of attention elsewhere at a critical time.

What then might lead to a concentration of activity and interest? The research studies are abundantly clear in this respect. The major source of immediate job satisfaction is the job itself. Jobs which increase the possibility of involvement, allow varied activities, utilize one's skills, offer greater space of free movement, and which have symbolic meaning in terms of their dramatic import and goals to be achieved have beneficial results. The quality of work is a vital and educational process for the worker himself, contributing both to his morale and his potentialities, his physical and mental health. What the job contributes to the worker may be just as important as what the worker contributes to the job.

Consider the aquanauts working on the ocean floor in a cylindrical capsule, subject to isolation and confinement, hyperbaric atmospheric pressures, exotic gas mixes, strained and garbled voice communications, physical dangers from other inhabitants of the ocean, danger of getting lost in the dark water, the possibilities of gas supply failures when outside the capsule, and the discomfort of the extremely cold water. Many of them described these stresses as the greatest experience of their lives! The stresses of the situation were clearly perceived and definitely appreciated as being genuine dangers. These dangers were offset by the pioneering nature of the work, the rewards inherent in being the first to accomplish something, and the fact that this experience might be important in each man's future career. The job implications are clear: stress on a job leads to accidents when the job is also monotonous and meaningless. Surely we can't send everyone into capsules to explore underwater space or the planets, but we can use these extreme examples to exemplify the human condition. People want meaningful, not simplified, jobs.

13. *Fun, Games and Jobs*

On the positive side, perhaps we can fashion jobs with the same criteria we use to design games. Games are fashioned to keep the participants entranced and engrossed. In games, all participants are usually equalized by some system of handicapping, "balancing terms," adjusting the betting limits, and so on. The outcome develops as the play goes along. But this is not enough. A maximally successful game (or job) would be one that allows the player to display attributes or satisfy motives otherwise unexhibited, such as dexterity, leadership, self-control, courage, intelligence and other characteristics which allow him to escape from the narrow definition of the job to a more complete expression of himself.

When the game is absorbing, external pulls can be held in check, as the game is a world in itself. Many worlds of work are not as involving and engrossing as games. As a consequence, workers find other worlds that are more absorbing, an overlapping situation that often leads to performance failure, accidents, and loss. A job needs to be a set of events in which the worker can easily become involved. This involvement organizes attention so that the rules for relevance, and what is in fact relevant, are the same. Our task is to construct jobs that are "playful" enough to allow the worker to take them seriously.

14. *Mysterious Accidents*

This analysis should help to illuminate one source of loss that would otherwise baffle us. Accidents often involve people who *nearly always* act safely but simply don't perform safely at *one particular time*. They are not accident-prone. They have the know-how. They know what they should do and why. They believe in safety. They have the skills and seem to be suffering no impairment. They seem to be motivated to perform safely. They are not taking any short cuts. They have not turned off the safety device in order to increase their productivity. They are not deliberately taking chances.

This is a "mystery" or catch-all category. Much safety literature is filled with descriptions of this kind of accident. They include descriptions of the man who was attempting to avoid bees or other insects, who was eating while operating, or was talking to colleagues, lighting a cigarette, reaching for a piece of equipment, thinking about a home problem, mad at his supervisor, or thinking about his vacation. What the safety literature doesn't tell us is *why* the demands of the job failed to rivet the worker's attention, *why* he was vulnerable to distraction.

WILL-DO

Is there a lack of commitment to safety?

With respect to human failure, our third category is WILL-DO: it refers to the readily observable fact that people vary in their safety-mindedness. That's why we need so many signs, signals and warnings to alert some people to the possibility of danger. How many persons are actually aware of the hazards facing them? Dr. Bernard Fox, safety specialist and psychologist, has classified people according to their risk-taking behavior. The predominant class consists of people who simply do not think about the hazard at all. The next most frequent class is uncertain about the hazard, but tends to equate the uncertainty with the essential existence of no real risk to themselves. The third group actually believes that no real hazard exists. The least frequent group deliberately appraises the hazard and the risk and acts accordingly.

What would account for this inability of most people to appreciate and respond appropriately to danger? Could an animal that has evolved and survived so many thousands of years have a capacity to turn off and ignore danger signals from the environment? It hardly seems likely.

Why Workers Won't Perform Safely

The motivational question of what makes people tick has usually been asked in terms of the causes of behavior — the events that precede it. Recently psychologists have reversed this trend. Instead of asking what goes on "in people's heads" before they act, they are looking at the motivational outcomes.

1. *Operant Conditioning*

   The basic idea behind this psychological principle is not new and you can easily see for yourself how it works. The fundamental principle called *operant conditioning* is that behavior (animal or human) is influenced by its effects. We all apply this basic rule in our everyday behavior, unwittingly, and sometimes deliberately. When someone tells us something we are interested in, we pay attention, and our attention increases the vigor and length with which the speaker holds forth. If we turn off our attention, most speakers turn off their conversation.

The theory is old, but its systematic application is new. Some students who had just learned this elementary notion decided to try it out on their psychology professor. They selected the behavior they would reinforce — standing on the left-hand side of the front of the room. When the professor stood in the selected area, the students were attentive, asked questions, nodded in approval, expressed their agreement, looked eager and alert. When the professor moved to the other side of the room, they were disinterested, slumped in their seats, let their attention wander. In no time at all the professor spent all of his time on the left "reinforced" side of the room.

Inspired by their success, the students then selected another behavior to reinforce — writing on the blackboard. Again, the students went into a slump until the professor went to the blackboard. At that point they woke up and provided their alertness, attention, interest, approval, and questions. After a short time the professor's lecture turned to a chalk talk. The students were convinced that the theory had some merit.

Readers can easily demonstrate the value of the basic notion by selecting some behavior by a friend or family member that occurs fairly frequently and reinforcing it when it occurs, as we have suggested above, with attention and expressions of approval and liking, nodding the head, and so forth. The behavior that is now reinforced will increase over its usual frequency of appearance.

The problems we are now encountering in influencing human performance in industry certainly indicate the need for new techniques to help us influence the quality and reliability of performance. Despite all of the efforts to improve human performance in industry, we are still encountering too-numerous cases of human fallibility.

2. *The Technology of Behavior*

Gaining new insight into human behavior presents one of the greatest challenges in the field of loss prevention. Although the basic idea behind operant conditioning is not new, its systematic application has led to a host of new technologies. One early fall-out has been of great benefit to school systems and industrial training programs: teaching machines and programmed learning. A methodical analysis of the behavior to be influenced when students learn (and the means of reinforcing these new behaviors) has led to a revolution in our ways of thinking about everyday human learning in schools, industrial plants, banks and offices.

The everyday world of school and work is not the only target of these new ideas about human behavior. They are now being used to restore the forgotten mentally-ill, now stored in remote human warehouses called mental hospitals. Articles describing the success of this approach to rehabilitation are moving from the technical journals to the mass media. One appeared in the July 11, 1969, issue of *Time*, titled "Reinforcement Therapy — Short Cut to Sanity." The article describes the amazing rescue of several hundred mental patients, previously diagnosed as incurable schizophrenics, from subhuman existence to the point where no signs or symptoms of their former mental illness existed. These dramatic improvements were achieved by the application of the relatively new technique of reinforcement therapy. Unlike the psychiatric techniques which seek to deal with the deep-seated causes of the patients' psychoses, reinforcement therapy concentrates on controlling and guiding everyday *normal* behavior. Its basic principle is that residual signs of normality should be encouraged, strengthened and widened by immediate rewards.

Unbelievably high rehabilitation rates are being recorded in most of the fifty-odd American institutions that are now using the reinforcement technique. Dr. Nathan Azrin, a reinforcement practitioner at the Anna State Hospital, bluntly states, "In the not-too-distant future, virtually all state mental hospital patients can be discharged into sheltered half-way house care."

At this time, operant conditioning is one of the hottest developments in the behavioral sciences, attracting faculty and graduate students as well as foundation and federal money. Applications to worldly problems are pouring forth at a rapid rate. Reinforcement therapists have reported success in working with autistic children, stutterers, obese persons, alcoholics, smokers, prisoners and even unhappily married couples.

3. *The Outcomes of Behavior*

Simplifying the situation, we give the worker two options — safe or unsafe performance. These behaviors can lead to any one of four outcomes, as the following table shows:

		Outcomes	
		Reward	Punishment
Performance	Safe Act	1	2
	Unsafe Act	3	4

The table generates four questions about the motivational outcomes of safe and unsafe acts:

a. Is the desired safe performance rewarding?

b. Is the desired safe performance punishing?

c. Is unsafe performance rewarding?

d. Is unsafe performance punishing?

Let us examine each of these in detail, beginning with the last.

Does Punishment Work?

If you can't easily persuade workers to behave safely, can you supress unsafe acts by discipline and punishment. Yes, but punishment has limited effectiveness, according to the operant reinforcers. It produces unwanted side effects. If you tried not paying attention to your wife in order to get her to stop talking, the side effects would be immediately obvious. If you are not convinced, try not answering, for at least ten minutes, an intimate friend or family member who is talking to you. The person will certainly become more quiet, but the frustration, resentment and hostility you create are typical side effects of punishment used to suppress behavior.

As applied to loss control, punishment can be expected to result in employees' behaving in the same way, but trying harder not to get caught! In addition, when the same unsafe act can lead to both reward and punishment, the employee will be in a state of conflict. He is then likely to take the resulting frustration out of his work by reduced output, substandard performance, damage, waste or fighting with other workers.

Abraham Lincoln must have understood how the negative approach stirs up resistance, for he said, "Assume to dictate to a man's reason, command his action or cause him to be shunned or despised and he will retreat within himself, close the avenues to his head and heart and, though your cause be naked truth itself, transformed to the heaviest lance, sharper and harder than steel can be made, you shall no more be able to pierce him than to penetrate the hard shell of a tortoise with a rye straw."

Of course, a supervisor has to take a firm stand when unsafe actions are repeated. He must show that he is ready to initiate or request disciplinary action. As a last resort, discipline can be a powerful deterrent. Yet there are a number of reasons it is inadequate as a general policy.

The occurrence of an unsafe act is a sure sign that something is wrong. Since unsafe acts do lead to accidents and loss, it is a natural tendency to attempt to suppress these acts through discipline. An unsafe act signals a potential accident, an event which is punishing for the supervisor. Therefore, he seeks to avoid this dissatisfaction.

It is true that one way of changing behavior is to make that behavior lead to an unpleasant situation. Aversive control or punishment involves suppression of behavior by penalizing it. Punishment acts the same way as a reward, but in the opposite direction. Unfortunately, punishment's side effects reduce its effectiveness.

Is Unsafe Performance Rewarding?

If punishment is of limited effectiveness, what can be done? The first step is to understand what maintains and supports unsafe behavior.

The worker who regularly performs unsafe acts when he "knows better" is probably following B. F. Skinner's law of operant conditioning. The unsafe act has been learned and is maintained *because it has been and continues to be reinforced by satisfying events.* Behavior that appears abnormal, because it courts possible injury, is as much a product of reinforcement as is "normal" behavior.

It is learned in exactly the same way. Children learn to throw tantrums because their parents provide the right reinforcements at the right time, say the operant conditioners. If parents refuse to give a child what he asks for when he cries, but give in if he cries harder, and each time they refuse to give in unless the child raises the intensity of his crying, the child will soon reach full-blown temper tantrums.

Unsafe behavior is also reinforced. There are many reasons for performing in a way that management would describe as an "unsafe act". Here are some of them:

- The *advantages and satisfaction* to be gained *at that time* seem greater than the disadvantages and dissatisfactions.

- The unsafe act *"makes sense"* to the employee. If he is challenged, he will explain to the foreman exactly why he thinks it is the most sensible way to do the job. The older employee will say he has been doing it that way for years.

- The unsafe act gives him *personal satisfaction*. It may attract the attention of co-workers, gain approval, admiration and the thrill of taking a chance, the satisfaction of bucking authority, pay back an imagined grudge, make him feel daring and offer many other personal incentives.

- The unsafe act may be seen as having job-related advantages, such as getting a job done sooner, increasing work output, avoiding extra effort, having more control over product quality, and so on.

This list of possible satisfactions produced by unsafe acts indicates that such acts may be strongly motivated. They have consequences for the employee that he enjoys and that therefore support and maintain his unsafe behavior.

1. *The Immediate Effects of Unsafe Acts*

   As viewed by the employee, one effect is the removal of work-related dissatisfactions. The unsafe act may take less time, energy, discomfort. The employee may also experience some *psychological gain*. He may feel independent (as a result of doing the job his way), get attention, obtain approval of co-workers, and enjoy the excitement of taking a chance. There may also be *job-related advantages* of getting the job done faster, higher work output, greater quality control.

In sum, unsafe acts may have the immediate effect of:

- removing work-related dissatisfactions,
- increasing job-related advantages,
- increasing immediate personal satisfactions.

2. *Unsafe Acts and Job Specialization*

Deliberately selected unsafe acts are more frequent in highly specialized jobs. The reasons for this are fairly subtle but easy to follow. It has to do with the amount of job satisfaction available in different kinds of jobs. Generally the more specialized the job, the less job satisfaction it yields. Mechanized industrial production jobs yield the least satisfaction. Satisfaction is higher in automated production, where jobs are less specialized. Satisfaction is highest in craft production systems where job specialization is less marked. *In highly specialized jobs, employees will seek alternate routes for job satisfaction, and one of them is the deliberate selection of unsafe acts.* Since their jobs are so narrowly constructed in terms of the actions permitted that their only personal contribution is "error", employees will search for ways of maintaining their individuality on the job. One of them is "unsafe performance." Providing other alternatives for individual participation is critical.

In summary, *the motivational consequences of unsafe acts may be equal to or greater than the motivational consequences of safe acts. The more specialized the job, the greater the potential for unsafe acts.*

3. *The Principle of Behavior Reinforcement*

The principle is clear: any behavior will sharply increase in probability if it is immediately followed by some event that satisfies a need. If this same satisfying state of affairs follows the same behavior with some consistency, its probability will increase to a high level. Unsafe behaviors may often have that characteristic. They are followed by need-satisfying events. These are both personal and job-related needs.

The purposeful control of behavior requires that *satisfying* events occur immediately following *desirable* behavior. If proper behavior can consistently be followed by need satisfaction, consistent control in desired directions can be achieved. As we have seen, however, safe behavior may not be followed by as wide a variety of satisfactions as unsafe behavior. Technically speaking, unsafe behavior is maintained and controlled by the satisfying events that follow it.

Summing up, it is difficult to suppress unsafe acts by disciplinary methods. Further, the unsafe acts may be rewarding. Put simply, unsafe performance may be highly motivated.

4. *What To Do About Unsafe Acts*

Every specialist knows that an unsafe act is "an accident about to happen." When we recognize these deviations from safe practices, we do our best to eliminate the satisfactions generated by unsafe behavior. This is an obvious solution to the problem, but it is only a small part of the solution. For there are many ways of behaving unsafely on the job, compared to the number of safe methods. The safety supervisor who spends his time trying to eliminate them all will soon run out of time, patience and energy.

An alternative, but less obvious, strategy is to increase the readiness for safe behavior. With these two strategies to choose from:

a. eliminate unsafe acts,

b. increase the probability of safe acts,

we will be better off emphasizing the second strategy. There are many ways the safety specialist will benefit from this second strategy:

a. He will *design the job* so that *unsafe acts are not rewarding.*

b. He will spend more time *recognizing* and *rewarding* safe performance compared to *disciplining* employees for unsafe performance.

c. He will *strengthen* and enhance the importance of the standard of safe performance.

d. He will *focus attention* on the importance of safe performance.

e. He will *remind* employees of the techniques of safe performance.

f. He will be seen by the men who work for him as *interested* in their welfare rather than as a disciplinarian or "nag".

g. *Accentuate the Positive*

Dr. Lloyd Homme of the Research Department of the Westinghouse Learning Corporation, Albuquerque, New Mexico, calls this application of reinforcement theory "contingency management." Importantly, he says a contingency manager can be taught in a short time. Parents, teachers, supervisors, and a variety of persons in positions of influence have been taught these techniques. While their application to safety and loss control problems is only beginning, there is no doubt that systematic application will change the picture.

In order to fully appreciate the dynamic reversal from our existing practices that the above program implies, let's review for a moment the motivational direction that many industrial programs have taken through the years.

Loss control personnel usually concentrate on unsafe practices and conditions, injury, doom, death, damage and other dismal topics. They are like the pediatrician who gives a repeated series of shots to a child. Soon the child cries at the sight of him. By highlighting unsafe practices and their effects in their contacts with workers, safety specialists may have wrapped the same mantle of doom around themselves.

Secondly, they may not be giving enough attention to Skinner's principle of behavior reinforcement. Regular safety inspections focus on unsafe practices and conditions. The safety specialist is more likely to deliver "punishment" than "reward." Seldom is safe behavior emphasized or even mentioned in a report.

The card that has been handed to the worker, stating that he has been committing an unsafe act that could have injured himself or a fellow worker, and that he ought to think of the consequences in the future, makes this point well. The little card goes further, pointing out that he should look for another fellow worker violating an established safety practice and pass the card along to him in the interest of safety.

Safety personnel in the field are generally trained to concentrate on *un*safety in their contacts and communications and observations. Even promotional materials reflect the overall negative approach of our safety motivation effort by giving major attention to what's *wrong* and little attention to what's *right* in behavior.

Is Proper Performance Punishing?

We have looked at the outcomes of unsafe performance — efforts to make unsafe acts punishing and to reduce the rewards connected with unsafe acts. We now face the possibility that the recommended safe performances may be avoided because they are in fact punishing.

As we have said previously, the presence of unsafe acts indicates that the employee sees the unsafe acts as having more immediate satisfactions than the safe act. There may indeed be obstacles in the way of safe performance.

Eliminating Obstacles To Working Safely

To reduce obstacles against working safely, the safety specialist can do a motivational analysis of the "punishments" that are the outcomes of safe acts. The following procedures can be adapted to his own experience.

1. Determine the obstacles. Time? Effort? Discomfort? Interference with production? Personal or group dissatisfactions?

Discussion with the workers involved should help identify these obstacles.

2. Reduce the obstacles. Use group participation techniques. Get ideas and suggestions from employees. Make it the topic of a safety meeting. Invite their ideas and reactions to any solutions you might come up with.

3. Evaluate any changes on a cost-benefit basis.

4. Discuss the change with the workers involved and get their agreement and cooperation before making the change.

Is Safe Performance Rewarding?

Motivational analysis suggests that we explore the balance of outcomes for the worker resulting from safe and unsafe acts. The most revolutionary innovation in the application of psychology to behavior is the concept that behavior can be controlled by increasing the rewarding outcomes that follow it. In psychology, this technique is called operant conditioning.

1. *Beginnings of the Technique*

   Techniques for the control of behavior were first worked out by Burrus Frederic Skinner at Harvard University just prior to World War II, and were applied in a manner so unbelievable that the generals in the Pentagon who saw the results of his work shook their heads in amazement and disbelief. Skinner applied his behavior control techniques to teaching pigeons to aim aerial bombs. Although Skinner was able to demonstrate the pigeons' accuracy, he wasn't able to cross the credibility gap of the generals' disbelief that any such feat was possible.

   After World War II, when *Life* magazine published pictures of pigeons he had taught to play ping-pong and peck out tunes on the piano, Skinner came to public attention.

   The Skinner animal laboratory spawned a set of psychologists who have perfected the art of animal training. Many of the current animal performances (the tricks performed by dolphins, pigs depositing coins in piggy banks, chickens playing baseball) are developed by students of Skinner.

Drug manufacturers were among the first to apply Skinner's operant conditioning to a serious problem, testing the effects of drugs on animals before they are tried on humans. Using Skinner's procedures, the animals are trained to repeat a stable performance over and over again. Then the effect of the drugs, including the time it takes for the effect to take place, the effects of the drug on performance, and time for disappearance of the effects, can be precisely measured against a stable base line.

While Skinner's students were busily working out the studies of animal behavior, Skinner went on to humans and soon spawned a whole new set of advanced technologies. A host of new business, organization, hardware and software applications sprang up behind him.

2. *From Animals To People*

We can find one of the roots of his theory in the work of Russia's Ivan Pavlov. Pavlov is the man who first made "conditioning" a familiar word, standing for the mysterious induction of changes in behavior. In fact, Pavlovian conditioning has not been as influential in human affairs as Skinner's experiments.

Pavlov's classic experiments with dogs first demonstrated the principles of what Skinner now calls "respondent conditioning." Previously neutral cues, like a buzzer which goes off just before the animal is fed, or a red light which goes on just before the animal is shocked electrically, are learned by the animal as signals of the onset of painful or pleasurable events. Advertisers who associate their products with the things we hold most dear in life, such as pretty girls, are using Pavlovian conditioning procedures.

This notion of conditioning was expanded by Skinner. He showed that dogs, rats, pigeons, and even men are conditioned not only to respond to signals of impending events, but conditioned by the *consequences* of their actions. The basic observation is that behavior is stronger in the future when it is followed by an event we ordinarily call a reward, and weaker in the future when it is followed by an event ordinarily called punishment.

This new principle is called *operant conditioning* by Skinner, because it deals with behavior that operates on and changes the environment in some way, as compared to Pavlov's *respondent* behavior which is responsive to cues and signals in the environment that have previously been associated with rewarding or punishing events. Skinner and his associates have systematically studied the details of the way in which this principle operates, and have made many applications to complex human behavior.

They emphasize desirable behavior or a first approximate of it. In another applied study with juvenile delinquents, Skinnerians had juvenile inmates learn by reading comic books. In working with stutterers, one researcher found he could make more progress by ignoring the stuttering and teaching a new mode of fluent speaking. Another Skinner student found that the road to reduction of obesity was to ignore how much was eaten and to teach a new method of eating. He noted that the cue used by some overweight people to ingest food was the fact that the fork or spoon was empty. He taught the obese to use a new cue, the food chewed and swallowed and an empty mouth, as the signal for lifting the next mouthful instead of the empty fork or spoon.

In addition to accentuating the positive, these experimenters found that the desired behavior had to be reinforced at the time it was demonstrated. Athletic coaches have used this principle of behavior reinforcement intuitively for years. They wait until they see the move they want or an approximation to it, and they congratulate the player. When that behavior is stabilized, they wait until they see an improvement before giving more praise and encouragement.

These experimenters were able to change behavior by focusing on the desirable behaviors and increasing their advantages. In mental hospital work, the researchers found that the patients sat and watched TV for most of their spare time. TV watching then became the "reward" for desirable behavior. Another experimenter, teaching four-year-old children, found he was able to exchange ten minutes of study for five minutes of running and screaming and whirling teacher around in his chair!

3. *Safe Behavior Reinforcement*

A reinforcement, these psychologists say, is a "high probability behavior," one that would occur if the person were left to his own devices and preferences, and can be used to reinforce the less probable ones, as in the illustrations above.

Most important from our point of view, the concept and technique has direct application to loss control . . . and we call it Safe Behavior Reinforcement. The Center for Programmed Learning for Business, in Ann Arbor, Michigan, has recently reported a successful program of reducing back injuries from improper lifting among supermarket employees in a large chain. Supervisors were taught the technique of positive reinforcement of correct lifting practices. Instead of responding mainly to errors, as was their usual practice, they concentrated on finding people who were lifting correctly and giving them attention, interest, praise and recognition. The procedure paid off far beyond expectations, according to Geary A. Rummler, the Director of the Center.

Certainly Safe Behavior Reinforcement is not entirely new. It has been the basis for prizes and contests, and has been included in supervisory training programs. It is practiced in industry and in everyday life. Yet an explicit and methodical application of the technology makes a difference. Safe Behavior Reinforcement requires two essential management procedures.

a. Organized and frequent confrontation by supervisors and staff personnel with employees, with immediate recognition of desired safe behavior, using attention, approval, praise, recognition and material rewards presented on the spot to the worker who is performing in the desired manner.

b. Re-design of jobs and equipment so that operator behavior produces a maximum of satisfying rewards and a minimum of job-related or personal punishment. In sum, Safe Behavior Reinforcement operates through both effective supervision and job design. The desired safe behavior produces rewards on the job itself and from the supervisory personnel.

Safe behavior reinforcement has many advantages:

a. It does not have the unwanted side effects of discipline, avoidance, conflict, frustration, and aggressive damage.

b. It increases the job satisfactions experienced by the employee.

c. It changes the nature of the relationship of foreman to his employees from watchdog and monitor to helpful resource.

d. It creates an atmosphere of mutual reciprocity between supervisor and employee.

e. It increases the probability of safe behavior, rather than reducing the probability of unsafe behavior. It is more direct in its effects.

How Can Management Use Safe Behavior Reinforcement?

Summarizing this chapter, the final objective is to reduce losses, injuries and property damage caused by unsafe acts (improper procedures, use of unsafe tools, equipment, materials, etc.) by increasing safe acts and reducing unsafe acts. Safety specialists can follow the following strategy:

1. *Increase the satisfactions associated with working safely:*

   - The ratio of time spent on recognition of proper behavior should increase compared with disciplining employees for improper behavior.
   - Recognition of safe behavior should be separated from discipline.
   - The ratio of recorded commendations to recorded reprimands should increase.
   - Safety observations and inspections should record an increasing ratio of safe to unsafe behavior.
   - Tangible recognition and awards should be given for safe behavior.
   - Proper behavior should be rewarded by increased opportunities for job satisfaction, work group satisfac-

tion, supervisory support and rewards from the company.

- Increased emphasis on the personal gains of working safely.
- Increased emphasis on the job gains of working safely.
- Increased individual and work-group participation in developing recommended safe procedures.
- Develop an operational procedure for identifying and rewarding safe practices. For example, each month a specific safe practice could be selected, or a set of safe practices for doing different jobs. If possible, an audit of the occurrence of these safe practices could be made to determine the effects of the program.

This strategy should be accompanied by the following strategies suggested by the motivational analysis discussed in this chapter.

2. Whenever possible, increase the overall satisfactions yielded by the job in order to increase the amount of energy and attention given to the job as compared to non-job activities.

3. Analyze the balance of rewards and punishments (satisfactions and dissatisfactions) associated with safe and unsafe acts, as *they are seen by the workers themselves*.

4. Analyze the job design and procedures for their contribution to undesired behavior by rewarding unsafe acts and punishing safe acts.

5. Reallocate some of the energies and attention paid to *unsafe acts* to safe behavior reinforcement.

6. Reduce the rewards and satisfactions that are the outcomes of unsafe acts.

7. Eliminate the obstacles and punishments associated with safe acts.

From Theory To Practice

This chapter has dealt with the strategy, not the tactics, of behavior motivation. Since behavior is influenced by its outcomes, there are several things safety specialists can do to maximize this relationship.

1. Make sure that the behavior learned in training programs and the behavior reinforced on the job are the same. There is often a considerable difference between the behavior acquired in training and that maintained or reinforced on the job. The know-how and skill learned in training need to be supported on the job.

2. Design jobs so that they are sufficiently rewarding to claim a man's attention. Some jobs are technologically frozen and don't permit enlargement. In that event, prizes, contests or games of any sort that will increase the amount of attention paid to the job are an asset.

3. Design jobs so that the rewards for safe performance are much greater than the rewards for unsafe performance. Many problems are designated as "training problems" because men continue to use unsafe acts although they know better. The safety specialist is often blamed because men "forget" their safety training on the job. He can now indicate what conditions must exist on the job before the safe behavior can be expected to occur.

4. Pinpoint the safe act so that the employee can see what behavior is being reinforced. Keep a count of the number of times it occurs. If possible, get employees to participate in selecting the safe act and keeping a count of occurrence. Of course, it is essential that the employee has the know-how and skills to perform the act.

5. Arrange the reinforcement so that it is given in time to be linked to the act. One problem with contests is that they don't reinforce any particular performance. An appraisal system is more effective when it is based on specific results.

6. Select reinforcers that are viewed as such by the employees. Every improvement, however slight, must be rewarded to get to the final goal. There are many rewards besides money that managers can dispense. These rewards include: measurable positive feedback on progress towards a goal; responses to requests made by the worker; social responses to his performance, including smiles, praise of performance, con-

versation, asking for suggestions; public recognition of his performance. There are no formulas to tell you what will, in fact, be most rewarding in each individual case.

BIBLIOGRAPHY

1. Bird, F. E. Jr. and Schlesinger, L. E., "Safe-Behavior Modification" *ASSE Journal*, American Society of Safety Engineers, Illinois, June 1970.

2. Goldiamond, I., "Stuttering and Fluency as Manipulatable Operant Response Classes" in Krasnas, L. and Ullman, L. P. (eds.) *Research in Behavior Modification.* New York: Holt, Rinehart, and Winston, 1969.

3. Harris, D. H. and Chaney, Frederick, B., *Human Factors in Quality Assurance.* New York: John Wiley and Sons, 1969.

4. Holland, J. G. and Skinner, B. F., *The Analysis of Behavior.* New York: McGraw Hill, 1961.

5. Homme, L. E., "Contingency Management," Albuquerque, New Mexico: Westinghouse Learning Corp., 1969.

6. Skinner, B. F., *Science and Human Behavior.* New York: MacMillan, 1953.

7. Staats, Arthur W. and Staats, Carolyn K., *Complex Human Behavior.* New York: Holt, Rinehart and Winston, 1963.

8. Ulrich, Roger, "Behavior Modification; Theory, Research, and Practice;" *Research Bulletin.* Volume 11, Number, 1, State of Michigan, Department of Mental Health.

X. Products Loss Control

The recent emergence of a new threat to corporate profits, in the form of products liability claims, is cause for substantial concern, not only among the insurance community, but for those involved in manufacturing and selling as well. Product liability claims occur when a product's consumers or users sustain injury, property damage or business interruption losses caused wholly or partly by the consumption or use of a faulty product. When this occurs, the user can, and with increasing frequency does, sue the product's manufacturer or seller to recover for damages and for the pain and suffering inflicted.

The heavy costs of these claims, whether covered by insurance or directly settled, constitute only a portion of the total losses incurred by industry through failure to manufacture and distribute safe and reliable products. Other costs or losses, not insurable and in some cases not directly measurable, may far exceed the costs of legally-imposed product liability losses. These include, but are not limited to, the costs of product warranty programs when the products do not perform as warranted, damage to product or corporate reputation growing out of product liability litigation, and the expense of product field modification and recall programs.

One has only to look at the automobile industry to see how defective and unreliable products can increase the cost of warranty programs and the damage to public confidence, and to witness the expenditure of millions of dollars annually to recall and modify unsafe products after they have reached the consumer.

Over the past 15 years, the number of product liability lawsuits has grown from a mere handful to approximately 700,000 annually. Awards of $100,000, to $200,000 are not unusual, and there are occasional instances of $1,000,000 and $2,000,000 awards. Why has the number of products liability claims increased so drastically?

One reason is the growing public awareness of the consumer's legal rights when injuries and damage caused by defective or unsafe products occur. This awareness is abetted by publicity given to large products suit awards, by legislation (such as the Consumer Product Safety Act and the Occupational Safety and Health Act) being passed on both the local and national levels, and by plaintiffs' agressive attorneys who encourage products liability suits. The law has been changed in most states to allow lawsuits by the injured party directly against the manufacturer. Previously, the manufacturer had considerable immunity against products liability suits because the injured party could sue only the person who sold him the product.

Another recent change in the law in many jurisdictions has been the trend to "strict liability." This legal theory does not require proof of the manufacturer's negligence. If the product is proven unsafe at the time it leaves the manufacturing plant, and if the unsafe condition is the cause of the injury, the injured party can recover in many states. In addition to the relatively new recovery concept of strict liability, an injured person has two other avenues of recovery for his injury. First, he may sue the manufacturer under tort law on the traditional allegation of negligence; secondly, he may sue under contract law on the basis of *a breach of express or implied warranty*. Future changes in the law allowing class-action suits by consumer groups will undoubtedly increase both the liability and non-liability problems arising from products.

It is not possible to separate the control of products non-liability losses from the control of products liability losses. The reason becomes obvious when we consider customer expectations for product purchase and use:

1. The customer wants products that have high utility at reasonable cost. Satisfaction of this need is the reason a company succeeds in business.

2. The customer wants products that can be used and maintained with reasonable safety.

3. The customer wants products that are reliable.

4. The customer wants products that are easy to maintain and to service.

All of the above items relate to the control of the products non-liability losses. Items 2 and 3 and, to a lesser extent, item 4 are the heart of a products liability control program. In the following pages, we will outline a program that will enable your company to maintain control in these areas. This kind of program, used effectively, will assist management in the prevention and control of the unnecessary costs of products accidents.

MANAGEMENT LEADERSHIP

The wholehearted backing of management must be given to the Loss Control Program if it is to succeed. Management *must* clearly communicate to all employees that the control of products losses is a key objective. Each department within the company has an important role to play in the program and must commit the time and resources required for the program's success.

A management decision to skimp on quality control procedures on materials needed to guarantee the safety or reliability of a product solely to remain competitive may prove the most expensive course of action in the long run. To meet the demands of a competitive market is not a viable defense in a court of law, should an accident occur because of an inferior quality product. The savings produced by a questionable "economy" measure have, on occasion, been lost through a single product liability lawsuit or recall program, to say nothing of the high warranty costs and the immeasurable loss of good will and future sales. Although no company can afford to spend unlimited amounts of money for product safety and reliability, most companies realize a more than adequate return on their investment in products loss control in terms of reduced costs from business and liability losses. The fact is that most companies today spend far too little on products loss control, and this is frequently reflected directly and indirectly in reduced profits.

It is top management, therefore, which must set the basic objectives and establish priorities for the program. Figure 1 is a sample letter which might be sent to all employees, informing them of the start of a formal products loss control program, emphasizing its extreme importance and indicating management's commitment to produce a safe product.

COORDINATION OF THE PROGRAM

The Products Loss Control Program will involve most departments in your company and will require the application of many diverse disciplines. Because of this, strong coordination will be required to assure a thorough and systematic approach. The program can be controlled and coordinated by the formation of a separate Products Loss Control Department, or by a group or committee composed of representatives from the various departments involved. It is suggested that, in order to provide capable, forceful leadership for the program, the chairman of the Products Loss Control Coordinating Group or Committee or the head of a

Belleview Electronics, Inc.
8900 Industrial Highway
Los Angeles, California

To All Employees:

The very essence of our company's continued profitability, growth, and security (and the security of all of us) depends on our ability to furnish to our customers products which are reliable and safe. When we manufacture and ship defective products we risk the possibility of serious injury or property damage to those we serve and also damage to the reputation of our products and our company.

We have always been proud of the reputation our products have had and the trust our customers have shown in our ability to produce reliable and safe products. Each and every one of our employees has contributed greatly to this reputation and trust by his job performance.

Present trends in our company make it imperative that we control the human and economic risks involved when our customers receive defective products. This can only be accomplished by the type of systematic, coordinated program which we are starting immediately. In the coming weeks you will receive further details on the program from your supervisor.

The management of Belleview Electronics is confident that all employees will contribute to the success of the program through their cooperation and continued high level of job performance.

Thomas G. Grebe

Thomas G. Grebe
President

Figure 1.

Products Loss Control Department be a top-level member of corporate management.

The loss control committee or members should have the ability and authority to speak for their departments. The various department heads, therefore, or their designated assistants, are likely candidates.

The following departments or their equivalents should be involved in the Products Loss Control Program to give it maximum effectiveness: Design Engineering or New Product Development, Manufacturing, Quality Control, Service and Installation, Legal, Insurance, Sales and Advertising, Purchasing, Personnel, Public Relations, and Safety.

Figure 2 is a sample memo confirming the appointment of one such department head to a coordinating committee and Figure 3 lists some suggestions on the effective use of coordinating groups.

COORDINATING GROUP OR DEPARTMENT

The coordinating group or department usually functions in a staff capacity to corporate management and to the departments involved in the Products Loss Control Program. It may have authority to initiate limited action, but usually major decisions must have the approval of top management. The coordinating unit will assist management in setting general policy for the overall program. One of its first duties should be to identify problem areas in the program and to set objectives and priorities. As priority objectives are set and met, the unit can establish long-range goals for the departments involved in the program. The primary areas of responsibility which the unit must coordinate include:

1. Prevention of product losses by:

    a. Design of safe and reliable products.

    b. Assurance that manufacturing and quality control standards are adequate to produce products to design specifications.

    c. Clear and accurate representation of products to customers in a way that avoids encouraging business and liability losses.

INTEROFFICE MEMO

ALLIED CASTINGS INTERSTATE HIGHWAY 45 HOUSTON, TEXAS

TO: Mr. Donald Allard, Manager

RE: Products Loss Control Group

Dear Don:

    I am sending this letter to confirm your appointment to the Products Loss Control Group which we discussed in our November meeting. Bob Hughes will serve as your Group chairman and will coordinate any problems between the departments.

    I feel that the excellent job you have done in the past in keeping rework at a minimum will place you in a position to give valuable advice to other committee members. As senior man in the production area, feel free to make whatever commitments that you think are prudent to maintain our reputation as a quality supplier of high pressure cylinders.

    The initial meeting will be held on Friday at one o'clock in the main conference room, at which time we will discuss the kick-off for the campaign. On Monday, all the employees will receive a letter from me explaining the program's objectives. This Friday's meeting should enable you to answer any questions your supervisors may have.

                                               Sincerely,

                                               James Bartlett
                                             President

Figure 2.

**MAKING EFFECTIVE USE OF
COORDINATING GROUPS**

1. Strictly define the purpose of the group. A written statement will achieve clarity and will eliminate the need for group members to spend time deciding exactly what they are supposed to do.

2. Make sure the authority and responsibilities of the group are clearly specified and understood by the members.

3. Keep the size of the group within manageable limits. Realizing that as the size increases it is more difficult to conduct an efficient meeting, it is recommended that the number not exceed ten persons and preferably should be kept to about six.

4. The chairman of the group should be able to conduct an efficient meeting.
    (a) He should prepare the agenda in advance so the members will have time to study the subjects and consider their views.
    (b) He must insure that all members are heard, encouraging the reticent and keeping the talkative in check.
    (c) When all the contributions of the members are in, the chairman should state the consensus of the meeting to be sure he has properly understood it.
    (d) He should see that minutes of the meeting are distributed for correction and review prior to their release.

Figure 3.

   d. Continual field monitoring to detect indications of customer dissatisfaction, unreliability, or unforeseen product hazards.

2. Legal defense of products liability claims (before and after the accident) by:

   a. Coordinating the defense of products liability claims.

   b. Documenting all efforts to produce safe and reliable products.

c. Periodically reviewing vendor relationships with emphasis on "hold harmless" agreements.

3. Maintaining good public relations by:

   a. Keeping in contact with federal and state governments with regard to laws, codes and standards affecting products.
   b. Participating in technical societies, trade associations, etc., in development and upgrading of product safety and reliability standards.
   c. Continuing contact with customers and the general public concerning acceptance of the products with regard to safety, reliability, ease of maintenance, and utility.
   d. Reviewing public statements and publications made on product accidents, product recall or field modifications.

Some of the more important duties of the coordinating unit are:

1. To coordinate the program activities of the involved departments to assure that they work together harmoniously.

2. To assure an adequate flow between the departments of information such as:

   a. Design test results.
   b. Complaint/incident/accident investigation reports and an explanation of their significance.
   c. Service Department reports on misuse.
   d. Quality control standards and their importance.
   e. Existing or new safety standards from technical societies, trade associations, governmental bodies, etc.
   f. Changes in the law or the legal climate pertaining to products.
   g. Changes in consumer attitudes toward product attributes such as safety, reliability, maintainability, etc.

3. To review complaint/incident/accident investigations to determine whether action taken or planned is adequate, and conduct analyses of these investigations to determine whether any trends are developing.

4. To recommend special action to top management when necessary concerning:

   a. Changes in product design, manufacturing, quality control, service and installation, advertising and sales, etc.
   b. Product recalls or field modifications.

DESIGN OR NEW PRODUCT DEVELOPMENT DEPARTMENT

American designers have the reputation of designing useful, practical products which can be produced at economical cost. These design efforts and our mass-production manufacturing methods are two of the reasons we enjoy such a high standard of living today. Unfortunately, product safety has too frequently been omitted from the designers' primary goals.

In the early 1960's, a chemical manufacturer produced and sold a masonry waterproofing compound that had a flash point of minus 50 degrees Fahrenheit. The material was literally so explosive that it was almost impossible for home-owners to use it without serious consequences. One death and a number of serious injuries resulted when the vapors of the flammable liquid ignited and exploded. The formula was reworked, and the company was able to remarket the product with a flash point of plus 70 degrees. It is now relatively safe to use. In another instance, the manufacturer of a large industrial machine neglected to put adequate guards on an in-running nip point. Two amputation accidents occurred which cost the manufacturer almost $300,000 in legal awards.

This condition still exists in American industry today; however, it is becoming increasingly expensive to the manufacturer who ignores this responsibility, as consumers win more and larger judgments in products liability law suits.

One of the primary functions of the Design or New Product Development Department should be to design reliable products which can be used with reasonable safety. It is more practical and usually much less costly to build reliability and safety into the product at the design stage than to suffer the consequences of catastrophic business and liability losses from failure to do so.

The level of safety required for a reasonably safe design is not static. The state of the art, court decisions, and public opinion necessitate periodic re-evaluations of product safety design criteria. What was adequate yesterday may not be acceptable today. Safety standards, as defined by the courts, will become even more stringent.

We have stated that product reliability is intimately related to product safety. In many cases an unreliable product is an unsafe product, irritating customers and making them less reasonable in settling products liability claims. Therefore, product reliability should be given utmost consideration in product design.

Products should be reasonably safe during normal use and also during normal service and maintenance. If there are any uses to which a consumer might put your product, even though you did not specifically intend it for those purposes, the law frequently requires that the product be safe during those uses also. The courts are saying today that the manufacturer has a responsibility to see that his products are safe for any "reasonably foreseeable uses" to which the consumer might put them.

You should know what types of accidents and malfunctions could occur with your products and how serious they could be. When analyzing potential types of accidents, include bodily injury, property damage, business interruption and extra expense losses. (If your product will be used in a large chemical plant and is necessary for the operation of that plant, or if it is used in an automated production line and the failure of your product will stop production completely, an accident involving a defect in your product could possibly result in business interruption losses involving thousands of dollars per day.) After potential accident types have been determined, the next step is to decide how to eliminate or to minimize their effects before they occur.

The following is a list of factors and techniques which the Design or New Product Development Department should consider in determining whether products are reasonably safe.

Codes and Standards

Does the product conform to all applicable safety standards (state and federal codes and regulations, industry standards, technical society standards, machine guarding standards)? These standards, in most cases, should be considered the minimum for safety and reliability. In many cases there will be *no* standards pertaining to the product or the standards which do exist will be *inadequate* to insure reasonable safety. In these cases, additional testing and development should be performed to develop new and better standards.

Human Factors Engineering

Have human factors been considered in the design or development of the product? Some of the human factors which should be considered are the physical, educational and mental limitations of people who may use the product and the inevitable tendency of people to attempt to use products for purposes other than those for which they were designed.

Critical Parts

Has the product been analyzed for "critical" parts or components? *Critical parts or components* may be defined as those whose failure could cause serious bodily injury, property damage, business interruption, or serious degradation of product performance. If there are parts or components in the product whose failure could cause a serious accident, then those parts should receive special attention in design, manufacturing, and representation of the product's qualities to the customer. Some of the techniques for analysis of critical parts or components are systems safety analyses, such as failure mode analysis and fault tree analysis, and analysis of the system for reliability. An analysis of product reliability is also necessary to estimate the cost of a product warranty program. In addition, where feasible, products should be field-tested under conditions *more severe* than they can be expected to undergo in actual use. Another technique for detecting critical parts or components is to review the existing failure experience of similar parts, components or products and relate this to the product.

After the critical parts or components are identified, it may be necessary to redesign the product so that critical parts or components will have increased reliability or will outlast the product itself. When this is impractical, product users must be warned of the possible hazards which could result from failure of critical parts or components. They must be instructed in inspection and maintenance techniques to detect and prevent potential failures. Depending on circumstances, these warnings and instructions should appear on the product itself and in the accompanying literature.

Packaging and Handling

Has product packaging and disposal been analyzed for hazards? For example, packaging products in aerosol containers usually adds to the hazards because the contents of the containers are under pressure. If a corrosive material is being sold, the container, of course, must be of a corrosion-resistant material. The packaging must also be adequate to protect the product from damage during storage and shipment.

Safety Audit

Is a safety audit made of all new designs in the blueprint and prototype stages? A safety audit is a systematic search for potential hazards which may have been overlooked by the product designers or developers. Many products are so complex that it is difficult for one man, the designer, to foresee and eliminate all the possible hazards. The safety audit should involve the Design and/or New Product Development, Manufacturing, Quality Control, and Service and Installation Departments and, in some cases, the Legal Department. The safety audit often has the additional benefit of detecting potential manufacturing, quality control, and service and installation problems before the design becomes final. In such cases, it may save thousands of dollars on the production line.

Warning Labels and Instructions

Have the operating, use and maintenance instructions been reviewed by the Design and Legal Departments to determine that potential product users will receive adequate instruction and warnings concerning potential hazards and proper product maintenance and use? Warning and instruction decals should also be placed on the product itself to comply with legal requirements for adequate hazard warnings. Many courts have held that for a warning to be adequate, it must (1) warn the user of the hazards involved; (2) instruct the user on how to avoid the hazards; and (3) where not obvious, state the possible consequences which could result if the user ignores the warnings.

Make certain that all sales brochures and product advertising have been reviewed by the Design and Legal Departments to be sure that they accurately depict the product's capabilities and show only safe operating and maintenance procedures, and that implied or express warranties are not inadvertently given.

Design and Changing Technology

Design engineers and technicians should participate in technical societies and trade associations and maintain contact with regulatory authorities, to keep abreast of new product safety standards as well as the upgrading of existing standards.

The Design or New Product Development Department should be periodically briefed by the Legal Department about current events in the business and legal climate related to products liability. Subjects which should be covered include:

a. The changing social and legal climate.

b. The upward trend in the number and size of products liability court awards.

c. The meaning and significance of negligence (including that of the designer), strict liability, "foreseeable uses," and express and implied warranties.

A design manual should be written and kept current to serve as a guide for experienced designers and as a training aid for new designers in product reliability, maintainability and safety. Job descriptions and job performance requirements should be written for product designers to include responsibility for safety as well as product reliability and maintainability. This should also be a help in developing and keeping qualified personnel.

Following is a partial checklist of hazards which can result from design errors:

A PARTIAL CHECKLIST OF POSSIBLE HAZARDS

- Poor biomechanical design (machine components and controls difficult to use because of poor design)
- Product component or environmental incompatibility

- Possible contamination (of or by the product)
- Possible corrosion (will it affect operation or properties of the product?)
- Insufficient or excessive ductility or hardness
- Electrical shock hazards
- Insufficient fatigue strength
- Flammable or explosive properties
- Toxicity (ingestion, inhalation, skin absorption)
- Possible irritant or sensitizer
- Inadequate machine guarding (points of operation, power transmission)
- Insufficient control
- Instability
- Product failure (or failure of critical components)
- Service, maintenance or adjustment hazards
- Structural weakness
- Unexpected movement

MANUFACTURING

After a reasonably safe and reliable product has been designed, the Manufacturing Department has the task of turning the design specifications into a finished product. Manufacturing errors can also mean that an unsafe and/or unreliable product will reach the consumer. Repetitive assembly-line work can become boring, causing the worker to lose pride in his work performance and in product quality. All Manufacturing Department employees make a vital contribution to the safety and reliability of the products they work on and they should all be reminded of that fact at periodic intervals.

There are many ways in which Manufacturing can contribute to the Products Loss Control program. One of the more important ones is to train, educate and motivate manufacturing employees by instituting job analyses and establishing specifications for each job. This is one of the most productive methods to establish efficient, low-error work performance.

Manufacturing should also insure that all workers — both older ones shifted to new jobs, and newer ones — are thoroughly instructed and trained in proper work procedures. One excellent method to help accomplish this is Proper Job Instruction.

Manufacturing should follow up with job observation to determine that work is being performed with maximum job efficiency. Total Job Observation serves as a check to insure that the worker is doing his task correctly and efficiently.

In addition, Manufacturing should be expected to work with the Personnel Department to develop education and training programs to upgrade skills; to seek to instill pride of workmanship, by insuring that each employee understands he is making a vital contribution to a quality product; and to initiate zero defects, error-free performance or other error elimination programs with the assistance of Personnel and Public Relations.

The Manufacturing Department can work with Quality Control to identify and eliminate trouble spots in the production lines, and can feed back to the Design Department any difficulties in maintaining specifications. In some isolated instances, Design may have set specifications which are more stringent than necessary, in which case a conference between the two departments will often result in more realistic requirements.

Manufacturing should avoid deviating, however, from design specifications without prior approval from the Design and Quality Control Departments. Design is in the final position to determine whether non-adherence to specifications in a particular case will affect product safety or reliability.

Maintaining manufacturing records for five years beyond the life of the product is vital, especially on critical parts and components (such records are necessary for product recall or field modification programs and for defense of liability suits). The records covering critical parts and components should be sufficiently complete to identify the batch or lot of raw mateirals from which they came or were made and the finished products in which they were used.

Manufacturing should serve on the Products Loss Control Group or as an adviser to the Products Loss Control Department, and participate in safety audits on new designs.

QUALITY CONTROL

We have seen in the preceding section that the Manufacturing Department has the job of translating design specifications into safe finished products. It is the job of the Quality Control

Department to assist in the production of quality products and to act as an effective check against faulty products leaving the plant. It is the function of Quality Control, using a number of testing procedures, to insure that the desired product quality levels are achieved and maintained and that scrap and rework is kept to a minimum by the Manufacturing Department.

The Quality Control staff should be free of undue influence from Manufacturing and should preferably report directly to top management. It must have the authority to stop the unapproved use and distribution of raw materials and finished products which do not meet specified standards. The distribution of non-specified raw materials or products may be permissible if approved by the Design Department, since its members are usually in the best position to determine whether product safety and reliability will be affected. The Quality Control effort should extend to all phases of product manufacturing, from the receipt of the raw materials to the completion of the finished product. In addition, Quality Control should monitor proper storage, packaging and shipping conditions. It must determine the type of tests that should be made on each part of a product and, in addition, the department should set up a quality control system to insure that minimum quality levels are met by the sampling and testing procedures. Since "critical" part or component failure can cause serious accidents, special attention should be given to those parts and components to make sure they are defect-free. (In some cases 100% testing may be necessary.)

Quality Control should be a preventive effort, instead of merely a process of sorting the bad from the good products. When the Quality Control and Manufacturing Departments work together, they can identify and control problem areas in the production process (for example, a machine which is gradually going out of specification or a worker who has difficulty producing to design specifications). Often these problems can be identified and corrected *before* the discrepant parts or products are produced.

One of the first places the Quality Control Department can save the company time and money is in the inspection of incoming raw materials and component parts. Without an incoming inspection, defective materials may not be detected until they have been incorporated in the manufacturing or assembly process. If they do show up later, during in-process work or through an inspection of the finished product, the cost of repairing or replacing them will obviously be much higher and possibly prohibitive. The manufacturer of the finished product is frequently held responsible for accident-producing product failure even when that failure is directly related to a product component which was both designed and manufactured by another company.

There are techniques other than inspection of incoming parts and materials that can be effectively used to control raw material quality. One is to require the vendor to certify that the material or part has been tested and meets the required specifications. This is often done with metals, particularly alloys, and with plastics.

Another method is called vendor rating and requires a cooperative effort between the Quality Control and Purchasing Departments. This involves the maintenance of records indicating the company's past experience with suppliers. An evaluation can then be made based on the supplier's past performance with the same type of materials.

Other practices would include a pre-shipment inspection by your Quality Control Department at the supplier's plant or a review of the vendor's own tests and quality control inspection. However, even though these vendor quality control plans can minimize the need for receiving inspections, they should never be dispensed with totally when "critical" parts are involved.

In-Process Testing

Sampling and testing which are carried out at various critical points during the manufacturing process are called in-process testing. The purpose is to implement quality control procedures during the manufacturing process, not only to detect the production of defective parts, but to work with production-line management to detect both human and machine errors and take corrective action. Here too, special attention should be given to critical parts or assemblies, including 100% sampling and testing if necessary. It is primarily during the production process that substantial time, money and manpower can be saved by eliminating needless scrap and rework of defective products.

Finished Product Testing

After the product is assembled, it should be tested finally for proper total performance. It should also be checked for proper installation and operation of guards or other safety devices. If warning signs or decals are required on the product, they too should be examined at this point. Usually, operating and maintenance instruction books or pamphlets, as well as warranties, are packed with the product and quality control should monitor this phase as well. A packing list for the packagers may be needed, depending on the nature of the product and its exposure to damage or degradation during handling and shipment.

Storage and Shipping

Some products are subject to damage through rough handling or deterioration from high humidity, heat, cold, etc. Where products may be subject to damage or deterioration after they have been produced, packaged, or even shipped to one's own or a distributor's warehouse, Quality Control monitoring should be extended to these areas.

The question is often raised whether the Quality Control Department or the Manufacturing Department should perform the in-process and finished-product inspection and testing. We believe this is best handled by the Quality Control Department because the Manufacturing Department, by its nature, may be less concerned than is the Quality Control Department in maintaining high quality standards. In the Manufacturing Department, test personnel might find themselves pressured to pass inferior materials in order to meet production schedules.

A less desirable, but acceptable, solution is to have the Manufacturing Department perform the in-process and finished-product inspection and testing, but under the overall supervision of the Quality Control Department. This procedure is usually satisfactory if the following conditions are met:

1. The Quality Control Department retains the overall responsibility for the inspection and testing.

2. The Quality Control Department gives the Manufacturing Department written testing procedures in regard to types of tests and the quantities of goods to be tested, and the instructions are followed rigorously.

3. The Quality Control Department conducts a continuous audit of the quality of the products being shipped.

4. The Quality Control Department continuously checks how closely the test procedures are being followed.

5. The Manufacturing Department inspectors are trained to perform the tests, and both tests and inspectors are kept up to date on new technology and job skills.

We have already mentioned that Quality Control should work with Manufacturing to identify and eliminate trouble areas in the manufacturing processes, and that they should work with the Purchasing Department on vendor rating plans.

Occasionally, the Design Department's product specifications may be unrealistic. Manufacturing may be unable to meet them. In those instances, Quality Control and Manufacturing should discuss those quality control problems with Design to see if more realistic standards which retain product safety and reliability can be set.

It is desirable for a company Quality Control Manual to be developed, spelling out general policies and setting out specific test and statistical sampling procedures for the various parts as well as the finished product. A Quality Control Manual formalizes the quality control procedure for the entire company and becomes a valuable management document, as well as a useful record of the company's effort to control quality in defense of product liability suits.

The Quality Control Department should supervise the calibration of test instruments. The importance of this function is evident, since these instruments are used to determine whether design specifications are being met.

Quality Control should keep records on *at least* the critical parts and components for the life of the product. These records should be complete enough so that they can be used to identify defective products after these have been sold. In addition, good quality control records can often be introduced as evidence in products liability law suits.

The Quality Control Department can be of great help in the products loss control program by preparing studies for management on the time and material costs of unnecessary scrap, failures, and rework. The results of such studies should establish and highlight the fact it is cheaper to "do it right the first time" and emphasize that a preventive quality control effort does save the company substantial money by preventing both business losses and liability losses.

SERVICE DEPARTMENT

In many companies, depending on the type of product sold, service personnel are required to maintain close contact with the customers and their employees. When this occurs, they are uniquely familiar with the people using their products; they are most likely to hear customer complaints and reactions to their products; they see the misuses and unsafe uses of the product; and they usually have some knowledge of incidents and accidents which have occurred. They are, in fact, usually the first company personnel to hear of product accidents.

In view of their unique and advantageous position, service personnel, when making a service call at a customer's plant, should use a check-off form to determine that products and safety devices are in good operating condition and being used properly. Unsafe conditions, such as guards removed from products, or unsafe practices should be reported, both orally and in writing, to the customer . . . along with written recommendations for correction and a statement of the likely consequences if such recommendations are ignored. A copy of the report should be sent to the Legal Department, where it will prove useful in the event of a later accident and products liability claim.

All customer complaints and reactions on products and product incidents and accidents should be reported to the appropriate department in your company. This could be the Design, Quality Control or even the Service Department itself. Customer complaints and reactions and product incidents should be screened by the department assigned, and all those situations involving possible products safety, reliability and maintainability should be forwarded to the Products Loss Control Unit.

The Service Department should serve on the Products Loss Control Committee or act as adviser to the Products Loss Control Department and also participate in the safety audits made on new designs.

SAFETY DEPARTMENT

Because of its past experience in developing and carrying out employee safety programs in your own plant, the Safety Department is often aware of both product hazards and misuse. Its personnel may have had experience in preventing employee accidents involving the very type of products your company is manufacturing. In addition, they are experienced in incident and accident investigation techniques and can be of assistance in training the department which will be investigating products incidents and accidents. Their surveillance of employee safety in production areas helps prevent production errors and costly employee accidents. Their experience may be valuable when safety audits are made on new designs.

PUBLIC RELATIONS DEPARTMENT

Product accidents, product recalls, and product field modification programs usually mean publicity and the news media often request statements for publication. Statements which are not carefully prepared can damage a company's reputation. However, if properly handled, these statements can result in fairer treatment by the news media and can help prevent public hostility toward your products and your company. In fact, skillful handling of publicity on product accidents, product recall, and product field modification programs can be used to demonstrate to the public your company's efforts to design, produce and sell safe, reliable products.

The Public Relations Department can also be of assistance in employee motivational programs, instilling pride in individual workmanship and reducing employee errors.

LEGAL DEPARTMENT

As we have already noted, the present legal climate in the nation is of great significance in products losses. Therefore, the Legal Department has a key role to play in the Products Loss Control Program. Some of its primary duties in the program are to educate and orient the various departments in the legal aspects of products loss control and in the legal techniques necessary for the reduction of legal exposures from product malfunctions. The department's personnel serve in an advisory capacity in specific matters such as advertising, instruction books, and recordkeeping, and they assist in the defense of products liability claims. They are, appropriately, the legal advisers in the program.

Early in the program the Legal Department should hold meetings with the Design, Quality Control, Manufacturing, Advertising, Sales and Service Departments to describe the legal climate and to point out hazards that these departments should work to avoid in conducting their company business. These meetings should include an analysis of court decisions on products similar to the company's own, the importance of designing to adequate safety standards, the importance of preventing defective products from leaving the plant, the dangers of overselling product capabilities and, finally, the hazards of giving undesirable express or implied warranties.

Records can be very useful in the defense of products liability claims and the Legal Department should advise the Design, Manufacturing, Quality Control, Sales and Service Departments of the records necessary for the legal defense of products claims.

In addition to providing orientation to the Advertising and Sales Departments, it is recommended that the Legal Department review advertising, sales brochures and instruction booklets for legal liability and warranty problems. It should check to see that product hazard warnings are given when necessary and that, when they are necessary, they contain the three essential ingredients: a warning of the hazard, instructions on how to avoid it, and an explanation of the possible consequences if the warning is disregarded. Because pictures of products in operation are sometimes held to constitute express or implied warranties, the Legal Department should verify that all illustrations and pictures used for advertising show only safe and correct operating procedures.

When product accidents resulting in damage or injury do occur, they should be reported promptly to the products liability insurance carrier. The investigation of such accidents may be done solely by the insurance company, but it must have the cooperation of your company. In such a cooperative effort, your Legal Department would probably play an important part. It is essential that accidents be reported and investigated promptly to determine your company's liability exposure. If this is not done promptly, essential evidence may be lost, witnesses may prove hard to locate, and the defense may be much more difficult. (Many companies have their own insurance departments which may perform these duties instead of the Legal Department.)

If product liability suits do go to trial, your Legal Department will probably help in the defense of the company's claims. Many times, the defense attorneys want to be, and should be, brought into the plant to observe first-hand all the design and manufacturing steps involved in the making of products similar to the allegedly defective one. This gives the defense attorneys valuable background and a more practical concept of the product and the steps involved in its production, as well as its operation. Those personnel expert in the design and manufacture of the particular product involved may be called upon to testify at the trial as expert witnesses.

The Legal Department should serve on the Products Loss Control Unit as legal adviser.

INSURANCE DEPARTMENT

The Insurance Department is too often considered merely the department which buys insurance and reports claims to the insurance company. However, because of their experience with product accidents and claims, its personnel usually have a background of valuable knowledge and experience in the products liability area.

Their primary duty is to recommend and to purchase at reasonable cost products liability insurance that is adequate to protect the company against the exposures involved. This is often quite difficult because products liability is one of the fastest growing areas of risk in business and industry today. In view of the rapidly-increasing number of products liability claims and the tremendous increase in the size of the awards, insurance that seems adequate today may be very inadequate tomorrow. New laws, new court decisions, revisions in insurance policies, new product uses . . . all must be continually followed and reviewed to insure, as far as possible, that your company insurance program meets your company needs. Your Insurance Department should have current information on trends in product safety and product liability through insurance risk management publications. Important developments should be conveyed to all involved departments, probably through the Products Loss Control Unit.

Through its contacts with the products liability insurance company, your Insurance Department has available to it safety engineers and products loss control specialists who have generally had wide experience with many different types of products and product accidents. These insurance representatives can often provide worthwhile suggestions on reducing products liability exposure from design, manufacturing, advertising and sale of products. Don't overlook this source of assistance.

Because the Insurance Department often has a knowledge of accident and claims experience on previous products, it can frequently be of help in analyzing product liability exposures from new products.

One of the axioms of business management is that management should know where it is spending its money. The Insurance Department should prepare cost studies for management and for the Products Loss Control Group, showing the impact of products liability losses on the company profits. These cost studies should include a cost accounting system, whereby products liability losses are charged directly to the responsible plant or department. This will help to emphasize the importance of products loss control efforts and also underscore to all departments the consequences of product failures.

The Insurance Department should serve on the Products Loss Control Unit.

The following duties which were previously discussed as being performed by the Legal Department may instead be handled by the Insurance Department:

1. Prompt reporting to the products liability insurance company of product accidents involving bodily injury or property damage.

2. Coordination of the investigation of product accidents involving bodily injury or property damage with the products liability insurance company.

3. Coordination of the legal defense of products liability claims with the products liability insurance company.

SALES, ADVERTISING AND INSTRUCTION BOOK WRITERS

Unfortunately, after "reasonably" safe and reliable products have been designed and manufactured, liability claims and business losses may be incurred through the manner in which the products are represented to users. Often the product user relies on your company sales personnel, advertising, sales brochures, and the operating and maintenance manuals for his knowledge of the capabilities and hazards of the product. If users are led to believe that the products have capabilities they do not possess, or if they are not adequately warned of hazards, injured persons may have grounds for a products liability suit based on breach of contract.

There are two types of warranties which can be involved in breach of contract suits. The first is an express warranty. The injured or damaged party alleges that a statement of fact or a promise made by the seller relating to the qualities of the product induced the purchaser to buy the product. Usually, the use of the words *warranty* or *guarantee* is held to be the giving of an express warranty. Other less forceful words or expressions may, in some cases, be considered an express warranty. It is usually unimportant that the actual intent to warrant was missing. Examples include: general statements as to the qualities of the product, such as a "safe" detergent or a "gentle" hair wave; negative representations, such as "harmless" product; and construction qualities, such as "shatter-proof" glass or "blow-out-proof" tires.

The second class of warranty is an implied warranty, of which there are two general types:

1. *A warranty of merchantability*; that is, that the goods sold are reasonably fit for the general purpose for which they are sold. This type of warranty accompanies all products sold.

2. *A warranty of fitness*; that is, that the goods sold are suitable for a special purpose of the buyer which is not satisfied by mere fitness for general purposes. For instance, if a product is sold with the knowledge that a customer intended to use it in an unusual way and if no instruction to the contrary accompanies the sale, this could be considered an implied warranty of fitness for the customer's stated purpose.

An illustration of how ill-advised advertising got one manufacturer into liability problems was the case of the back-hoe manufacturer who showed in his advertising literature a picture of a back-hoe lifting a large section of pipe. A customer attempted to use one of the back-hoes for a similar purpose. The back-hoe collapsed and the operator was seriously injured. The operator successfully sued the manufacturer for a breach of warranty.

While we are on the subject of warranties, we should mention the serious business losses that can result from giving warranties which the products cannot meet. Consider very carefully whether the products are reliable enough to fulfill the warranty conditions without imposing a severe financial drain on the company.

Because sales personnel, advertising, and instruction book writers have the greatest impact on how the products are represented to the customers, they must be involved in the Products Loss Control Program to minimize the business and liability exposures from undesirable express and implied warranties. The first step is to have the Legal Department advise the Sales and Advertising Departments on express and implied warranties and their importance in controlling liability exposures.

All advertising and sales materials should be reviewed by the Legal and Design Departments to determine:

1. Are they clear and accurate? Do they overstate the product's capabilities?

2. Do they encourage the customers to believe the products have uses for which they were not designed?

3. Can the products safely and reliably do all that the advertising and sales materials say they can?

4. Are only safe operating procedures shown?

All operating service and maintenance instructions should be reviewed by the Legal and Design Departments to determine:

1. Are all the hazards in the use, servicing and maintenance of the products clearly indicated?

2. Are safe procedures given so that the users can avoid the hazards?

3. Are the consequences of ignoring the hazards and safe operating procedures clearly indicated?

Finally, another important duty of the Sales Department is to keep sales records which can be used to identify product purchasers. These records are essential if product recall or field modification programs are to have any chance of success. Sales records should be retained for the life of the product.

PERSONNEL DEPARTMENT

The caliber of the employees in the company will have a great influence on the quality of your products. Employees who lack the proper job skills or who are not interested in their work can greatly increase the number of defective products leaving the plant. Therefore, it is essential that the Personnel Department be as selective as finances allow in the selection of new and transferred employees and that it continually upgrade the performance goals for established employees.

Some suggested duties for the Personnel Department include assisting the various departments in setting up job classifications, screening prospective employees for suitability for the position, assisting in the development of educational and training programs to improve employee performance, and participating in employee motivational programs to instill pride in individual work and reduce work errors.

PURCHASING DEPARTMENT

The Purchasing Department is charged with buying quality raw materials and components at a competitive cost. This requires a knowledge of the requirements set by the Design Department and also a knowledge of the capabilities and reliability of the suppliers. Purchasing should work with Design and Manufacturing to determine what specifications have been set and what delivery schedule is required. Purchasing should not deviate from specifications without the written permission of the Design Department.

Together Purchasing and Quality Control can evaluate the capabilities and reliability of suppliers through vendor rating plans.

The list of reliable suppliers compiled by Purchasing can be of great assistance in producing reliable and reasonably safe products.

We have discussed the role and duties of each department in the company in the Products Loss Control Program. The importance of recordkeeping and field monitoring as part of the program has been mentioned briefly; now these aspects will be covered in more detail.

RECORDKEEPING FOR PRODUCTS LOSS CONTROL

Recordkeeping is a very important part of a products loss control program. A primary reason is to identify and locate products which might reach the customer in a defective condition. It can be costly, frustrating and sometimes embarrassing to be unable to do so when it becomes necessary. Design, manufacturing, quality control and, in some cases, service records, identify particular batches of products which turn out to be defective. Adequate sales and distribution records must be kept, in addition, to identify the customers who purchase your products. Should a field modification or product recall program become necessary, these records will make it possible to locate and notify customers.

Moreover, these records are a prerequisite to any successful legal defense in any products liability suits. A partial list of the required records includes design, design test and qualification, production, quality control, sales, distribution, accidents, failure analyses, letters to customers/users, recall, warranties, advertising, sales brochures, and operating and maintenance instructions.

Since liability extends to the life of the product (plus the time limit of the statute of limitations) these records should be kept at least for the life of the product plus an additional five years. This will ensure the records are available for most lawsuits.

FIELD MONITORING OF PRODUCT COMPLAINTS, INCIDENTS, ACCIDENTS

After the products are sold, the field monitoring system feeds information about customer complaints, incidents and accidents back to the company. This information should be received and reviewed by a central source in the company, to determine if product safety or reliability is involved. In many cases, a preliminary screening of customer complaints, accidents and incidents (occurrences which do not cause property damage or bodily injury but have the potential to do so) will indicate potential product hazards.

Where preliminary screening indicates potential hazards, prompt product failure analysis and/or field investigations to determine their sources (design,, manufacturing, quality control, advertising, service or whatever, should be made. After the field investigation, the causes of the problems should be analyzed, to determine if there are trends developing or if the problems are isolated ones. This information should be fed to the Products Loss Control Unit or Department for further action, which might include changes in design, manufacturing or quality control, or even a field modification or product recall program.

In addition to preventing products accidents, field monitoring assists in the continual task of improving customer satisfaction with your product.

SUMMARY

Products Loss Control is a critical and integral part of the commercial enterprise. Its success, as you have seen, depends upon all departments working together in an effectively-coordinated program to exert stringent and continuous control over all phases of production, from the initial product design to the eventual sale . . . and even beyond.

The catalyst that initiates such a program — the ingredient which provides impetus, continuity and effectiveness — is the wholehearted commitment and support of company management. A few loss control programs fail even *with* such support. *None* succeeds without it.

A program similar to that outlined in this chapter, if effectively used, can mean the difference between profit and loss for your company.

On the following pages you will find sample documents to aid you in your products loss control program.

Included are a model telegram (Figure 4) used for urgent recall of a contaminated product, a letter asking permission to field-modify a marginal safety interlock system on a molding machine (Figure 5), and a follow-up letter (Figure 6) to that

```
216P EST JAN 9 76 AB076

AAA125 (1312) (1-134106G 160) PD 01/09/76 1313      FY ASK 466-2355

PMS DES MOINES IA                                   MR GARFIELD PRICE

N J TELEXED FROM PLAINFIELD 1-609-935-8210

ALLWEATHER FINISHES
2520 GRAND AVE
DES MOINES IA

ATTENTION MR GARFIELD PRICE MANAGER    OUR SHIPMENT FAST DRY NO L2-5876

CAR NO GTX-10071 POSSIBLY CONTAMINATED AND COULD AFFECT NON-

FLAMMABILITY OF YOUR PRODUCT STOP CALL COLLECT FOR DISPOSAL

INFORMATION STOP  . J W CARPER GENERAL MANAGER  COLORANT INC  8200
                    BRUNSWICK AVE PLAINFIELD N J
```

Figure 4.

AMERICAN PLASTICS COMPANY
3780 HUDSON AVE.
NYACK, NEW YORK 10960

February 10, 1976

Mr. Edwin McCue
Plant Manager
Edlin Molding Company
2700 Windsor Street
Greenville, South Carolina 29610

Dear Mr. McCue:

Our records indicate that your firm purchased a model D-111 30 oz. injection molding machine from us on October 30 of last year.

As part of our continuing effort to provide the safest possible equipment for our customers, we would like to install a new safety limit switch on your machine.

We are currently using this improved safety device on our new models, and we would like all our customers to have this added protection. The switch circuit can be installed, free of charge, in 45 minutes, so that your downtime can be held to minimum.

Our service representative will be in your area the week of March 23-27 and will contact you at that time for an appointment. If this week is unsatisfactory, please feel free to contact me for other arrangements.

Thank you for your cooperation.

Sincerely,

Albert T. White
Albert T. White
Vice President of Engineering

ATW:nb

Figure 5.

AMERICAN PLASTICS COMPANY
3780 HUDSON AVE.
NYACK, NEW YORK 10960

February 28, 1976

Mr. Edwin McCue
Plant Manager
Edlin Molding Company
2700 Windsor Street
Greenville, South Carolina 29610

Dear Mr. McCue:

Our service representative has informed us that he was unable to make arrangements with you regarding modifications of your D-111 molding machine. We regret that you were unable to see our representative; however, we remain willing to modify your machine free of charge, at any time convenient to you.

We would appreciate your sending us a letter stating a convenient time so that we may continue our policy of upgrading our customers' equipment to the highest safety standards of the industry.

However, unless we hear from you within two weeks, we will assume that you do not wish to have the alterations made.

Sincerely,

Albert T. White
Vice President of Engineering

ATW:nb

Figure 6.

request. It is suggested you make use of similar instruments where circumstances require.

In addition there are four model forms: a design checklist for use in product safety/reliability evaluation (Figure 7); a product suggestion/reaction form, (Figure 8) useful in finding customer complaints and correcting product hazards; a notice of incident or accident form (Figure 9); and a follow-up investigation report form (Figure 10). You are encouraged to make use of these forms as is, or to modify them to meet your particular information requirements.

ORGANIZATIONS FOR DEVELOPING PRODUCT SAFETY STANDARDS

American Conference of Governmental Industrial Hygienists

American Gas Association

American National Standards Institute, Inc.

American Petroleum Institute

American Society of Automotive Engineers

American Society of Mechanical Engineers

American Society for Testing and Materials

Chemical Specialities Manufacturers Association, Inc.

Compressed Gas Association

Illuminating Engineering Society

Manufacturing Chemists Association

National Safety Council

National Fire Protection Association

Trade Associations

Cities and States

Underwriters' Laboratories, Inc.

U. S. Government

    Atomic Energy Commission

    Bureau of Mines

    Department of Agriculture

Department of Health, Education and Welfare
Department of Housing and Urban Development
Department of Labor
Department of Transportation
National Bureau of Standards

DESIGN DEPARTMENT
SAFETY/RELIABILITY EVALUATION REPORT

PRODUCT _____
    (NAME)   (TYPE)   (MODEL NO.)   (RESEARCH CODE NO.)   (PROJ. RELEASE DATE)

1. LIST ALL CRITICAL COMPONENTS* ON THE CRITICAL COMPONENTS FORM. LIST THE METHOD USED TO DETERMINE CRITICAL PARTS, AND THE SAFETY STANDARDS APPLICABLE TO THEM.
    *(CRITICAL PARTS OR COMPONENTS ARE THOSE WHOSE FAILURE COULD CAUSE SERIOUS BODILY INJURY, PROPERTY DAMAGE, BUSINESS INTERRUPTION, OR SERIOUS DEGRADATION OF PRODUCT PERFORMANCE.)

2. LIST HERE ANY CRITICAL PARTS NOT COVERED BY ADEQUATE SAFETY STANDARDS.
    (A) _____
    _____
    _____

    (B) HAS ADEQUATE TESTING AND RESEARCH BEEN DONE TO ESTABLISH THE SAFETY AND RELIABILITY OF THESE CRITICAL PARTS? YES ☐ NO ○ IF NO, WHAT DO YOU RECOMMEND? _____
    _____
    _____

    (C) HAVE MINIMUM QUALITY CONTROL STANDARDS BEEN SPECIFIED FOR CRITICAL PARTS? YES ☐ NO ○
    FOR NON-CRITICAL PARTS? YES ☐ NO ○

3. HAVE ADEQUATE GUARDS BEEN PROVIDED FOR ALL DANGER POINTS (TRANSMISSION POINTS, NIP POINTS, AND ALL LOCATIONS WHERE HAZARD THROUGH CONTACT EXISTS)? YES ☐ NO ○ IF NO, WHAT DO YOU RECOMMEND?
    _____
    _____

4. HAVE WARNINGS BEEN REVIEWED WITH LEGAL DEPARTMENT –
    (A) ON PRODUCT OR CONTAINER? YES ☐ NO ○
    (B) IN INSTRUCTION MANUAL? YES ☐ NO ○
    (C) RECOMMENDATIONS: _____

5. HAVE SALES AND ADVERTISING MATERIAL BEEN REVIEWED BY THE DESIGN DEPT.? YES ☐ NO ○
    RECOMMENDATIONS: _____
    _____

6. LIST THE TYPES OF ACCIDENT WHICH COULD OCCUR DURING –
    (A) NORMAL OPERATION _____
    (B) MAINTENANCE OR ADJUSTMENT _____
    (C) PRODUCT FAILURE OR MALFUNCTION _____
    (D) MISUSE _____
    (E) UNINTENDED FORESEEABLE USES _____

7. HOW CAN THE PRODUCT BE MODIFIED TO REDUCE ACCIDENT HAZARDS, IMPROVE RELIABILITY, OR REDUCE MAINTENANCE PROBLEMS? _____
    _____
    _____

EVALUATED BY	DATE	CHECKED BY	DATE

Figure 7.

PRODUCT SUGGESTION/REACTION REPORT

PERSON REPORTING SHOULD FILL OUT SECTION "A" OR "B"

DATE:

SECTION A: CUSTOMER SUGGESTION OR REACTION

CUSTOMER _____ ADDRESS _____

PRODUCT _____ _____ _____ _____
(TYPE OR NAME) (MODEL NO.) (SERIAL NO.) (PURCHASE DATE)

REACTION OR SUGGESTION _____

_____

COMMENTS OF REPORTER _____

REPORTER _____ DEPT. OR DEALERSHIP _____

DATE:

SECTION B: DEALER (OR EMPLOYEE) SUGGESTION

NAME _____ DEPT. OR DEALERSHIP _____

PRODUCT _____ _____ _____
(TYPE OR NAME) (MODEL NO.) (OTHER IDENTIFYING INFORMATION)

SUGGESTION _____

_____

FORWARD TO ENGINEERING DEPT., HEAD OFFICE, FOR REVIEW

Figure 8.

NOTICE OF REPORT INCIDENT/ACCIDENT

NOTE: USE THIS FORM (1) TO REPORT ANY ACCIDENT WHICH CAUSED BODILY INJURY OR PROPERTY DAMAGE OR (2) TO REPORT ANY INCIDENT WHICH HAS THE POTENTIAL TO CAUSE BODILY INJURY OR PROPERTY DAMAGE.

PRODUCT NAME _____

(MODEL NO.) (SERIAL NO.) (LOT NO.)

CUSTOMER'S NAME AND ADDRESS _____

REPORTED LOCATION OF INCIDENT/ACCIDENT

DATE OF INCIDENT/ACCIDENT | TIME AM/PM | DATE REPORTED

PERSONAL INJURY

NAME AND ADDRESS OF INJURED PERSON _____

OCCUPATION _____ AGE _____
NATURE OF INJURY _____
PART OF BODY INJURED _____

PROPERTY DAMAGE

PROPERTY DAMAGED _____
NATURE OF DAMAGE _____
EXTENT OF DAMAGE _____

WITNESSES:
NAMES
AND
ADDRESSES
(1) _____
(2) _____
(3) _____
(4) _____
(5) _____

DESCRIBE THE INCIDENT OR ACCIDENT IN ALL THE DETAIL YOU CAN _____

SOURCE OF INFORMATION: CHECK ONE OR EXPLAIN
CUSTOMER ☐   DEALER ☐   DISTRIBUTOR ☐   NEWSPAPER ☐   OTHER _____

REPORTED BY | DATE | REVIEWED BY | DATE

SEND TO: XYZ DEPARTMENT, HEAD OFFICE

Figure 9.

Figure 10.

BIBLIOGRAPHY

1. *Accident Prevention Manual for Industrial Operations*, 6th Edition, Chicago, Ill.: National Safety Council.

2. Coccia, Dondanville, and Nelson, *Product Liability Trends and Implications*. New York, N.Y.: American Management Association, Inc., 1970.

3. Guida, Anthony, "In Search of the Prowling Oops — A Blueprint for Liability Prevention," *Standardization News*, Vol. 1, No. 3, American Society for Testing and Materials.

4. Hopkins, Stuart K., "Voluntary Standards at the Crossroads," *National Safety News*, April, 1967.

5. Jacobs, Richard M., "Measuring Consumer Product Reliability," *Quality Progress*, September, 1969.

6. MacCollum, David V., "Reliability as a Quantitative Safety Factor," *Journal of the American Society of Safety Engineers*, May, 1969.

7. Peters, George A., *Product Liability and Safety*. Coiner Publications, Ltd., 1971.

8. Powers, Francis E., and Keating, William E., "Changing Liability Challenges Business Practices in Safety," *Environmental Control Management*. January, 1970.

9. Proceedings of Product Liability Prevention Conferences, 1970, 1971, 1972, 1973. Newark, N.J.: Newark College of Engineering.

10. "Products Liability and Reliability — Some Management Considerations." Washington, D.C.: Machinery and Allied Products Institute, 1967.

11. "Products Liability: Guides for the Corporate Manufacturing Executive," and "Products Liability: Guides for the Corporate Executive." Milwaukee, Wisc.: Defense Research Institute.

12. "Products Loss Control Guide." Philadelphia, Pa.: Insurance Company of North America, 1970.

13. Robb, Dean A., and Philo, Harry M., *Lawyers Desk Reference, A Source Guide To Safety Information, What To Find, How To Find It.* Rochester, N.Y.: The Lawyers Cooperative Publishing Co., 1966.

14. Shankula, Robert E., "An Insurance Engineer Looks at Products Liability," *Materials Research and Standards*, December, 1969.

15. "Zero Defects, Doing It Right The First Time," Bulletin 71. New York, N.Y.: American Management Association, 1965.

Environmental Health

XI.

INTRODUCTION

In a modern Loss Control program, the safety professional must consider the health hazards associated with the job environment. In the past, efforts have been made to prevent accidents by focusing attention on man and his machine. One method was to issue personal protective equipment and to install machine guards. Any action, however, that would protect the worker from toxic process chemicals and cleaning solvents was almost completely ignored. For example, until recent years, few efforts were made to reduce the worker's exposure to noxious air contaminants, such as lead fumes, asbestos fibers and beryllium dust.

Activity in this area has been new to most safety people. For the industrial hygienist, however, whose full-time activity is devoted to the recognition, evaluation and control of occupational hazards, it has been a hard and long battle since the Industrial Revolution. Diseases normally associated with certain industries have been eliminated with the aid of these specialists. Mercurialism, a disease common to the hat industry, was eradicated primarily through the efforts of the Industrial Hygiene staff of the U. S. Public Health Service. Carbon tetrachloride, a once common industrial solvent, caused serious damage to the liver and kidneys from single or repeated over-exposure to its vapors.[1] Successful substitution of a less toxic solvent reduced the threat of injury from this type of exposure. Silicosis, the scourge of miners and foundry workers, has been brought under control by the use of personal protective equipment, proper ventilation, and early detection. These and similar accomplishments indicate the progress of industrial hygienists in their continued attempts to improve the work environment.

These efforts, however, were few and far between because of the lack of legislation and compensation. Prior to the adoption of the compensation principle, workers injured on the job had to sue their employers for remuneration. Little relief was offered to workers disabled as a result of exposure to toxic air contaminants on the job. Several years later, however, workmen's compensation laws provided relief to workers for accidental injury, and today they provide coverage for occupational disease. These laws were an important factor in prodding management to initiate safety programs designed to curb on-the-job accidents and to help reduce personal injuries. Occupational health hazards, on the other hand, were ignored primarily because:

1. The presence of toxic agents could not be detected by ordinary senses and the instrumentation designed for this purpose was inadequate.

2. The effects of long-term exposure were not known, and the disease quite often did not manifest itself until after many years of exposure.

3. Occupational diseases went undetected because they often resembled non-occupational-related conditions.

Today, there is little improvement in the occupational health picture. A major portion of industry still does not receive the benefits of a well-organized safety and health program. Moreover, many consider it a soft core item that can be eliminated when a reduction in force or cost is necessary. Mr. Eugene L. Newman, Director of the Board of Certified Safety Professionals of the Americas, reflected management's position on safety and health when he said, "Management considers its responsibility for safety fulfilled with payment of a workmen's compensation premium. After that, it's the insurance carrier's problem." Similarly, the intrinsic subtleties of typical occupational health hazards, compared with more conventional safety aspects of industry, can detract from their importance in budget considerations. There is often a tendency at management levels to rationalize or procrastinate, instead of giving such exposures their proper attention. One reason for management's indifference stems from the fact that if governmental health authorities aren't concerned, then health hazards are considered to be under proper control. Because of this, management, in most cases, has been reluctant to upgrade — or even properly maintain — desirable protective programs. Since most of the activity of the health service is being directed to the worst risks and little attention is given to widespread borderline health conditions, minimum official assistance has been provided towards rectifying this problem.

Recent trends in legislation regarding occupational health, and an increased awareness of health hazards in the job environment by employees and the public in general, indicate a need for reappraisal of company safety and health programs.[1]

The safety professional will no longer concentrate solely upon old-time problems, like loose wires tripping people, but he will assume the responsibility of instituting preventive programs for health hazards generated by a booming industry, technological changes and increased work hazards. He must know how the health of the employee and his working environment interact, as well as how to measure and control the deleterious factors of the job.[2] It is, therefore, important to discuss what the safety man must know in order to upgrade his role in industrial safety.

This section is not designed to train industrial hygienists, but it will attempt to expose individuals responsible for safety programs to various industrial health hazards that exist and to develop within them a professional awareness.

Generally speaking, loss control measures in industrial health are *not* adequately employed at this time. Unprotected workers are still being gradually deafened from excessive job-related noise levels. Employees continue to work with materials that emit dangerous dusts or vapors; in most cases, it is only a matter of time before they become seriously, and perhaps irreversibly, impaired. Many such eventualities can be predicted, and adequate safeguards can be recommended. It takes an enlightened safety professional to suspect and evaluate potential or actual health hazards so that he can avail himself of the professional advice of specialists.

CLASSIFICATION OF OCCUPATIONAL HEALTH HAZARDS

The type of contaminant which occurs in a workplace depends upon the process used. However, for simplicity contaminants in the industrial environment, are divided into groups, according to their physical characteristics. These are defined by the American National Standards Institute as follows:[3]

Dusts: Solid particles generated by handling, crashing, grinding, rapid impact, detonation, and decrepitation of organic materials, such as rock, ore, metal, coal, wood and grain. Dusts do not floccuate, except under electrostatic forces; they do not diffuse in air, but settle under the influence of gravity. Dusts are generated by such operations as mixing, grinding, crushing or screening. Air concentrations are measured in millions of particles per cubic foot of air (m.p.p.c.f.)

Fumes: Solid particles generated by condensation from the gaseous state, generally after volatilization from molten metals, etc., and often accompanied by a chemical reaction, such as oxidation. Fumes result when volatilized solids, such as metal, condense in cool air. Examples are fumes from lead melting and burning, welding, brass melting and zinc galvanizing. Air concentrations are measured in milligrams per cubic meter (mg/m^3).

Gases: Normally formless fluids which occupy the space of enclosure and which can be changed to the liquid or solid state only by the combined effect of increased pressure and decreased temperature. Gases diffuse. Concentrations of toxic gases, such as carbon monoxide, hydrogen sulfide, chlorine, sulfur dioxide and ozone are measured in parts of gas per million parts of air by volume (ppm).

Vapors: The gaseous form of substances which are normally in the solid or liquid state and which can be changed to these states by either increasing the pressure or decreasing the temperature alone. Vapors diffuse. Vapor concentrations may be found at operations using cleaning agents, paint thinners, spot removers and drying agents. These are also measured in parts per million parts of air by volume (ppm).

Mists: Suspended liquid droplets generated by condensation from the gaseous to the liquid state or by breaking up a liquid into a dispersed state, such as by splashing, foaming, and atomizing. Some examples are sulfuric acid mist (during steel pickling) and chrome acid mist (during chrome plating). Spraying of solutions for rust removal or surface treating will form a mist. Mist concentrations are measured on a weight basis in milligrams per cubic meter (mg/m^3).

Smoke: Particulate matter resulting from the combustion of materials that are organic in nature.

Living agents, such as bacteria, molds and other parasites are not included in this classification, but should be considered as causative agents of many occupational diseases. Their relative importance has not been as great as chemical and physical agents because they cause illness-exhibiting symptoms related to those of non-occupational origin. Furthermore, environmental sanitation and anti-infection drugs have played a major role in reducing the risk of exposure to biological agents. It should be noted, however, that the possibility of contracting a disease via these agents is not remote. Machinists have experienced extensive skin inflammations, or dermatitis, as a result of cutting oils in contact with the broken skin. Although mineral oils do not act as a good medium for germ growth, germs may be picked up by oil from contaminated storage tanks and oil supplying systems. In many cases, workers discard food into oil pans or drains. This practice may carry disease-producing germs into the oil.[4]

The recent use of enzymes, substances capable of breaking down proteins, in the detergent industry has generated considerable concern among industrial hygienists. These enzymes are added to washing powders to increase their activity on stains produced by protein matter. Exposure to these enzymes during the manufacturing process has resulted in provoking complex allergic reactions in the lungs, thus impairing a worker's efficiency. However, evidence is available indicating that irreversible lung damage will occur. Rigorous preventive measures that were recommended have produced favorable results.

Safeguarding a man from acute and chronic exposure to hazards that may arise in the pursuit of his occupation requires knowledge of his physical and mental performance capacity and his tolerance limits to the stresses of the physical environment.[5] Ergonomics, a relatively new discipline, can offer most of this information by applying principles designed to reduce monotony and fatigue in performing a job task.

MODE OF ENTRY

In order to protect workers from toxic materials in industry, it is essential that we have a clear understanding of *how* these materials enter the body and how they effect types and degrees of physiological response. There are three ways in which contaminants are absorbed into the body:

1. *Inhalation* — The great majority of occupational poisonings that affect the internal structures of the body result from breathing airborne substances.[6]

2. *Skin Contact* — Serious and even fatal poisonings can occur from short exposures of skin areas to strong concentrations of extremely toxic substances.

3. *Ingestion* — This route plays a minor role in the absorption of most toxic materials.

Inhalation

The easiest and most common route that toxic substances use to enter the living body is the respiratory tract. More occupational poisonings affecting the internal structures of the body result from breathing airborne contaminants than any other mode of entry, because of the efficiency of the respiratory tract in assimilating materials quickly into the body system. Particles lodged in the lungs may have a direct effect on them or may pass directly to other organs by way of blood or lymph.

The type and severity of the action of inhaled substances will depend on the nature of the substances, the amounts absorbed, the rate of absorption, and individual susceptibility. Different types of dusts, fumes, and mist will elicit different physiological responses.[7] Some of these reactions include:

1. The fibrous hardening of lungs, known as pneumoconiosis, caused by irritation created by the inhalation of dust. Silicosis and asbestosis are examples of this condition, resulting from exposure to silica and asbestos respectively. Both conditions are serious and incurable.

2. The systemic reactions caused by toxic dusts of such elements as lead, manganese, cadmium, and mercury, by their compounds and by certain organic compounds.

3. Metal fume fever, resulting from the inhalation of generated fumes of zinc or magnesium, or of their oxides. This condition is transient.

4. Allergic and sensitization reactions, caused by inhalation or skin contact with organic dusts, such as flour, grains or some woods, and dusts of a few organic and inorganic chemicals.

5. Bacterial and fungal infections, caused by inhalation of dusts containing active organisms.

6. Irritation of the nose and throat, caused by acids, alkalies or other irritating dusts or mists. Chromic acid mists may cause ulceration of the nasal passages.

7. Damage to internal tissues, resulting from inhaled radioactive materials.

Skin Contact Absorption

Although the emphasis in industrial toxicology is upon substances that represent inhalation hazards to workers, there are a great many substances that can cause injury by skin contact.

Some materials, such as strong acids and alkalies, may damage the skin itself; others will permeate the skin and become absorbed into tissue. Aniline and carbon disulfide, for example, can cause death upon contact over a large area of the skin. In the latter case, the substance attacks a body function, such as the nervous system or the circulatory system, and may result in paralysis or anemia. In most cases, toxic chemical agents are not absorbed through the skin if it is intact. Skin abrasions, however, favor absorption.[9]

Ingestion

Ingestion of toxic substances may result from inhalation. Particles caught by the ciliated (hair-like) cells in the lower part of the respiratory tract propel mucous and foreign material towards the oral cavity, where they are expectorated or swallowed. (8) Ingestion of material usually results, however, from contaminated food, tobacco or beverages, or from putting fingers into the mouth, or from licking the lips. Poisoning by this route is far less common than by inhalation, because the frequency and degree of contact with toxic agents from material on hands, food and cigarettes are far less than by inhalation.

MODE OF ACTION

Industrial poisonings do not manifest themselves in identical fashion. Their severity is described as acute, sub-acute, or chronic.

An *acute poisoning* is characterized by a single heavy dose, analogous to an industrial accident in that the exposure is sudden and severe and can usually be pin-pointed as to time and place. These poisonings are rare but they may result from an accidental break in a pipeline carrying toxic materials, or from similar accidents which release materials suddenly into the work atmosphere. Some of these result in unconsciousness, shock or collapse, severe inflammation of the lungs, or even sudden death.[10]

Chronic Poisonings refer to those occupational diseases which develop slowly from the continued absorption of small doses of toxic materials over a long period of time. Chronic poisoning does not occur with all poisons, because the body eventually removes most of the material through destruction, change, coupling or elimination, so that no accumulation of the material or permanent injury of the organ functions occurs.[11] Harmful materials which produce *irreversible* damage, so that the injury accumulates, produce most of the chronic conditions in industry. Benzene, once a common solvent, used in the printing industry, is one of these agents which, in low concentrations, has been known to produce irreversible chronic effects.

SEVERITY OF ACTION OF TOXIC AGENTS

The degree of toxicity depends on three factors. First, the *nature* and *concentration* of the material determines to a large extent the toxic response. Rates of solubility, detoxification and excretion are primarily dependent on the chemical properties of the harmful material. Consideration should be given to the chemical properties of a material before its use, so that safeguards can be applied. Moreover, a close watch should be maintained to keep air concentration at a safe level. Second, the intensity of action depends on the *length of exposure.* Single exposures to large doses of toxic material may produce greater damage than small doses administered over a long period of time. The reverse is also true. Small doses of a substance, such as carbon monoxide, will not cause any harm on single exposure but may produce lasting damage when administered over a long period of time. Third, *individual susceptibility* to certain substances or to extended exposures largely determines the manner in which an individual will respond. Some people may show no evidence of exposure, while others may become severely ill. Individual biochemical responses, *age, sex* and *previous illness* are believed to play a significant role. Additional factors, such as *rates* of *working speed* and *nutrition*, suggest a relationship to individual susceptibility, but most of these are not supported by scientific evidence.

EXPOSURE LIMITS FOR DANGEROUS SUBSTANCES

Threshold Limit Values

In order to determine the severity of exposure to toxic agents, it is necessary to use guidelines indicating air concentrations of contaminants to which a person may be exposed without ill effect or discomfort. These limiting values are commonly referred to as the Threshold Limit Values (TLV), as adopted by the American Conference of Governmental and Industrial Hygienists (ACGIH), and are extensively used throughout industry. Many of these have been incorporated in the Walsh-Healey Public Contracts Act and recently in the "Occupational Safety & Health Act of 1970."[1,2] The TLV's are recommended values of time-weighted average concentrations for an eight-hour working day. These concentrations can be exceeded only if equivalent time is spent below the recommended level during the work day.

Certain substances, however, have been assigned a "C" designation. This is a ceiling limit which should not be exceeded. Those substances which are followed by the notation "skin", may be absorbed by the skin, mucous membranes and the eyes, and preventive measures should be taken to avoid this type of contact. Caution is to be used when applying these limits. The preface to the 1971 TLV list states that "these limits are intended for use in the field of industrial hygiene, and should be interpreted and applied only by persons trained in this field. They are not intended for use, or for modification for use, (1) as a relative index of toxicity, by making a ratio of two limits, (2) in the evaluation or control of community air pollution or air pollution nuisances, (3) in estimating the toxic potential of continuous uninterrupted exposures, and (4) as proof or disproof of an existing disease or physical condition . . ."

Acceptable Concentrations

The 237 Committee of the American National Standards Institute has adopted and published a series of guides for a number of substances, (13) which include the following:

1. An acceptable ceiling concentration for protection of health, assuming an eight-hour work day.

2. An acceptable time weighted average for protection of health, assuming an eight-hour work day.

3. An acceptable maximum for peaks above the ceiling for an eight-hour work day.

Like the TLV's, the ANSI guides should be used only as intended.

Emergency Exposure Limits

The Toxicology Committee of the American Industrial Hygiene Association has published emergency exposure limits (EEL) for short-term accidental exposures to airborne contaminants that can be tolerated without permanent toxic effects.[14] These limits can be used in disaster planning, to predict what type of emergency action to take in the event of an accidental spill, and to recommend appropriate personal protective equipment.

Criteria Used for Establishing Standards

Standards used for limiting occupational exposures are based on the best available information from actual industrial exposures, from experimental animal studies, and occasionally from human studies. Toxicity data obtained from human subjects are favored over animal data in establishing exposure limits. Epidemiological studies aid in determining the incidence of occupational disease with long latent periods, such as cancer.[1,5]

With this information available, exposure levels may be decreased or increased accordingly. A safety factor is usually applied to the limits, to minimize the degree of risk. This factor is usually set at one-tenth of the minimum value of hazardous substances which will produce an adverse effect. Since the limits are regarded as guides useful in the control of health hazards, they are not to be considered as fine lines between safe and dangerous concentrations, and are subject to revision with additions to the basic available knowledge.

PHYSICAL HAZARDS

Employees are often subjected to harmful energies which produce tissue injury. These environmental conditions cause injury which in many cases is *cumulatively damaging* rather than disabling. Typical stresses encountered in the work environment are:

1. Ionizing radiation

2. Ultra-violet rays

3. Visible light

4. Infra-red rays

5. Microwaves and lasers

6. Excessive noise

7. Shock and vibration

8. Temperature and humidity

Ionizing Radiation

Ionizing radiation has progressively acquired a firm stronghold in industry. Its uses, once limited to medical research and treatment, now encompass a wide range of useful activity in the manufacturing process. Non-destructive testing of pipelines and castings is commonly done by x-ray devices. Radioactive gauges are a useful tool in quality control, where critical tolerances must be maintained on the thickness of various materials. Radioactive tracer techniques have helped in analytical chemistry.

Depending upon the type of radiation, length of exposure, and individual susceptibility, body damage may be slight or severe; it may even result in death. A health physicist is better qualified to assess radiation exposure and to recommend controls; however, safety engineers and industrial hygienists should be aware of radiation problems and be able to request expert counsel.

Ultra-Violet Radiation

Exposure to ultra-violet radiation causes mild to deep burns of the skin, depending on the exposure. Sunburn is a common reaction to these rays. Flash burns (producing a feeling of "sand in the eyes") are a common affliction among welders. The most common industrial sources are arc welding, ultra-violet lamps used for inspecting flaws in metal castings, and sterilizing lamps used in hospitals. Eye and skin protection should be offered, to decrease the likelihood of exposure.

Visible Light

Good lighting is important to the performance of work tasks. A large number of accidents may be attributed to poor lighting conditions. Minimum illumination levels[16] recommended for a specific task have been documented and should be consulted in correcting unsafe conditions.

Infra-Red Radiation

This type of electromagnetic energy, similar to visible light but of longer wave length, gives off radiant heat, which is potentially injurious. Typical sources of exposure are hot metal surfaces and glowing molten glass. Steel rolling operations and glass-blowers are particularly involved.

Radiant heat does not heat the air through which it passes, but heats any object upon which it falls. Ventilation cannot be applied as a control measure. Shielding of both the source and the worker is the most practical method.

Infra-red energy is absorbed by the lens of the eye. The lens cannot lose heat readily. Progressive exposure to these rays has resulted in cataracts among glass-blowers, who frequently look at molten material in furnaces.

Microwaves and Lasers

Microwaves comprise that portion of the electromagnetic spectrum which has been used in radar and other communication systems. The heating effect of microwave radiation has been recently used for cooking and drying purposes in commercial installations. Since microwave energy is capable of producing heat inside a material, it presents potential dangers to human beings. Extensive tissue destruction can occur through overheating, and cataracts have been produced by exposing the eye to excessive amounts of microwave energy.

The electron tubes which produce microwave energy are known to produce x-rays because of the higher voltages needed for their operation. Manufacturers are aware of this problem and have incorporated adequate shielding to prevent any exposure.

Like any other form of radiation, microwave energy can be controlled with proper shielding, minimal exposure time and recommended safe distances. Exposure criteria for microwave energy have been outlined in the American Conference of Governmental Industrial Hygienists' 1973 publication, "Threshold Limit Values of Physical Agents."

A laser (Light Amplification by Stimulated Emission of Radiation) is a device for generating coherent electromagnetic waves of such intensities that, if not controlled, they may result in permanent injury to employees. The eye is the organ most vulnerable to injury because of the ability of the cornea and lens to focus the parallel laser beams on the small spot on the retina. Reflections from laser beams are just as intense as direct beams. These reflections are difficult to predict and can make off-axis viewing just as dangerous as on-axis viewing. Not all exposure to laser energy results in eye injury. Certain factors have a bearing on its degree. These are:

1. Pupil size

2. Distance from the source of energy

3. Energy and wavelength of the laser

4. Divergency of the laser light

5. Presence of scattering media

6. Place on the retina where the light is focused

Tentative permissible exposure limits have been suggested[18,19,20] but animal experimentation has not been sufficient to permit industrial health experts to establish any safe exposure limits. Until these limits become more definitive, *no exposure* to laser energy should be the rule.

Meanwhile, lasers have been taken out of the research institutions and are finding useful applications in industry, as in communications, surveying, mechanical measurements (flow rates and seismology), welding and cutting, three-dimensional holography, as well as in bloodless surgery and classroom demonstrations. It is impossible at the moment to foresee the numerous applications that lasers may have in the future, but we should bear in mind that the dangers inherent in this energy source are insidious and very much dependent on the power outputs produced. A recommended protection program for laser installations includes the following:

1. Personnel protection, in the form of protective devices for the eyes and the use of drapes and protective devices for the eyes and the use of drapes and protective black felt coverings to minimize damage resulting from scattered laser light.

2. Protective respiratory equipment designed to prevent the inhalation of vaporized material.

3. Ventilation effective in removing vaporized material.

4. Proper shielding to avoid electrical shock.

Noise

Prolonged exposure to high noise levels can produce a permanent hearing loss in employees. Early damage is often unnoticed because the hearing loss usually occurs first in the high frequencies or tones above those important for understanding speech. After a prolonged exposure to the high noise levels, the hearing loss can progress until the speech frequencies are involved — this can ultimately result in the loss of the ability to hear speech even though the person is shouting from a short distance. Many states have compensation laws which permit hearing loss claims to be made. If an audiometric test shows an appreciable hearing loss in the most important speech frequencies (usually at the frequencies of 500, 1000, 2000 Hertz) the employee has a compensable hearing loss claim. The cost of these claims can be considerable when a number of employees have hearing losses.

The problem of hearing loss due to noise is complex. In this section, it will suffice to say that the most important factors are the noise levels and the length of exposure. Employees exposed to moderately high noise levels for 40 hours a week will probably have as much hearing loss as employees exposed to higher noise levels for short periods each day, over a long period of time. The frequency of the noise is also important; high noise levels which are primarily in the frequency range of 300 to 2,000 Hz cause the greatest hearing loss in the speech frequencies. Not only does noise produce hearing loss and sizable compensation claims; it also reduces employees' output and efficiency, induces excessive fatigue, causes equipment failures and, finally, increases the occupational accident rate. All of these factors indicate that a definite hearing conservation program should be initiated and its efforts directed towards the following objectives:

1. Preventing noise from interfering with adequate speech or other audible communications.

2. Preventing hearing loss to employees exposed to excessive noise levels.

3. Preventing industrial noise sources from disturbing members of the community.

Industry has been generating noise for several hundred years. The decibel level has been increasing at an alarming rate. Improperly lubricated equipment is often the cause, indicating a lack of routine maintenance. Lack of foresight on the part of the manufacturer may create a poorly designed unit; the machine housing is often inadequate to withstand the impacts and vibrations during normal operations; the operators are frequently allowed to operate equipment without proper instruction and, finally, the operation itself may be such that noise is a direct by-product.

All these factors contribute to unbearable noise. Typical examples of such operations are listed below, but this list is, of course, not all-inclusive.

- High-speed machinery
- Power saws, shapers, planers
- Pneumatic tools
- Impact tools
- Nail-making and screw-making machines
- Vibrating or shaking devices
- Metal-forming equipment
- High-speed grinders
- Plasma-torches

Potentially-hazardous noise levels indicate the advisability of a noise survey. The following guidelines usually determine when noise levels will become hazardous:

- During plant inspection, it becomes difficult to understand shouted speech at close distances.
- A review of accident experience reveals a number of hearing-loss claims.
- Employees complain of head pains and ringing in the ears after a work day.
- Loss of hearing develops, which has the effect of muffling speech and certain other sounds after several hours' exposure to the noise.

The absence of pain is no proof that hearing loss is not occurring. This loss may be produced by levels well below the threshold of pain. Once it is determined from the above guidelines that a noise problem may exist, a hearing conservation program should begin. This program should include:

a) Noise measurement

b) Reduction of noise exposure

    1) engineering methods

    2) ear protection

c) Measurement of hearing ability

Noise measurement

To determine whether noise is a potential hazard to hearing, it is usually necessary to measure its overall level and to perform a sound analysis. Knowing the frequency of noise, the sound pressure levels of the various frequencies and the exposure time to the noise is important to the overall analysis of the noise problem. The two most common instruments used to measure noise are the sound level meter and the octave band analyzer. The sound level meter incorporates three weighting networks (A, B, C) to measure overall noise level.

The three networks are necessary to approximate the response of the human ear over the entire frequency and intensity range. The A-weighted sound level, expressed as dB(A), is usually sufficient for illustrating the intensity of noise exposure in employee work areas, particularly where the exposure is determined to be within acceptable levels. The simplicity of a single numerical value is appealing when presenting such information to non-technical individuals in management. The dB(A) is the unit of sound measurement in many hearing conservation criteria.

The octave band noise analyzer measures sound as a series of noise levels, expressed in decibels (dB) in successive individual octave bands from about 20 to 20,000 Hz. This form of analysis is useful for determining the noise control methods to be tried. Such analysis is also advisable where dB(A) results indicate levels slightly excessive for hearing conservation, in order to establish whether the noise is in the frequencies considered most likely to produce compensable hearing loss. In the case of a hearing loss claim, attainment of an octave band analysis of the alleged offending noise may facilitate medical testimony that the hearing loss is unrelated to the characteristics of the noise in question.

Procedure for Noise Survey

Noise surveys may be done for screening purposes — this is a rapid evaluation of the problem to determine if a more detailed study is necessary. Surveys may also serve to determine the distribution of the sound pressure level throughout the frequency range. Sound measurements are often made to evaluate noise-control measures.

In a screening survey, a large number of measurements is usually taken with the sound survey meter or sound level meter throughout the area, with particular attention to the noise sources.

In making the survey, the area should be covered completely in some orderly fashion. A walk-through procedure is usually sufficient, as long as time is taken to record the following essential information:

1. Equipment used for the measurements.

2. Location in plant (supplement with diagrams).

3. Sound pressure levels.

4. Machines which are operating.

5. Number of employees exposed and duration of exposure.

6. Characteristics of the noise.

On the basis of the selected criteria, it is possible to decide on the need for more detailed surveys.

In our case, the criteria to be used are those recommended by the ACGIH whose standards have been incorporated in the Walsh-Healey Act and, recently, in the Occupational Safety & Health Act of 1970. The general noise survey is made to determine the characteristics of a noise and its potential effect on individuals. In the evaluation of noise exposure to individuals, the microphone would be placed as near the individual as possible without interfering with his work. Several measurements should also be made with the sound level meter in the area to determine if the sound pressure level is relatively uniform and to ascertain if the measurements being made represent the individual's exposure.

It may be necessary to make a number of measurements to determine accurately the noise levels and individual exposure. This will be the case if the machine is operating only part of the time or is cyclic in operation. Also if an individual works in a number of locations, it will be necessary to make measurements at each.

Basic information that should be obtained during the survey in almost all cases includes:

1. Equipment used for the measurements

2. The time and date of the survey

3. Description of the location in which the measurements are made

4. Description of the noise source under study

5. Description of secondary noise sources

6. Noise-control measure in use

7. Overall and octave band levels

8. The meter speed and weighting network used.

Interpretation of Data

Both the ACGIH and the U. S. Bureau of Labor Standards have adopted Threshold Limit Values for Noise. Sound pressure levels and employees' exposures should be compared to the values in Figure 1.

Control of Noise

The control of noise is an exact science which is best left in the hands of expert acoustical engineers. Its complexity, however, should not discourage the safety professional from understanding the general principles applied in the control of noise. The application of these principles has often resulted in a homemade approach to eliminating noise which has produced admirable results.

As in the case of other health hazards, the approach to controlling noise should make the worker inherently safe on the job, no matter what his own action or attitude may be. One method used to accomplish this is to persuade management to

Permissible Exposures	
Duration per Day Hours	Sound Level dB(A)*
8	90
6	92
4	95
3	97
2	100
1 1/2	102
1	105
3/4	107
1/2	110
1/4	115 – C**

* Sound level in decibels, as measured on a standard level meter operating on the A-weighting network with slow meter response.
** Ceiling value.

Figure 1.

reduce exposure by initiating a task-rotation system or reducing the hours of work. Caution should be exercised here to prevent the shifting of a worker from an area of low noise to an area where noise level exceeds permissible limits. The other approach is to place the noisy machinery in a low-traffic area.

When these administrative measures are not feasible, then it is necessary to control noise by considering three methods commonly used: reduce noise at the source, reduce the amount of noise transmitted, and protect the operator.

A reduction of noise at the source is often accomplished by substituting a quieter process, such as welding instead of riveting.

Additional noise reduction can be achieved through proper maintenance at regular intervals and the installation of vibration dampers.

Transmitted noise is easily reduced by enclosing the process, by absorption with suitable materials, and by the use of well-fitted mufflers.

If the noise cannot be attenuated and it reaches the operator, then we should provide personal protective equipment such as ear plugs or ear muffs. If the operator remains stationary throughout his work day, then it may be wise to enclose his work station.

Abnormal Temperature and Humidity

Many industrial operations expose the worker to temperature and humidity extremes which cause adverse stress conditions.

Excessive heat together with increased humidity and minimal ventilation has produced heat exhaustion, heat cramps and heat stroke. Abnormal cold temperatures, on the other hand, are known to induce frostbite. Both of these environmental conditions impair the health of employees and seriously affect production efficiency. Temperature extremes may be found in any working environment. Even in a downtown office building, an excessive amount of humidity can affect the performance of daily tasks.

The worker should be properly protected in extreme temperature environments. Suitable clothing and good ventilation may often solve the problem. In many cases, it will be necessary to enclose the work station and provide the enclosure with conditioned air.

Shock and Vibration

The effects of vibration on man will depend on its frequency, acceleration and the duration of the exposure. Based on these factors, human responses to vibrations have included changes in respiration and peripheral circulation. The most common effects of vibration are exemplified by the use of hand-held power equipment. Prolonged use of pneumatic chisels and drills has resulted in numbness of fingers and hands. Clinical findings indicate that these local changes can lead to pathologic changes in tissues, bones and the nervous system.[2][1]

CONTROL OF HAZARDS

Identification of Exposures

So far this chapter has discussed the types of hazards found in industry and listed their various effects on the human organism. To achieve any control over these hazards, it is best to begin with a systematic procedure for recognizing them in the industrial environment.

A common approach used by many governmental agencies is to consider the raw materials used in the manufacturing process. Once these items are listed, their toxic potential is evaluated, using a source such as Irving N. Sax's *Dangerous Properties of Industrial Materials*. Then the manufacturing process is observed, to determine what happens to the raw materials and the final product. Any operation which may cause any of the raw materials to come in contact with the operator should be noted, so that a follow-up check can evaluate the worker's exposure.

A checklist (as shown in Figure 2) can be useful in locating problem areas.

HAZARDS SURVEY SHEET			
Location	**Operation**	**Hazard**	**Controls**
Screen Room	Silk screening	Toluol vapors	Local exhaust ventilation
Casting area	Pouring	Lead fumes	General vent. respirators
Finishing	Grinding	Lead dust	Local exhaust vent.

Figure 2

It should be remembered that any of the operations can cause conditions which could affect workers in other areas of the plant. For example, vapors from spilled mercury can be picked up by an air-conditioning system and circulated throughout the plant. Therefore, consideration should be given to the effect of any operation on the total environmental picture.

A visual inspection of the plant is a must in order to locate hazards. Even though process flow sheets indicate the operations are well controlled, they do not indicate whether a breakdown in the system has occurred.

As soon as the problem areas have been defined, a quantitative measurement of the hazard should be made. Direct-reading instruments and laboratory analysis are frequently used in determining quantities of air contaminants. A complete discussion of the methods and specific instruments employed is too great for the scope of this section, and the reader may instead consult the sources listed at the end of this chapter.

Measurement of the hazards in the environment is an easy task for an experienced individual, but evaluation of the hazard's potential to inflict damage on the worker is an art requiring the teamwork of several disciplines. A reasonable attempt can be made to assess the hazard, by comparing concentrations with recommended TLVs, obtaining additional exposure data from biochemical monitoring of the man (both on-the-job and off-the-job), and measuring the efficiency of control systems in effect.

Control of Exposures

Some control measures have already been discussed in the section dealing with occupational hazards. This section will cover the basic techniques frequently used to prevent and control occupational hazards. It should be noted that engineering control efforts are definitely preferred to personal protection efforts where health hazards are significant, and should be the ultimate goal in the treatment of industrial hygiene exposures. Some advantages of engineering control are:

1. The hazard is satisfactorily reduced without dependence upon human behavior.

2. Disciplinary problems associated with dependence upon personal protective equipment are absent.

3. "Hazardous" operations become "non-hazardous" operations, with consequent reduction in employee concern, "risk pay" considerations, etc.

4. The additional fatigue and pre-occupation with maintenance activity from the use of protective apparel is converted to additional man-hour productivity.

These methods are also preferred because they free the employee from the exposure in a manner that requires minimum effort on his part.

1. *Isolation of the exposure from employee work areas.* Operations not requiring continuous attendance, such as ball mills, can be completely walled off from employee work areas. The isolated area should have self-closing doors and no windows that can be opened. It should be maintained at a slightly lower air pressure than that of the adjacent employee work areas by an appropriate exhaust system, so contaminants will not infiltrate these areas.

2. *Complete enclosure of employee operation.* Often employed in permanent sand-blasting installations, this method consists of a sealed booth, completely containing the operation. The operator may control the operation from outside the booth or he may work inside the booth with the necessary personal protective equipment. Some grinding operations are similarly enclosed. A lower air pressure should be maintained inside the booth by an appropriate exhaust system.

3. *Local exhaust collection.* If complete enclosure of the operation is impractical, an effective collection system should be mounted as close as possible to the point of generation. This is probably the most widely used means of control. Airborne contaminants must be drawn away from the face of the employee and captured by the system.

   A large fan near a dusty operation, drawing without the aid of a duct, is usually not a suitable substitute for a ducted system. Absence of ductwork to direct the air flow admits the possibility that turbulence and cross-drafts will seriously impede the efficiency of this system. Dust may be inadvertently directed into other work areas by such an arrangement.

   It is inadvisable to filter or similarly process the air used for capturing contaminants and to recirculate this processed air into the work area. If the process becomes inefficient for any reason, such a recirculation system may present a significant health hazard to the workers for a long duration before it is corrected.

4. *General ventilation.* A significant health hazard exposure is rarely controlled efficiently by a general ventilation system. The contaminants tend to circulate in breathing zones of the workers, creating an undesirable condition. Opening windows to "relieve" hazardous conditions often intensifies them instead; strong drafts and gusts will agitate settled dust and keep airborne dust in suspension.

The three preceeding control methods are much preferred.

5. *Wetting methods.* The wetting down of dusty operations with water or other liquids is a common practice in industry, and is used frequently in mixing, cutting, grinding and drilling techniques. Foundry sand is often handled in a moist condition but wetting is often not completely effective as a dust-eliminating measure.

6. *Substitutions.* The replacement of toxic materials with relatively innocuous substances is occasionally feasible. Abrasive blasting, for example, need not be done with sand if steel particles can be satisfactorily substituted. Low silica or non-silica parting compounds are available for foundries. Benzene has been successfully replaced with a less toxic solvent (such as methylene chloride) to perform many cleaning operations.

Air Cleaning

Any time significant contaminated air is exhausted from a work area or a process, it should be cleaned of significant contaminants prior to being released. There are a number of methods for accomplishing this. Common methods are described in F. A. Patty's *Industrial Hygiene and Toxicology*, Volume I, by W. N. Witheridge, in his chapter entitled "Air Cleaning."

Respirators are not suitable substitutes for engineering control methods of air contaminants, but they must sometimes be used under conditions where better controls cannot be applied. A major complaint concerning respirators is that they are inconvenient or uncomfortable, and enforcement of their use is often difficult and inadequate. These devices also require careful fitting to the individual, and a respirator cleaning and maintenance program is necessary. A failure in enforcement, fitting, or maintenance may result in a return of the injurious exposure to the worker.

If respirators are necessary, they must be approved for their particular application by the U. S. Bureau of Mines. Use of unapproved equipment by employees in hazardous concentrations of toxic contaminants is considered unsatisfactory protection.

Medical Control Program

Industrial medical programs have developed into sophisticated techniques which play an important role in controlling occupational diseases. No longer is a physician confined to the undemanding medical services which were once limited to pre-placement and periodic physical examinations. Today the industrial physician and his staff perform a multitude of services which serve to protect and maintain the health of the employee. A good medical program should include the following:

- Pre-placement examinations: These aid in the placement of the employee in a suitable job.

- Periodic examinations: These determinations confirm or disprove the suitability of the employee to his job and help uncover any occupationally-induced illness.

Any occupational-related injury or disease should receive prompt treatment, and efforts should be made towards the prompt rehabilitation of the employee. Non-occupational injuries or diseases should receive the same attention. In those cases, the employee should be made aware of his condition and advised to consult his physician.

MAINTENANCE OF CONTROL PROCEDURES

The responsibility of the safety professional in preventing industrial health hazards does not end with the installation of a control system, whether it is engineering control or simply a change in management attitudes. If the system is to continue to work, it must be properly maintained to achieve its intended results.

Often, employees' complaints to their supervisors or to the medical department provide concrete evidence that the system is breaking down. But reliance on this method may not help prevent serious conditions; the recommended method is to conduct periodic surveys of the system, which will be designed to provide data on the entire system's performance. A complete evaluation should include:

- *Air sampling.* This may be accomplished either with a direct reading instrument designed to measure a known contaminant, or by a personal sampling technique which will measure concentrations in the breathing zone over carefully selected intervals and then average the results according to the length of exposure.

- *Visual inspection.* A periodic visual check of the control system will, in many cases, uncover any mechanical failures or breakdown in supervision. Leaks in ventilation systems, poor housekeeping, the failure of employees to wear protective equipment, and improper labeling are all clues easily detected by a visual check.

- *Process materials change.* A frequent cause of a breakdown in the control system is a change in the materials processed. Although the control system employed is functioning properly, a change in materials used in the manufacturing process may cause the process to generate contaminants of such a nature that the control system is inadequate to handle it. For example, a work station designed to remove toluol vapors may not be sufficient to remove silica sand. Employees should be advised when new materials are introduced in the process, so that proper handling procedures can be established.

- *Process change.* Any modification of the manufacturing operation can alter the environmental picture. A change that calls for new temperatures and larger quantities can produce new and larger quantities of contaminants, which the present system cannot handle. New processes should always be reviewed for their hazard potential before they are adopted.

- *Record-keeping.* Survey data and inspection notes should be kept for planning and design. These records also serve as a measuring tool in the success of the control methods.

THE NEED FOR AN OCCUPATIONAL HEALTH PROGRAM

The largest single expenditure a business faces is salaries. This is an investment which yields greater returns once a product is marketed and sold at a profit. The health of the employees is just as important when business is trying to maximize profits. Lost time has slowed production lines and proved costly in terms of increased insurance premiums. Therefore, it is good policy to insure that the work environment does not affect the employees adversely. A working industrial hygiene program can insure employees' health and help business by:[22]

- *Reducing the cost of workmen's compensation insurance.*
  The smaller the number of accidents, the lower the rates.

- *Reducing the cost of hospital and surgical insurance claims.*
  A consistent reduction in the number of claims will result in a lower premium.

- *Reducing absenteeism.*
  A better working environment can prevent any disruption of production schedules.

- *Reducing labor turnover.*
  A safe and pleasant working environment contributes to maintaining a steady work force.

A good occupational health program helps business in other, less tangible ways. It creates a good working atmosphere for the employee, which influences his attitude toward management; it promotes a mutual respect between employees and management; it increases the period of good earning power for the worker and guarantees a longer period of usefulness to the employer.

ORGANIZING AND MANAGING AN ENVIRONMENTAL HEALTH PROGRAM

An industrial hygiene and health program that fits the needs of both management and employee can be organized and managed to produce results for either large or small industrial plants. Regardless of the plant's size, however, there should be wide company representation by both staff and operating management. This organization will indicate the breadth of company involvement in the abatement of health hazards. A task force or committee represented by department heads and chaired by a top-management member is entrusted with the responsibility of formulating, establishing and guiding the administration of company policies, programs and procedures relating to environmental conditions, including water pollution control, air pollution control and solid waste disposal.

The committee recommendations will direct specific duties and responsibilities requiring the cooperative effort of the engineering, industrial hygiene, safety, medical and purchasing organizations, as well as the responsibilities of the supervisors and employees.

The objectives and responsibilities of the committee include the following:

1. Provide expert information on all matters pertaining to environmental control and recommend projects to be undertaken.

2. Approve news releases on selected environmental control matters.

3. Review appropriation requests dealing with new facilities or additions to existing facilities, to insure compliance with environmental health standards.

4. Be responsible for liaison with public regulatory agencies.

5. Recommend environmental control consultants and coordinate their activities on company projects.

6. Maintain representation on trade and technical organizations, to keep abreast of developments in environmental control affecting the industry.

Cynics have often said that if a project is to be scrapped, it is turned over to a committee. This may be true if the committee's recommendations are not implemented by management and line personnel. It is the responsibility of every individual in the company to insure that the committee's programs and recommendations concerning the maintenance of a safe and healthy work environment are carried out diligently.

The following program which describes the responsibilities of plant personnel has been adapted for use in this chapter from a similar program used by the Western Electric Company in its Industrial Health Bulletin entitled "Guidelines for the Safe Use of Lasers."

Responsibilities of the Engineering Organization

Since new manufacturing processes and operations are introduced in this organization, the control of environmental hazards begins here. Their responsibilities are:

1. To plan all operations so that personnel are not exposed to any health hazard.

2. To locate much of the automated equipment away from heavily populated or traveled areas.

3. To notify the medical, industrial hygiene and safety organizations whenever it is planned to introduce new processes and operations, and to insure safety by requesting an industrial hygiene survey.

Responsibilities of the Medical Organization

The responsibilities of this group are:

1. To recommend the placement of only those employees and prospective employees whose physical and emotional health capacities meet the minimum requirements specified for the job.

2. To cooperate in the development of effective measures to prevent exposure to occupational health hazards.

3. To examine periodically those employees who are working in harmful environments.

4. To remove employees from a hazardous job whenever periodic examinations indicate excessive exposure.

Responsibilities of the Industrial Hygiene Organizations

The main function of the industrial hygiene organization is to recognize, evaluate and control occupational health hazards. It has the responsibility:

1. To advise other departments of the potential hazard arising out of any current or proposed process.

2. To adopt nationally-recognized standards and to make appropriate tests periodically to ensure that the standards are met.

3. To specify the design and quality of all types of personal protective equipment and to prescribe standards for their use.

4. To recommend appropriate controls to minimize employee exposure to occupational health hazards.

5. To assist the supervisor in educating employees on practices, precautions and procedures established to control their exposure to health hazards.

6. To review present and proposed practices and to insure that they are in accordance with established standards.

Responsibilities of the Safety Organization

The safety organization acts as the coordinating body for all safety and health activities. Its responsibilities are:

1. To coordinate the educational, engineering, supervisory and enforcement activities of the safety program.

2. To provide the necessary educational material and assistance to supervisors in teaching employees safety rules, regulations and procedures.

3. To conduct safety surveys with the industrial hygiene staff to insure that proper practices and procedures are being followed.

4. To recommend any changes in safety rules, regulations and procedures to keep pace with technological advancements.

Responsibilities of the Purchasing Organization

The responsibility of this organization is to ensure that the equipment and materials purchased meet recommended safety and health requirements designed to prevent hazardous exposures to employees.

Supervisor's Responsibilities

The supervisor is responsible for maintaining safe working conditions within his department and for implementing the safety program. His responsibilities are:

1. To insure that the work environment is safe for his employees.

2. To instruct employees periodically on precautions, procedures and practices to prevent possible exposure to health hazards.

3. To enforce proper housekeeping at work stations.

4. To prevent the consumption of food, beverages and tobacco in the work area.

5. To inform appropriate departments of any operation or condition which appears to present a hazard to employees.

6. To inform the medical department of any accidental exposure and to send the employee(s) involved to the medical department for examination.

7. To provide his employees with the proper personal protective equipment, instruct them in its correct use, and enforce the wearing of such equipment.

8. To administer appropriate disciplinary action when safety rules are violated.

Employee's Responsibilities

The success of any program relies on the cooperation of the individuals involved. The employee will benefit only if:

1. He observes all safety and environmental health rules, including wearing appropriate personal respiratory equipment.

2. He notifies his supervisor when a condition develops which may cause personal injury or property damage.

3. He reports to his supervisor any accidental exposures.

4. He practices good personal hygiene.

5. He reports to the medical department for an examination at prescribed intervals.

MEASURING THE PERFORMANCE OF AN ENVIRONMENTAL HEALTH PROGRAM

An industrial hygiene and health program must be examined from the standpoint of profit and loss the same way any other project in industry would be evaluated. Industry should not only compensate its workers adequately for their labor and provide a safe healthful place to work, but should also provide a reasonable and adequate return to the shareholders who have invested their money in the undertaking. Therefore, industry must look at costs. Every investment must be able to justify its costs in terms of results and achievement.[2,3]

Industrial hygiene and health activities, however, are difficult to measure in terms of cost-benefit thinking, because of the many intangibles involved. Attempts to develop a measure for the effectiveness of performance of an industrial hygiene and health program have been suggested in the report of the Industrial Hygiene Workshop of the "Symposium on Industrial Safety Performance", sponsored by the National Safety Council in May, 1966.

REFERENCES

1. "For Management." A.I.J.A. 1962 Publication.

2. ORC Management Memo, "New Dimensions for Management in Occupational Safety and Health." Organization Resources Counselor, Inc.

3. ANSI, 1430 Broadway, New York, N.Y. 10018. Standards developed for the handling of specific chemicals.

4. Key, Marcus M., Ritter, Edmond J., and Arndt, Kenneth A., "Cutting and Grinding Fluids and Their Effects on the Skin." *American Industrial Hygiene Association Journal*, 27:423 (1966).

5. Francis N. Dukes-Dubos, M.D., "The Place of Ergonomics in Science and Industry", *Journal of the American Industrial Hygiene Association*, 31:565, 1970.

6. Stokinger, Herbert E., "Occupational Diseases — A Guide to Their Recognition." U. S. Department of Health, Education and Welfare, Public Health Service.

7. Data Sheet 531, National Safety Council, 1963. "Dusts, Fumes, and Mists in Industry."

8. Patty, F. A., *Industrial Hygiene and Toxicology, Volume 1.* Interscience Publications, Page 161. New York, N.Y.: John Wiley & Sons, Inc.

9. Mayers, May R., *Occupational Health, Hazards of the Work Environment.* Baltimore, Md.: The Williams & Wilkins Company, 1969.

10. Sax., N. Irving, *Dangerous Properties of Industrial Materials.* New York: Reinhold Book Corporation, 1968.

11. Lehmann, B. K. and Flury, F., *Toxicology and Hygiene of Industrial Solvents.* Baltimore, Md.: Williams & Williams Company, 1943.

12. Walsh-Healey Public Contracts Act.

13. Acceptable Concentrations. Z-37 Series. American National Standards Institute, 1430 Broadway, New York, N.Y. 10018.

14. "Emergency Exposure Limits," *American Industrial Hygiene Association Journal*, 25:578, November-December, 1964; 27:193, March-April, 1966.

15. AMA Publication, "Guide to the Significance of Occupational Exposure Limits."

16. Practice for Industrial Lighting; ANSI A11.1 (1965).

17. Daniels, R. G., and Galdstein, B., "Lasers and Masers-Health Hazards and Their Control," *Federation Proceedings*, Supplement 14, Volume 24, No. 1, Part III, January-February, 1965, Pages 25-27.

18. Martin Marietta, "What You Should Know About Laser Safety," Orlando Division, Orlando, Fla., 1965.

19. "Laser Systems-Code of Practice," British Ministry of Aviation of the United Kingdom, London, England, 1965.

20. Goldman, L. and Hornby, P., "Personnel Protection from High-Energy Lasers," *American Industrial Hygiene Association Journal*, November-December, 1965, Page 553.

21. Stockinger, op. cit. (6).

22. American Medical Association, Council on Occupational Health, "A Management Guide For Occupational Health Programs."

23. N. H. Collison, "Management and an Occupational Health Program," *Archives of Environmental Health*, February, 1961, Volume 2, Pages 116-123.

BIBLIOGRAPHY

1. "The Industrial Environment — Its Evaluation and Control," Publication No. 614. U. S. Department of Health, Education, and Welfare. Available from U. S. Government Printing Office, Washington, D.C. 20402. (1965).

2. *Industrial Hygiene and Toxicology, Vol. I, General Principles*, 2nd Ed., edited by F. A. Patty, Interscience Publications, p. 830, New York, N.Y.: John Wiley & Sons, Inc., 1958.

3. "Occupational Health Hazards, Their Evaluation and Control", Bul. 198. U. S. Department of Labor, Washington, D.C. 20210 (1968).

4. *Air Sampling Instruments Manual*, 3rd Ed., American Conference of Governmental Industrial Hygienists, Cincinnati, Ohio 45202. (1969)

5. *Industrial Noise Manual*, 2nd Ed., American Industrial Hygiene Association, Southfield, Michigan 48075 (1966).

6. "Industrial Noise — A Guide to Its Evaluation and Control," Publication No. 1572. U. S. Public Health Service. Available from U. S. Government Printing Office, Washington, D.C. 20402 (1967).

7. Peterson, A. P. G. and Gross, E. E. Jr., *Handbook of Noise Measurement*. West Concord, Mass: General Radio Co.

8. "Documentation of Threshold Limit Values," American Conference of Governmental Industrial Hygienists, Cincinnati, Ohio 45202 (1966).

9. Gafafer, W. M., "Occupational Diseases — A Guide to Their Recognition," Public Health Service Bulletin 1097, U. S. Government Printing Office, Washington, D.C. 20402 (1964).

10. *Industrial Hygiene and Toxicology, Vol, II, "Toxicology,"* 2nd Ed., F. A. Patty. Interscience Publications. New York, N.Y.: John Wiley & Sons, Inc., 1963.

11. *The Chemistry of Industrial Toxicology*, 2nd Ed., H. B. Elkins. New York, N.Y.: John Wiley & Sons, Inc.

12. *Industrial Ventilation*, 10th Ed. Lansing, Mich.: Committee on Industrial Ventilation, American Conference of Governmental Industrial Hygienists, 1968.

13. *Respiratory Protective Devices Manual*. Lansing, Mich.: Committee on Respirators, 1963.

XII.
Fire LossControl

On Sunday evening, January 15, 1967, final arrangements were being completed for the opening of the National Housewares Manufacturers Association's 46th semi-annual exhibit. Instead of an opening, the citizens of Chicago were shocked on Monday to learn that McCormick Place, the largest exhibition hall in the United States, had suffered substantial fire damage. According to later reports, a major portion of the approximately 1,250 exhibits, involving 3,700 booth spaces which had been erected for the show, was destroyed. Fire damage was estimated at 25 million dollars. Structural damage to the building itself was an additional 20 million dollars. McCormick Place, a windowless, non-combustible concrete and steel exhibition hall built in 1960, took its place with others on a long list of "fireproof" buildings that could not burn ... but unfortunately did!

How can a modern, so-called "fireproof" building be destroyed so quickly by a fire that was discovered almost in its incipiency? "To be accurate," said R. E. Gaudet, N.F.P.A., "McCormick Place did not burn; it merely fell in. Fortunately there were not thousands of people in the building at the time.

After the fire, Chicago's Mayor Daley appointed a committee of engineers to investigate the fire and evaluate the findings. Their primary function was to propose improvements in the Chicago Building Code. The committee's report contained a section titled "Recommendations for McCormick Place", in which the findings were summarized. Most of the fifteen recommendations in this section were aimed at correcting conditions brought about by the lack of professional management planning and control. Let's examine several excerpts from the recommendations:

1. "... *adopt a resolution* making all future construction and operations subject to full compliance with the Chicago Building Code and the Chicago Fire Prevention Code."

2. "A *complete study* ... of the private water supply system to insure adequate design ..."

3. "*Make* necessary *repairs* to pumping station ..."

4. "*Review* the operating *controls* in the pumping station . . ."

5. "*Train* maintenance *employees* in methods of operating the pumping station . . ."

9. "*Initiate* a *regular maintenance and inspection program* (with log) to assure that all system components are operable . . ."

10. "*Appoint* a qualified full-time *fire marshal* . . ."

11. "*Train* all *employees* in methods of extinguishing small fires . . ."

15. "The storage, handling and use of flammable solids, liquids and gases should be closely *regulated*."

    Look at the key words (highlighted) in these recommendations. The engineers who prepared them would probably agree that their investigation did not reveal anything particularly complicated or mysterious (or new, for that matter) about the proximate or contributing causes of the fire damage. A professional manager would readily recognize them as a form of a management deficiency analysis. That is the primary lesson to be learned from this destruction. Because of the structural and protection deficiencies known to exist in McCormick Place prior to the fire, *better management planning* and *tighter management controls* were necessary to either prevent the occurrence of fire or to make certain it would be contained and extinguished during its incipient stage. It is often the case that company management remains complacent for too long. These same people often assume that fire loss control is merely a matter of a physical inspection once or twice a year by an insurance company or a rating bureau's technical representative, whose function it is to point out the most obvious hazards resulting from their failure to plan and establish controls.

    The National Fire Protection Association clearly points out that dealing with the unavoidable consequences of a destructive fire challenges the resourcefulness of management. The origin of fire, however is not unavoidable. The *real* challenge to management is to remove, where practical, the potential causes of fire in plants and operations, and then to establish the necessary emergency procedures which provide reasonable assurance that minor fires which do occur cannot grow to disastrous size.

Although new processes and materials have introduced new hazards, and will continue to do so at greater rates in the future, there is nothing new or mysterious about the causes of the great majority of industrial plant fires. Since the National Fire Protection Association began recording and analyzing fires in 1896, the causes most frequently listed range from faulty electrical wiring to careless smoking and poor housekeeping practices. The remedies for most of the causes on this list are also uncomplicated and may be as simple as conducting frequent, thorough facility inspections and implementing modern maintenance programs. Success of these programs hinges upon the establishment of a critical-item inventory of all plant facilities and equipment.

The frequency of large loss industrial fires indicates a failure to use the protective measures and equipment which have been developed and standardized, largely under National Fire Protection Association auspices, to limit the amount of damage a fire may cause. Here again there is nothing complicated or mysterious about the reasons, in most instances, why small fires too often grow to disastrous size. NFPA studies of such fires consistently pinpoint the principal reasons as the absence of division walls and vertical opening enclosures to confine the fire near its area of origin, the lack of either operational automatic sprinkler systems to check the fire in its initial stage or detection and alarm systems for quickly alerting fire-fighting forces. Major fires in industry, with relatively few exceptions, are simply the results of the planning that was never done, the designs that were incomplete and the equipment that was never installed or that was poorly maintained!

The design of automatic protection and detection systems and other needed safeguards requires competent engineering or architectural advice. It is the business owner or manager who bears the ultimate responsibility for seeing that the threat of fire is considered in every phase of his operation, and that every practical preventive and protective measure is taken. The most significant development of the past half century in fire protection is the vast increase in the knowledge of how fire behaves and how it can be contained. There has been and continues to be progress made in the effective use of this accumulated knowledge.

TOTAL LOSS CONTROL

The discipline of fire loss control and industrial safety share the same prime objective: conservation of the company's assets, property and human life alike. Each approaches the goal of providing a safe physical environment by the same methods (i.e., elimination, segregation and/or protection of hazards) and both utilize the same or similar yardsticks as a measure of successful performance. The perfect score in meeting objectives is attained by the absence of accidental occurrence, whether fire or industrial!

Considering the similarity and overlap of objectives, methodology and measurement of performance, structuring Fire Prevention and Protection and Industrial Safety as two separate departments can lead to serious management problems. It is virtually impossible to establish clear-cut lines of demarcation between these functions. Despite common interest in loss control, their overlapping fields of responsibility and duplication of effort often seem to lead to conflict between the administrative heads of the Fire Protection and the Plant Safety Departments, rather than teamwork.

It would appear to be logical (and, in fact, mandatory) that the loss control efforts of a corporation be administered and coordinated by a single head of a combined department.

This same reasoning should be applied to include the responsibility of plant security in the Loss Control Department. This permits management utilization of the security force as a cadre for company fire brigades, for continuing facility inspections and for trained first aid personnel.

PROFESSIONAL ASSISTANCE

A wealth of fire loss control information and assistance may be obtained from insurance companies who maintain technical departments, as well as the National Fire Protection Association, the American Insurance Association, Society of Fire Protection Engineers, Underwriters' Laboratories, and architectural, engineering and trade associations. Independent research laboratories and consulting firms also provide a source of competent technical and engineering advice for those loss control administrators who cannot afford and do not need full time specialists on their own staff.

The National Fire Protection Association (NFPA) is a non-profit voluntary membership association, founded in 1896 to "promote the science and improve the methods of fire protection and prevention; to obtain information on these subjects and to secure the cooperation of its members and the public in establishing proper safeguards against loss of life and property by fire."

NFPA's present membership includes representation from industry and commerce; fire departments; federal, state and local agencies; architects and engineers; building inspectors and other officials; electrical inspectors; building materials makers; hospitals, schools and other institutions; and insurance companies.

The basic NFPA function is the preparation of standards and codes which are widely utilized as the basis of laws and good practice. Of the more than 200 standards developed to date, some 40 relate to building construction and design and to building services. The Association does not promulgate a building code as such, but adaptations of many NFPA standards and codes are used in building codes. The technical committees which prepare NFPA standards are made up of individuals especially qualified in the subject area. Currently some 1700 individuals serve on these committees, which are balanced to have all interested parties represented. In the process of development and adoption, standards go through a democratic legislative procedure, allowing proponents and opponents to be freely heard.

In addition to standards and codes, NFPA publishes a wide range of technical and informative material on fire protection and prevention, including the authoritative 2,100-page *Fire Protection Handbook*, now in its 13th Edition. It also issues three periodicals: *Fire Journal, Fire Technology,* and *Fire Command.*

NFPA carries on an extensive public education activity in fire safety and operates field services in flammable liquids, gases, electricity, marine fire hazards, and fire extinguishing systems. Affiliated with NFPA as sections are the Fire Marshals Association of North America, the Society of Fire Protection Engineers, the Industrial Fire Protection Section, the Electrical Section, and the Railroad Section.

The American Insurance Association (AIA) is a non-profit insurance industry organization, whose purposes are to: (1) provide a forum for the discussion of problems which are of common concern to member companies; (2) promote the interests of such companies in every legitimate manner consistent with the public welfare; and (3) serve the public interest through appropriate activities, including the promotion of the safety and security of persons and property.

The Engineering and Safety Service is dedicated to providing the research and facilities the member companies must have to keep pace with new processes, materials and procedures produced by advancing technology. The service assists the underwriting and engineering efforts of the companies in a common objective of maintaining high standards of service in the face of complex developments affecting fire protection and industrial safety.

The Municipal Survey Service provides authentic information to assist subscribing companies in their underwriting, and promotes effective municipal fire protection throughout the nation.

Periodically during the year, the AIA sponsors instructional courses in Fire Safety and Loss Control, centered at various key geographic locations.

Insurance Rating Boards and Bureaus provide statistics, rates and advisory assistance to the insurance industry. The individual or corporation with an insurable interest may have access to these services when involved in problems having a direct bearing on the insurance rates assessed against his property.

Factory Insurance Association, (FIA), is an organization of capital stock fire insurance companies. Their organizational objective is to provide stock fire insurance, together with superior fire engineering, supervisory inspection and loss adjustment services, to large manufacturing risks enjoying superior management, maintenance, construction, automatic sprinklers and other private protection. The Association fosters an engineering council, which develops and coordinates fire protection practices and standards as dictated by the changing requirements of industrial America.

Factory Mutual Inspection Department and Laboratories, (F.M.), is an organization maintained by and for manufacturers; it specializes in industrial fire protection. Its membership consists of leading manufacturers. Inspection services are operated jointly by the Associated Factory Mutual Fire Insurance Companies for the benefit of its policyholders, to minimize likelihood of heavy property losses and to safeguard production. This organization engages in much the same activity as the FIA, in respect to development and coordination of fire protection practices and standards.

Business and industry have found that individual insurance carriers usually provide substantially the same type of services offered by the FIA and F.M.

Underwriters' Laboratories, Inc., (U.L.), is a non-profit organization engaged in the testing of devices, systems and materials to determine their relative fire prevention and fire protection characteristics. The results of these tests determine whether a company's name will be included in the several lists published by U.L.

These lists contain the names of companies which have qualified to use the U.L. "Listing Mark" on their products and which have agreed that the Mark will be used only on products which meet the U.L. requirements.

In order to promote continuing quality of tested and approved products, U.L. engages in a "follow-up service." Under this service, representatives of U.L. perform periodic re-examinations and/or tests of the products at U.L. testing stations in order to determine continuing compliance with the Laboratories' requirements.

Reasons for using U.L. lists include:

1. To obtain the names of those firms authorized to use the Listing Mark on specified products.

2. To determine information regarding the form of the Listing Mark to be used for a specific product class.

3. To obtain information pertaining to limitations or special conditions applying to the product.

The published U.L. lists, together with the research and testing effort they represent, can be a valuable tool to the Corporate Fire Loss Control Manager and his staff when purchasing equipment. The wide recognition afforded Underwriters' Laboratories, Inc. should not be overlooked in the corporate decision-making process.

A partial list of other service organizations providing services in the fire protection area includes:

Inland Marine Insurance Bureau

American Mutual Insurance Alliance

Fire Insurance Research and Actuarial Association

National Insurance Actuarial and Statistical Association

National Safety Council

Nuclear Energy Property Insurance Association

Colleges and universities also are expanding into the field of fire protection engineering. Degree or certificate programs are offered by many state and private schools, such as the University of Maryland, the Illinois Institute of Technology, the University of Southern California and Oklahoma State University.

Governmental agencies at the local, state and national levels also provide various materials which can assist the fire loss control manager. At the national level, numerous publications relating to a wide range of topics are available by addressing an inquiry to the Superintendent of Documents, in Washington, D.C.

The U.S. Department of Labor, Workplace Standards Administration of the Bureau of Labor Standards, holds instructional and training seminars at various times and places. State and municipal agencies responsible for fire protection and safety also sponsor periodic training sessions and seminars on fire loss control topics.

Fire protection equipment manufacturers and their distributors also play a major role in fire loss control. This is particularly true of those industries engaged in the manufacture, supply and distribution of equipment designed to reduce or eliminate fire losses. Their programs consist of publications, engineering services (frequently at no cost to customers), demonstrations and seminars. The areas of fire protection covered in such programs include: fire detection and alarm systems, automatic sprinklers, portable extinguishers and sophisticated automatic extinguishing systems for specific classes of hazard.

In addition to the materials and services provided by the manufacturers individually, valuable information is disseminated by them collectively through their trade associations.

From this brief overview of the available sources of professional assistance, it is evident that even a neophyte fire loss control manager need not find himself at a loss to solve a specific problem. By the judicious use of the sources of information available to him, he should not fail to arrive at a solution for the correction of any undesirable or unsafe condition with which he may be confronted.

C.O.P.E.

Fire Loss Control is the conservation of assets by reducing the incidence and severity of undesired and unexpected fires. Effective administration necessarily requires some knowledge of the basic factors which must be considered to intelligently plan and implement a program that will eliminate or minimize the occurrence of fire and its resultant destruction. The essential areas which must receive prime attention can be summarized by the acronym "COPE" — Construction, Occupancy, Protection and Exposure.

CONSTRUCTION

To illustrate the importance of building construction as a factor in fire loss control, you need only consider the tremendous range of possibility of loss inherent in various types of buildings. Compare the combustibility and susceptibility to fire damage of a building constructed of wood, for example, with one built of reinforced concrete. It is obvious that the type of construction has a direct bearing on both the incidence and extent of fire damage. Moreover, this conclusion dictates that those charged with the corporate responsibility of administering a fire loss control program be familiar with the four basic, most prevalent classes of construction: (1) Fire Resistive, (2) Non-Combustible, (3) Masonry, (4) Frame.

Fire Resistive Class Buildings are constructed entirely of non-combustible materials such as concrete and steel. All structural members — columns, beams, girders, floors, roofs and exterior walls — must be constructed to withstand for three hours or more, without failure, the most severe fire to be expected within the building.

> NOTE: The fire resistance ratings of building materials and assemblies are determined by testing organizations such as the Underwriters' Laboratories and the American Society for Testing Materials through the use of a standard fire test procedure. The test, using a controlled time temperature method, shows how well a specific material or construction assembly can be expected to control or limit the spread of fire to its area of origin. Therefore, a property should be termed "fire resistive" only when the building structural members have a fire resistance capability to withstand, without failure, the total amount of heat (BTU's) produced by burning its combustibles in a pre-determined period of time (liberation rate.)

Non-Combustible Class Buildings are constructed of non-combustible materials, such as steel, iron, concrete, brick tile, etc., which fail to meet the standards for fire-resistive construction. Since 1946, there has been a large increase in building construction classed as noncombustible. Buildings which have metal walls and roofs on an unprotected steel framework constitute the largest number of non-combustible buildings. Many all-metal buildings are pre-fabricated in a steel shop and then assembled at the actual location after the foundation and concrete floor have been poured.

Masonry, Brick or Ordinary Class Buildings have exterior and other bearing walls of brick, stone or concrete and the remaining structural members — columns, beams, floors and roofs — are entirely or partly of wood. Most of the older commercial buildings have brick bearing walls with wood floors and wood roof deck. In constructing these buildings over three stories, it was the practice to have the thickness of the walls vary from the lower floors to the top floors to support the load.

Frame Class Buildings are those in which bearing members of walls and interior construction are wood. This category includes stucco on wood studding, metal on wood studding, asbestos shingles on wood studding, and up to four inches of brick or stone as a veneer over wood studding. Most frame buildings have wood floors and the roof covering may be wood shingles, asphalt shingles, built-up tar and gravel, or metal applied over a wood deck, or a wooden framework.

How Fires Spread

In order to more clearly understand the importance of building construction, let's examine how fires spread. Fires spread both vertically and horizontally. Building design should take these factors into consideration and provide appropriate protection to control the spread of fire. A study should be made of the building under consideration, to determine the extent to which such protection is necessary and the relation between protection of openings and general features of building construction, occupancy, and automatic or manual fire protection systems.

It is important to remember that technical compliance with building codes or insurance rating schedules will not always produce the best design from the point of view of fire safety. Proper protection of openings in walls, floors, and partitions is essential to prevent the spread of fire and smoke and to safeguard life.

Construction openings are classified as either vertical or horizontal, which refers to the way fire may spread through them. Vertical openings include not only openings in floors, but also openings into stair enclosures, elevators, dumbwaiters, chutes, and any other opening which extends through one or more floors.

Horizontal openings are those in walls or partitions which sub-divide a story of a building into fire areas, rooms, or corridors. This classification also includes openings in exterior walls, which may be exposed to possible exterior fires, such as those in other buildings or in outdoor storage.

Standard practice for protecting against vertical spread is to enclose all vertical openings with walls or partitions of substantial fire resistive construction, and to provide fire doors or other protection for all interior openings in such walls. To protect against horizontal spread, solid walls are preferable to those with openings, no matter how effectively the openings are protected. Therefore, the first thought should be to reduce both the number and size of openings. Such action, however, should not be to the point where fire department access is impaired. Fire doors, shutters, and fire windows are designed to protect openings under normal use, with clear spaces on both sides of the openings.

Vertical Fire Protection. Fires spread vertically from floor to floor in the following ways:

1. Through unprotected vertical openings, such as elevators, shafts, stairways, escalators, dumbwaiters, ducts, conveyors, chutes and pipes, and through light walls.

2. Through the floors themselves.

3. Through failure of protection devices on vertical openings.

4. Through external wall openings.

5. Through unburned gases, in the form of smoke, passing into upper floors with the possibility of resultant fire or explosion.

Each floor of a building should form a barrier to the vertical spread of fire. Unprotected floor openings permit drafts which fan a fire and carry heat and smoke to upper stories. The lack of enclosure for vertical openings is usually responsible for the ignition of combustible contents on floors above the fire and for cutting off the means of escape for occupants.

Horizontal Fire Protection. Fire spreads horizontally to separate sections of the same building or from building to building in the following ways:

1. Lack of fire walls.

2. Unprotected wall openings.

3. Failure of fire doors or other protection for wall openings.

4. Lack of parapets for fire walls where required.

5. Passage of the fire around the end of fire walls.

6. Exposure of openings in outside walls at angles to each other.

7. Heat passing through walls, especially in bearing walls, at points where timbers and joists are built into them.

8. Unburned gases in the form of smoke passing through wall openings, resulting in explosions.

9. Failure of walls.

Preventing the horizontal spread of fire is particularly important in buildings of large areas, housing large quantities of combustibles or containing hazardous processes. This is accomplished by dividing the interior floor areas into smaller areas, separated by fire walls or fire partitions; fire doors, which help confine any fires to a single section, are used to connect the areas. When a building is close to others and subject to exposure fires, windows and other exterior openings should be protected by wire glass and metal sash, fire doors, or other means.

OCCUPANCY

The second factor considered in "COPE" is occupancy, which is the use to which the premises are put by the occupants. Some types of operations or processes create severe and even unique exposures to loss, referred to as special hazards, such as hazardous chemicals, flammable liquids and gases, combustible dusts and fibers, explosives, etc. Other exposures to loss can exist in most occupancies, and are called common hazards; examples of these are smoking and matches, heating and electrical equipment and poor maintenance or housekeeping practices.

The degree of the hazard presented by the occupancy is of major importance in determining the proper construction, size and even the location of buildings. Storage of stock, whether in large, high piles, or racks with limited aisle spaces, or located in smaller areas in low piles protected by automatic sprinklers, influences the fire hazard significantly.

Unfavorable conditions, such as congested, combustible operations or large unbroken areas without fire cut-offs or automatic protection, affect not only the probable frequency, but the severity potential of fire as well.

The loss control manager's basic responsibility encompasses proper analysis of the operations and processes, taking into consideration the degree of fire hazard they introduce in terms of the presence or absence of ignition sources, the cause of ignition and the total assets subject to one fire. The need for professional consulting services is frequently a must in carrying out this management function, particularly when dealing with the problems associated with such special hazards as new plastics, high-piled rack storage, cryogenic liquids, radiation, and explosive atmospheres.

PROTECTION

The third factor to be considered in "COPE" relates to protection, both public and private. The importance of this area cannot be minimized; both aspects should be of prime concern to a corporation in the planning and implementation of a fire loss control program.

A well-managed fire loss control program uses available public protection, together with a carefully-selected program of private protection, to accomplish its goal. This program must be reviewed continually for changing hazards and to be certain that the systems consist of the most modern components available.

Public Fire Protection

Public fire protection refers to the facilities provided by municipalities, including fire departments, water supplies, water mains and hydrants, fire alarm systems and other related services. While this aspect cannot be controlled by any individual or corporation, the degree of protection afforded can be influenced by responsible community action.

As a taxpayer and a responsible member of the community, the corporate citizen has a vital interest in the fire protection facilities in the area. The most obvious reason for this concern is assurance that adequate fire-fighting facilities are available, to minimize loss in event of a fire. A less obvious economic factor for management's consideration is the fact that the quality of municipality protection directly affects the amount of insurance premiums paid for fire and allied coverages. Public protection in every city and town in the United States and Canada is examined and graded by either the American Insurance Association or the state fire insurance rating organization having jurisdiction. The resulting published "grade" of the individual locality bears a direct result on all fire rates promulgated for that rating jurisdiction.

The municipal grading evaluates the water system, including pumping facilities, supply, distribution system and hydrants, building codes, fire alarm system, police department, conflagration potential, fire department and other factors. Statistical studies have shown that the annual percent of fire loss compared to the total value of buildings and contents runs consistently from an annual 2.47% of loss in Grade 2 localities to 18.44% of loss in Grade 10 localities.

The fire department is one of the most important factors in this grading. The criteria for measuring its effectiveness are rigidly scheduled and cover the organization, membership, fire stations, apparatus and equipment.

Organization ranges from the large city fire department, a professional force headed by a Chief, who is directly responsible to the administrative head of the municipality, through intermediate types of organizations such as county fire departments, fire districts and fire protection districts, down to the volunteer fire company that enjoys a quasi-public position, but raises its own funds and maintains complete independence in regard to responsibility and recruitment.

Regardless of the membership of the department, professional or volunteer, the personnel must be thoroughly trained, both initially, on entrance to the department, and on a continuing basis. Leadership must be competent; meetings must be regular; and drills must be frequent.

Fire stations must be centrally located, heated (if subjected to cold weather), and preferably used only for fire department purposes. The number of stations required is dependent on the area covered, but should be of sufficient number and location to insure that all buildings are within three miles of a fire station.

Apparatus, in order to qualify, must be of a specified size and capacity, must be certified by Underwriters' Laboratories for fire service, and must carry minimum equipment, including hose, portable extinguishers, nozzles, ladders and a host of other incidental, but necessary, items.

Private Fire Protection

Private protection consists of facilities provided by individual firms, and is made up of such items as sprinkler systems, fire brigades, detection and alarm systems, portable fire extinguishers and other fixed automatic extinguishing systems. Some of these systems can be used in the area of "spot protection," i.e., in the protection of specific operations.

TYPES OF EXTINGUISHING AGENTS

Carbon Dioxide

Carbon dioxide protects flammable and combustible liquid hazards, electrical equipment, water damageable materials, and other special hazards. Areas can be reoccupied immediately after extinguishment, usually without the need of extensive cleanup.

A nontoxic, nonflammable, noncorrosive and nonconductive gas, carbon dioxide reduces the oxygen content of the air to starve a fire of this essential element. Carbon dioxide's three-dimensional quality enables the fire-killing gas to penetrate all areas of the protected equipment or spaces; it can be used in vaults, engine test cells, transformer rooms, on electrical raceways or to protect individual hazards in computers, dip tanks, spray booths or printing presses. Carbon dioxide can extinguish a fire fast.

Carbon dioxide is stored as a liquid under high pressure (in cylinders), or under low pressure (refrigerated-insulated containers), so that it is practical for almost any size of hazard. It expands to a gas upon discharge.

There are three methods of automated carbon dioxide applications:

1. Total Flooding: providing protection to an enclosed or semi-enclosed space.

2. Local Application: discharged to blanket an individual hazard or piece of equipment effectively.

3. Delayed or Prolonged Discharge: engineered to start discharge at a high rate of application, supplemented by a prolonged discharge at a lower rate. This technique may be used on either total flooding or local applications, and is designed to maintain a nonhazardous fire condition following initial extinguishment.

Dry Chemical Interrupts Combustion Reaction.

Fast-acting dry chemical has a wide variety of uses in automated fire protection. Its nontoxic qualities make it popular for hotel and restaurant kitchens, where it quickly extinguishes fires in ranges, grills, fryers, broilers, hoods and ducts. As a nonconductor, it can be used on electrical fires, such as motors, generators and similar equipment.

Dry chemical is especially noted for its sensational speed of extinguishment on flammable and combustible liquid fires. Certain types of dry chemical can also extinguish fires in ordinary combustibles (wood, paper, fabrics).

Although the fire department equipment industry is continuing its research into new types of automatic systems, there are four basic dry chemical formulations now available. The type of dry chemical selected depends on the hazard being protected, with system components tailored to the specific usage.

Total application will give protection to large enclosed areas, while local application will protect specific hazards. Its efficiency minimizes fire damage and reduces production delays that can be costly in lost time or process shutdown.

Stored in metal cylinders under pressure, systems can be designed to tolerate extreme variations in temperature, from $120°F$ to $-65°F$, without reduced effectiveness or efficiency.

Foams Provide Blanketing and Cooling Effectiveness.

Fire fighting foams are an aggregate of gas-filled bubbles formed from aqueous solutions. The great extinguishing efficiency of these foams results from excluding air (oxygen) and cooling. Forming fluids of low density, high heat-absorption capacity, and film coalescence, foams have wide application in automatic systems.

On flammable and combustible liquids, foam systems form a protective covering that extinguishes fire by starving it of its air supply. Hazardous vapors are suppressed after extinguishment. As foam contains a high percentage of water, the cooling effect repels radiant heat during the extinguishment process.

Conventional automatic foam systems are used to protect such hazards as oil storage tanks, aircraft hangars, solvent and paint processing areas. These systems are of the mechanical (air) foam type or of the chemical type with expansion ratios from about 4 to 1 to 16 to 1.

High-expansion foam systems can be used in several different and unique ways, such as basement and vault areas, specialized fire problems like mine shafts, and some industrial processes and storage situations. High-expansion foams are generated from special aqueous surface-active compounds which are aspirated or "blown" through fine mesh screens and have expansions of from 100 to 1 to 1,000 to 1.

Halons are Used for Explosion Suppression.

Halons are compounds containing one or more atoms of an element from the halogen series such as: fluorine, chlorine, bromine and iodine. Four compounds are now used: bromotrifluoromethane (1301), dibromodifluoromethane (1202), bromochlorodifluoromethane (1211), and dibromotetrafluoromethane (2402). The numbers are to simplify reference to the actual chemicals and are called "Halone Numbers."

The chemicals are either of the liquefied gaseous types (liquids under pressure, released as a gas) or vaporizing liquid type (liquids at atmospheric temperatures, which vaporize when applied to a fire).

Explosion suppression effectiveness is achieved by inhibiting the combustion process through instant release of the selected Halon in milliseconds of time. Fire extinguishing effectiveness is largely due to the chemical interruption of the combustion process and to dilution of the oxygen content in the air.

Automatic systems are designed for multiple applications. Their initial use was in aircraft power-plants and to protect against fires in the engine compartments of boats, military tanks, and racing cars. Systems are now being used to protect computer and data process systems, bank and storage vaults, and for flammable liquid and vapor hazards.

Sprinklers Do the Humanly Impossible By Being Constantly on Guard Against Fire.

Automatic sprinkler systems provide around-the-clock, proven protection for all types of buildings and are well established as sentinels against loss of life. As individual sprinklers operate, water is discharged directly on the fire. The water flow sets off an automatic alarm to alert occupants and/or protective agencies.

In standard systems, heat-sensitive elements cause each sprinkler to function independently and release water. (Deluge and preaction systems are operated by separate heat-sensitive devices responsive to fire conditions.) Sprinkler equipment can be installed to blend aesthetically with the ceiling decor.

Sprinkler systems are the most widely used means of automatic protection and have achieved an efficiency ratio of 96% satisfactory performance in hundreds of thousands of actual fires. Without sprinklers, business and commerce simply would not survive. The life-saving quality of sprinklers has been especially recognized in custodial-care facilities and in schools, hospitals, hotels, and similar properties.

Since most fires in buildings initially involve the contents, sprinklers are particularly effective, because they actuate before the structure is threatened and effectively extinguish or control the fire. The efficiency of sprinkler protection has been recognized by substantial insurance premium rate reductions in properties protected by complete systems which are properly supervised.

Water Spray Systems Control Burning, Extinguish Fires, and Protect Exposures.

Fixed nozzle systems discharging water spray (sometimes called "water fog") are designed to cool vessels and piping subject to fire exposure and to extinguish fires in heavy oils.

By engineering the pattern of water discharge and by controlling outlet size to achieve the desired density, maximum efficiency is obtained from each gallon of water used. These systems are commonly used to protect storage tanks of flammable liquids, gases, piping and process equipment, electrical transformers, oil-filled switches, and rotating machinery.

Each installation is tailored to the particular hazard, and the piping is hydraulically calculated and balanced for the intended usage. Systems may employ wetting agents, additives, or viscosity-controlling materials for special situations.

Water spray systems can be designed solely to protect tankages of volatile and heat-reactive materials from exposure fires, spontaneous heating, or the heat of polymerization. Ultra-high-speed systems, which operate in less than 50 milliseconds, protect such hazards as solid propellant machinery and gunpowder manufacturing operations.

CLASSIFICATION OF FIRES AND RECOMMENDED PROTECTIVE EQUIPMENT

In addition to sophisticated fire protection systems to cope with potential major fires, all properties should be provided with facilities to cope with the small fire in its incipient stage. The great preponderance of fires can be controlled and/or extinguished by the prompt use of appropriate portable fire extinguishing equipment. This equipment, however, must never be considered as a substitute for major fire fighting facilities.

Fires are grouped into four classes, ("A" through "D"), and identified as follows:

Class "A" identifies fires occurring in ordinary combustible materials such as wood, paper, and cloth. This type of fire is most effectively fought by means of quenching or cooling through the medium of water or aqueous solutions. The temperature of the burning material is reduced by this cooling process to a point below its ignition temperature and the fire is extinguished.

Class "B" identifies fires occurring in materials such as greases, flammable liquids, gases and petroleum derivatives. Fires affecting these materials are normally extinguished by the smothering or blanketing action of chemical-dispensing extinguishers which provide an oxygen-excluding agent.

Class "C" identifies fires occurring in electrical equipment such as panel boxes, motors, etc. An extinguishing medium having an electrical non-conductivity characteristic must be the first consideration.

Class "D" is a relatively new classification which identifies fires in combustible metals such as magnesium, sodium, potassium, etc. Fires in this group react violently with water and, therefore, a device which dispenses an appropriate dry chemical agent is recommended for this class.

Portable Extinguishers. Several factors must be considered to obtain maximum effectiveness from the use of portable fire extinguishers:

a) Suitability for controlling the type of fire which might be anticipated

b) Adequacy of the capacity and number of the extinguishers selected

c) Proper distribution of extinguishers to provide prompt access and use

KNOW YOUR FIRE EXTINGUISHERS

	WATER TYPE				FOAM	CARBON DIOXIDE	DRY CHEMICAL	
	STORED PRESSURE	CARTRIDGE OPERATED	WATER PUMP TANK	SODA ACID	FOAM	CO2	CARTRIDGE OPERATED	STORED PRESSURE
CLASS A FIRES — WOOD, PAPER, TRASH HAVING GLOWING EMBERS	YES	YES	YES	YES	YES	NO (BUT WILL CONTROL SMALL SURFACE FIRES)	NO (BUT WILL CONTROL SMALL SURFACE FIRES)	NO (BUT WILL CONTROL SMALL SURFACE FIRES)
CLASS B FIRES — FLAMMABLE LIQUIDS, GASOLINE, OIL, PAINTS, GREASE, ETC.	NO	NO	NO	NO	YES	YES	YES	YES
CLASS C FIRES — ELECTRICAL EQUIPMENT	NO	NO	NO	NO	NO	YES	YES	YES
CLASS D FIRES — COMBUSTIBLE METALS	SPECIAL EXTINGUISHING AGENTS APPROVED BY RECOGNIZED TESTING LABORATORIES							
METHOD OF OPERATION	SQUEEZE HANDLE OR TURN VALVE	TURN UPSIDE DOWN AND BUMP	PUMP HANDLE	TURN UPSIDE DOWN	TURN UPSIDE DOWN	PULL PIN - SQUEEZE LEVER	RUPTURE CARTRIDGE - SQUEEZE LEVER	PULL PIN - SQUEEZE LEVER
RANGE	30'-40'	30'-40'	30'-40'	30'-40'	30'-40'	3'-8'	5'-20'	5'-20'
MAINTENANCE	CHECK AIR PRESSURE	WEIGH GAS CARTRIDGE ADD WATER IF REQUIRED	DISCHARGE AND FILL WITH WATER ANNUALLY	DISCHARGE ANNUALLY - RECHARGE	DISCHARGE ANNUALLY - RECHARGE	WEIGH SEMI-ANNUALLY	WEIGH GAS CARTRIDGE-CHECK CONDITION OF DRY CHEMICAL	CHECK PRESSURE GAUGE AND CONDITION OF DRY CHEMICAL

NOTE — There is also a multi-purpose dry chemical extinguisher suitable for A, B, & C. classifications.

d) Establishment of an effective maintenance program to assure that extinguishing equipment is operative at all times. The maintenance program must provide a definite schedule for periodic cleaning, repair and recharging of individual units.

e) Training programs must be instituted to assure the presence of personnel capable of effectively using the extinguishers.

EXPOSURE

The fourth and final factor to be considered in the acronym "COPE" is *exposure*. For effective fire loss control, exposure must be considered both from the point of view of the area surrounding the building and the occupancies and operations within the building. These are commonly referred to as exterior exposures and interior exposures.

1. *Exterior Exposure*

   a. *Building Construction*. An adjacent building of fire-resistive construction would obviously present a lesser degree of exterior exposure than would one of masonry and wood construction, or one of frame. Fire can spread from one building to another by one of several means. The ways in which the spread of fires occurs are "Conduction", "Radiation," and "Connection." These terms are clearly defined at the end of this chapter.

   b. *Building Occupancy or Operations*. This refers to the use of surrounding (exposing) buildings or structures such as an office building, garment factory, machine shop, paint manufacturing warehouse, oil tank, farm, etc. The manner in which these operations are conducted and in which they are safeguarded should never be overlooked in designing or assessing a protection program. A fire developing from this direction could be just as damaging as one which starts within the subject's premises.

Internal exposures are those created by operations with a greater degree of loss potential than the remainder of the occupancy. These exposures may be illustrated by the storage, handling and use of volatile flammable materials in a metal working operation. The methodology of protection from these hazardous operations can be summarized by *ESP — Eliminate, Segregate* and *Protect*. Recognizing and taking corrective action to eliminate, segregate and protect them requires judgment and awareness of the potential hazards, a thorough inspection, and experience.

Elimination can be accomplished by various means. It may be feasible to replace an entire process with a safer one, or it may be possible to substitute less hazardous materials in either construction or manufacturing processes. In certain cases, it may be necessary to remove the material or process entirely from the premises and place it in a separate smaller building designed solely for this purpose.

Segregation of hazards can be effective in reducing loss potential, by isolating hazardous operations to areas affording safest conditions. A process creating an exposure to loss could be segregated from the remainder of the plant by enclosing the area with a masonry wall and providing vertical and/or horizontal openings with self-closing or automatic-closing standard fire doors. Another method of segregation would be the provision of special-purpose vaults and/or cabinets.

Protection — those hazards that cannot be eliminated or segregated without adversely affecting production or degrading the process should be protected. The type and degree of protection must be designed so that production downtime is kept to the desired minimum. The protection afforded may be automatic sprinklers or one of the special protection systems mentioned earlier in this chapter under automatic protection — or both. Decisions must be made whether to control or extinguish the fire and whether to risk the extent of the damage likely to be caused by the extinguishing agent. It should be obvious that when manual, semi-automatic or automatic protection is the desired solution to the control of fire hazard, professional guidance from a fire protection engineer should be sought.

DEFINITIONS

Conduction:

Conduction is the transfer of heat from one body to another, by direct contact or through an intervening solid, liquid, or gas heat-conducting medium. The amount of heat transferred by conduction depends upon the thermal conductivity of the materials through which the heat is passing, and the area and thickness of the conducting path. The rate of heat transfer through any material is in direct proportion to the temperature differential between the points of entrance and departure. The transmission of heat by conduction through air or other gases is also directly proportional to the absolute pressure of the gas. There is no conduction transfer through a perfect vacuum. In the normal pressure range, solids are better heat conductors than gases. The base commercial insulators consist of fine particles or fibres of solid substances to give them strength and rigidity, but the insulating property is achieved by the spaces between the particles which are filled with air.

The transmission of heat cannot be completely stopped by any insulating material. In this respect, the flow of heat is unlike the flow of water, which can be stopped by a solid barrier. Heat insulating materials have a low heat conductivity. Heat flows through them slowly, but no amount of insulating material can actually stop the flow. This fact should be remembered when inspecting protection against stoves or other sources of heat which might ignite nearby woodwork.

Convection:

Convection is the transfer of heat by a circulating medium, which may be either a gas or a liquid. Heat generated in a stove is distributed throughout a room, heating the air by *conduction*, and the circulation of heated air through the room to distant objects is heat transferred by *convection*. Heat is transferred from the air to the objects in the room by *conduction*. Heated air expands and rises, and, therefore, heat transfer by *convection* occurs naturally in an *upward direction*, even though air currents can be made to carry heat by *convection* in *any direction*.

Explosive or Flammable Limits:

The vapors or gases given off by flammable liquids must be mixed with air or oxygen in a proper proportion for fire to occur. For each substance there is a lower limit, or minimum concentration of the gas with air, below which it will not burn, even though the temperature has reached the ignition point. This is usually referred to as a mixture that is too lean. There is also an upper limit or maximum concentration above which combustion will not take place. In this case the mixture is too rich. These concentrations are usually expressed as a percentage of gas or vapor in the air by volume. Since most vapors of flammable liquids burn at extremely rapid or explosive rates, the percentage between the lower and upper limits is called the "explosive range" of the substance. As an illustration, the explosive range of gasoline is from 1.4% to about 7.6%, depending upon the grade of gasoline. This means that combustion cannot occur if there is less than 1.4% or more than 7.6% of gasoline vapor in the air on a volume basis.

Fire:

There are many definitions of fire. Webster defines fire as "the principle of combustion manifested by light and heat, especially flame." Funk & Wagnall's defines fire as "the evolution of heat and light by combustion." Some scientists consider fire an occurrence which produces visible fire combustion. While these are all proper definitions of fire, they do not quite meet the needs of the fire protection engineer. Authorities on fire prevention and protection consider the definition of fire to be "Rapid oxidation, producing light and heat."

Air contains approximately 20% oxygen and 80% nitrogen by volume. Generally, nitrogen does not enter the chemical reaction, but serves to dilute the oxygen, which will effectively reduce the intensity of combustion. An increased percentage of oxygen produces increased intensity of combustion. Oxidation of a material takes place continuously as long as an oxidizing agent, such as oxygen, is present. At relatively low temperatures, however, the reaction is usually so slow that it is not perceptible. The rusting of iron is an example of slow oxidation. At substantially higher temperatures, the oxidation rate may become rapid and generate large quantities of heat. If the heat released is sufficient to keep the reaction going at a rapid rate, and, if flames appear, ignition is said to have taken place.

The burning of materials in the presence of a normal oxygen atmosphere is generally accompanied by a luminosity called flame. Visible flame is rarely separated by any appreciable distance from the burning materials. However, heat, smoke, and gas can develop in certain types of smouldering fires without readily visible evidence of flame, and air currents can carry these elements far in advance of the fire.

In the burning of most fuels, combustion can only take place after the solid or liquid fuel has been vaporized or decomposed by heat to produce a gas. There are a few fuels which do not evaporate or decompose to form gases at ordinary fire temperatures; combustion takes place by the direct combination of the fuel with oxygen, such as in the burning of charcoal. This point may be illustrated by a relatively simple experiment. Put wood shavings into a test tube and heat the bottom of it. Then, place a source of ignition at the top of the test tube and you will observe a flame. This flame will be the visual evidence of the burning of the gas, which was produced by the decomposition of the wood shavings, caused by the application of heat at the bottom of the test tube.

Flash Point:

It has already been mentioned that liquid fuel will not burn until it has been vaporized by heat to produce a gas. The temperature at which vapors or gases are produced in sufficient quantities to be ignited is called the flash point. Each liquid fuel has its own unique flash point. There are several types of apparatus used to determine flash point. Two basic types are commonly called the closed cup tester and the open cup tester. Open cup flash points represent conditions with the liquid in the open and are approximately 10 to 20% higher than the closed cup flash point figure for the same substance.

Heat Transfer:

The transfer of heat is responsible for the start and extinguishment of most fires. Heat is transferred from one material to another by one or more of three methods: conduction, radiation, or convection.

Ignition:

Ignition of combustible material can only occur when the temperature is high enough and the quality of heat is adequate enough to initiate self-sustained combustion. An example is the fact that an ordinary match will not ignite a large log. The heat is dissipated through the log and thus a quantity of heat sufficient to initiate combustion of the log is not achieved at any one point. A match, however, will ignite wood shavings, as the heat absorbed by the shavings is not dissipated, but is concentrated in each small shaving. The same is true in the case of a spark which often will not ignite combustible materials in their solid state, but will ignite finely divided particles, such as combustible dust or flammable vapors.

Oxygen:

Excluding oxygen from a fire usually involves the use of agents that will blanket the fire and smother it. Chemical agents, such as carbon dioxide, foam, vaporizing liquids and dry chemicals, are used in most instances, as these substances are heavier than air and will settle on the fire, eliminating its oxygen supply.

Products of Combustion:

Most ordinary combustible materials are compounds composed of carbon and hydrogen. When these materials burn in a free air supply, the principal products of combustion will be carbon dioxide and water in the form of steam. When they burn in a restricted air supply, they may produce carbon monoxide and steam. Under certain conditions, including an additional supply of oxygen, the carbon monoxide will burn again to form carbon dioxide. The products of combustion can be divided into four categories: fire gases, flame, heat and smoke. These four products produce a variety of physiological effects on people (such as burns and the toxic effects of inhalation of heated air and gases).

The term fire gases refers to the gaseous products of combustion. Many variables determine what fire gases are formed by fire, the principle ones being the chemical composition of the burning material, the amount of oxygen available for combustion, and the temperature. Accurate statistics on actual causes of fire deaths due to the toxicity of fire gases are not available, but it is recognized that fire fatalities from the inhalation of hot fire gases and hot air are far more common than fire deaths from all other causes combined.

Heat is largely responsible for the spread of fire in buildings. Smoke is small particles of unburned carbon and tarry particles that make the fire gases visible. It is true that there are certain combustion conditions under which materials can burn without producing visible products of combustion, but, in general, smoke accompanies fire, and like flame, is visible evidence of fire.

Properties of Flammable Liquids and Gases:

There is no sharp line of demarcation between flammable liquids and gases. Liquids become gases at either low pressures or high temperatures, whereas gases become liquids at either high pressures or low temperatures. It is the vapor from evaporation of flammable liquids which burns or explodes. Explosion is nothing more than extremely rapid burning.

Radiation:

Radiation is the transfer of heat from one body to another by heat rays through intervening space. Radiated heat will pass freely through a vacuum and through gases such as hydrogen, oxygen and nitrogen. Since air is primarily a mixture of oxygen and nitrogen, there is no absorption of radiated heat by the air, except as it may contain water vapor or other contaminants. As does light, radiated heat will travel through space in a straight line until it encounters an opaque object, where it is absorbed and then proceeds through the object by conduction. Like light, radiated heat is reflected from bright surfaces and will pass through glass. Heat radiation is a two-way process, in that heat radiates from a stove to the wall, and the wall, in turn, radiates heat in all directions when heated above the temperature of other objects in the room. The rays from a heat source spread out in all directions and, therefore, the further an exposed object is from the source of heat, the less is the concentration that will reach it.

Removing Fuel:

Eliminating fuel to extinguish a fire is usually difficult while the fire is in progress. In forested sections, large areas or lanes are stripped of vegetation, creating "fire breaks." In urban areas, during the time of a conflagration, the fire marshal may order sections of building razed in order to create the same kind of "fire break."

Specific Gravity:

Specific gravity is the ratio of the density of any substance to the density of some other substance taken as standard. Water is used as the standard for liquids and solids, and hydrogen or air is the standard for gases. As the specific gravity of water is one, a liquid with a specific gravity of less than one will float on water, whereas a specific gravity greater than one indicates that water will float on the liquid. This is an important consideration in handling a flammable or combustible liquid fire. Generally speaking, water can be used as an extinguishing agent for any liquid with a specific gravity of more than one but not for a liquid with a specific gravity less than one.

Temperature:

Water is the most common agent used to remove the heat and, hence, lower the temperature of fuels below their ignition-point. In many instances, it is necessary to use large quantities of water to remove the heat and to keep the fuel below its ignition temperature, even after the fire has been extinguished.

Theory of Fire Extinguishment:

A fire can only occur by the proper combination of *fuel*, *oxygen*, and *heat*, and it can be extinguished by eliminating any one of these three ingredients.

Vapor Density:

The ratio of the mass of a given volume of vapor to the same volume of air is its "vapor density." If the vapor density is greater than air, which is considered one, the vapor will tend to settle to the floor. However, if it is substantially less than one, it will tend to rise. Most flammable liquid vapors such as gasoline, naphtha, turpentine and alcohol are heavier than air and will settle to the lowest surrounding point. However, natural gas is lighter than air and, where present (such as from a leaking heater or stove), will rise to the highest surrounding point.

Volatility:

The volatility of a liquid is the rate at which it will evaporate in the open air at a given temperature. Evaporation rate is the rate of changing from the liquid to the vapor state at a temperature below the boiling point. All materials evaporate, and it is the difference in the rates of evaporation of mixtures that is of primary concern for fire protection. Although there are many exceptions, vapor pressure, evaporation rates and boiling points vary. In general, as the boiling point goes down, the vapor pressure and the evaporation rate go up. As a general rule, the faster a liquid evaporates, the more quickly its vapors will reach the lower explosive or flammable limit in any given situation. Flammable liquids, such as gasoline, are hazardous, as they have the characteristic of a very high evaporation rate, or volatility, coupled with a very low flash point.

SUMMARY

In summary, let's briefly review the key points presented in this chapter. First of all, better management planning and tighter management controls are necessary to help prevent the occurrence of fire or to assure that it is contained and extinguished during its incipient stage. To accomplish this, management should remove, where practical, the potential causes of fire and establish emergency procedures to control minor ones.

Since fire loss control and industrial safety share the same prime objective (the conservation of the company's resources, both, property and human by eliminating, segregating and protecting against hazards), all loss control efforts should be coordinated and administered by a combined department.

To assist the loss control manager in meeting this objective, a wealth of fire loss control information and assistance is available from a variety of sources, such as The National Fire Protection Association, The American Insurance Association, Factory Insurance Association, Factory Mutual Inspection Department and Laboratories and Underwriter's Laboratories. By using these sources in a judicious way, even the inexperienced fire loss control manager can find solutions to the most difficult problems.

Effective administration of a fire loss control program requires some knowledge of the basic factors which must be considered to intelligently plan and implement a program that will eliminate or minimize the occurrence of fire and its resultant destruction. Critical attention, therefore, must be given to the following areas: construction, occupancy, protection and exposure.

1. *Construction*: Those responsible for corporate administration of fire loss control programs must be familiar with the most prevalent classes of construction.

2. *Occupancy*: They should also be familiar with the intended use of the premises by the occupants.

3. *Protection*: Both public and private fire protection of resources, a third major consideration for implementing a successful fire loss control program, must also be considered.

4. *Exposure*: For effective fire loss control, exposure must be weighed both from the point of view of the area surrounding the building and from the occupancies and operations within

the building. These are commonly referred to as exterior and interior exposures.

In essence, then, management must be concerned, involved and committed to designing and implementing an effective fire loss control program within its operations. It is no longer possible to gamble on the chance that a plant or office will be spared from the ravages of fire. If management fails to initiate such a program, the results can be costly in terms of dollars and cents, but even more so in terms of human lives. The lessons that were learned following the destruction of McCormick Place were costly. Fire and structural damage alone accounted for approximately 45 million dollars. Today, there is a new McCormick Place standing on the same site. This multimillion dollar exhibition hall, however, contains the most modern fire protection facilities, geared specifically to the occupancy or contents of each area and to ease of access for the fire department. To avoid problems of heat and flame spread, the 2 1/2 million square foot building has been divided into 12 separate fire areas; within these areas there are approximately 20,000 sprinklers, 225 hose cabinets, 150 manual stations, 450 fire detectors, pumps with 8,000 gpm flow rates and 270,000 gallons of water storage. The new building has been designed with fire protection in mind. In severe contrast to the former exhibition hall, which was a rectangular building with solid walls and no windows or openings for hoses and other equipment, the new McCormick Place has a more than adequate number of exterior access routes.

Although the experience gained from the holocaust at McCormick Place was indeed costly, it has shown how vital an effective fire loss control program is, and why it should be a definite concern to management.

BIBLIOGRAPHY

Approved Equipment and Materials for Industrial Fire Safety. Norwood, Massachusetts: Factory Mutual Engineering Corporation.

"Carbon Dioxide Extinguishing Systems," Pamphlet No. 12. Boston, Massachusetts: National Fire Protection Association.

"Central Station Protective Signaling Systems," Pamphlet No. 71. Boston, Massachusetts: National Fire Protection Association.

"Dry Chemical Extinguishing Systems," Pamphlet No. 17. Boston, Massachusetts: National Fire Protection Association.

"Fire Brigades," Industrial Data Sheet 588. Chicago, Illinois: National Safety Council.

"Fire Extinguishing Appliances," Pamphlets 10-19. New York, New York: American Insurance Association.

"Fire Prevention and Control on Construction Sites," Industrial Data Sheet 491. Chicago, Illinois: National Safety Council.

Fire Protection Equipment List. Chicago, Illinois: Underwriters' Laboratories, Inc.

Fire Protection Handbook, 13th Edition. Boston, Massachusetts: National Fire Protection Association, 1969.

"Foam Extinguishing Systems," Bulletin No. 11. Washington, D.C.: U.S. Department of Interior, Bureau of Mines.

Gaudio, Joseph A., "The Fire Protection Engineer in Europe," *Fire Journal*, Vol. 66, No. 4, July, 1972.

Handbook of Industrial Loss Prevention. New York, New York: McGraw-Hill Book Co., 1967.

"Hazardous Chemicals Data," Bulletin No. 49. Washington, D.C.: U.S. Department of Interior, Bureau of Mines.

"Inerting for Fire Prevention," Bulletin No. 69. Washington, D.C.: U.S. Department of Interior, Bureau of Mines.

"Inspecting, Recharging and Maintaining Portable Fire Extinguishers." Pittsburgh, Pennsylvania: Fire Equipment Manufacturers Association, Inc.

"Judging the Fire Risk." Chicago, Illinois: American Mutual Insurance Alliance.

"Management Control of Fire Emergencies," Pamphlet No. 7. Boston, Massachusetts: National Fire Protection Association.

National Fire Codes. Boston, Massachusetts: National Fire Protection Association, published annually.

Portable Fire Extinguisher Guide. Evanston, Illinois: National Association of Fire Equipment Distributors, Fire Equipment Manufacturers' Association, Inc.

Smith, Edwin E., "An Experimental Determination of Combustibility", *Fire Technology*, Vol. 7, No. 2, May, 1971.

"Spontaneous Ignition and Its Prevention," Research Bulletin No. 2. Chicago, Illinois: Underwriters' Laboratories, Inc.

"Sprinkler Systems," No. 13 and 13A. Washington, D.C.: U.S. Department of Interior, Bureau of Mines.

Stop Fires — Save Jobs. New York, New York: American Insurance Association.

Your Plant's Fire Protection. New York, New York: American Insurance Association.

Air Pollution

XIII.

INTRODUCTION

Serious air pollution incidents in London, England; Donora, Pa.; New York City and other cities throughout the world began to intensify man's interest in what effect his polluted environment has on his ability to survive. The first identifiable federal interest in air pollution was the development of a program in 1955 that involved a modest research program and offered technical assistance to states. This program was followed by the Clean Air Act in 1963, with subsequent amendments in 1965, 1966, 1967 and 1970. These amendments listed levels of air quality for various pollutants that should be met throughout the United States within a designated time period. Public interest and research generated by these acts has made man more cognizant of the effects which the various air pollutants have had, not only on human health, but also on vegetation, animals, and items such as building materials. As man gains more knowledge, he will become less willing to live and work under potentially dangerous conditions. The following discussion of man's *general and working environment* will serve as an introduction to the types of pollutants which exist, their effects, and the means of control needed to satisfy the demand for clean air.

GENERAL ENVIRONMENT

In the United States today, over 200 million tons of man-generated pollutants are poured annually into the atmosphere. If we had the ability to disperse these pollutants evenly throughout the atmosphere, as occurs with most naturally-formed pollutants, air pollution would be of little concern. However, as a result of industrial development and high population densities, most man-generated pollutants are emitted from less than 1% of the land area, containing 50% of the population. When winds are slack, especially during a temperature inversion, prohibiting vertical mixing as a result of a warm layer of air aloft, stopping pollution from escaping, contamination levels in these areas can reach very high concentrations.

In October, 1948, Donora, Pennsylvania, was covered by a smog caused by concentrated emissions from heavy industrial production and stable weather conditions. During four days of slack winds and temperature inversion, fifteen people, more than the statistical mean, died.[2] However, the most catastrophic incident occurred in England in December, 1952. Heating systems burning coal poured out sulfur dioxide and soot into an atmosphere already heavily polluted by industrial emissions. A temperature inversion prevented escape of these pollutants and caused a particularly heavy smog to cover the London area from December 5th through December 9th. When medical authorities studied its effect, they found that 4,000 people, more than the statistical average, had died.[1]

WORKING ENVIRONMENT

It is very difficult to define responsibility for poor health which might be partially caused by man's occupational exposure to particulate or gaseous emissions. Difficulty arises because of variable conditions, such as pollution exposure time and concentration found in the working environment.

In the past, only those pollutants that had identifiable detrimental effects were diagnosed as the culprits. As pollutant effects become better understood, new pollution control standards and stronger enforcement have followed. Since *all* emissions of pollution will come under observation, no matter how difficult to control, the upgrading of existing control systems and the design of new systems should be fully investigated. Many manufacturing processes require very costly control systems that usually present a serious economic obstacle to the solution of the problem. Qualified consultants may maximize the probability of arriving at a satisfactory solution.

Although control efficiency and the initial cost of control systems may be the major factors in determining the system to be employed, operating costs should also be carefully considered. In order to maintain a system's performance at its high initial efficiency, constant preventive maintenance is required, and this cost must also be taken into account. In some cases, the collected contaminant can be used or sold, offsetting some of the control system cost. This, however, is not usually the case, and the contaminant must be disposed of in some other acceptable manner.

There is no question about the need for compliance with pollution code requirements, in order to protect man in his environment. One of the major problems facing enforcement agencies, however, is that enforcement of these codes may financially destroy companies working on small profit margins. States, as well as the federal government, are developing tax incentives to partially offset the financial burdens associated with the purchase of control systems.

POLLUTION APPRAISAL

When evaluating air pollution, consideration must be given to determining the nature of the contaminant emitted, recognizing the pollutant's effects, assessing these effects, measuring the pollution concentration to determine pollution code compliance, and selecting the proper pollution control equipment.

TYPES OF POLLUTANTS AND THEIR SOURCES

Air pollution is defined by the Public Health Service as: "the presence in the outdoor atmosphere of one or more air contaminants, or combination thereof, in such quantities and of such duration that they are, or may tend to be, injurious to human, plant, or animal life or to property, or which unreasonably interferes with the comfortable enjoyment of life or property."

Pollutants can be classified as either (1) aerosols (also called particulate matter) or (2) gases, including vapors. A brief summary of some of the contaminants and their principal sources follows:

    A.   AEROSOLS — These include dusts, smokes, mists, and fumes which account for about 28 million tons of pollutants emitted annually to the atmosphere. These air suspensions differ widely in terms of size, density, and importance as a pollutant (toxicity).

        1.   *Dusts:* These are solid particles, normally greater than one micron in size. (One Micron (u) is one millionth of a meter or one twenty-five thousandth of an inch.) Principal sources of generation are by crushing, impaction, or frictional tearing. Fly ash, unburned carbon formed through the combustion of coal or oil, is also an important dust contaminant. Dusts do not diffuse in air and normally settle under the influence of gravity. Particles under 2 microns in size tend to remain suspended, and can be detrimental to man's respiratory system.

2. *Fumes:* These are solid particles, generally smaller than one micron in size, usually resulting from chemical reactions other than combustion. Fumes can also result from condensation of vapors from the gaseous state, generally after volatilization of minerals being processed. Chemical process plants, and roasting and heating processes required in smelting operations are primarily responsible for their generation.

3. *Mists:* These particles result from condensation of gases or vapors to a liquid state or by the atomization of liquid by spraying, foaming, or splashing.

4. *Smoke:* These particles arise from incomplete combustion and contain many particles in the 0.1 to 1.0 micron range. Smoke consists mainly of carbon and other combustible materials.

An example of the seriousness of the particulate concentration problem facing urban areas in the United States is cited by the following excerpt from the May 14, 1970 issue of the *Birmingham Post Herald:*

> Here is Wednesday's air pollution count in the number of micrograms of solid pollution particles per cubic meter of atmosphere sampled:
>
> Norwood Park 150
> Wahouma 235.16
> Huffman 127.46
>
> Number of Jefferson County respiratory deaths last week was two. The national urban average particulate count is 97, while the non-urban average is 35. A count 80 or more has an adverse effect on health, according to the U.S. Public Health Service, and a count of 200 is considered a critical level.[3]

Particulate concentrations causing effects on man, visibility, and materials were listed in the *NAPCA Air Quality Criteria for Particulate Matter* as follows: "adverse health effects were noted, in areas where studies were conducted, when the annual mean level of particulate matter exceeded 80 mg/m^3; visibility reduction to about 5 miles was observed at 150 mg/m^3; and adverse effects on materials were observed at an annual mean exceeding 60 mg/m^3."[4]

B. GASEOUS MATERIALS — These contaminants are formed from combustion processes or, as in the case of vapors, are derived from materials usually solid or liquid, such as gasoline and solvents. The combustion of fuel (industrial, commercial, and domestic) is the principal source of gaseous generation. Gaseous materials to be discussed are carbon monoxide, sulfur oxides, hydrocarbons, nitrogen oxides, and interaction products.

1. *Carbon Monoxide:* This colorless, invisible, and odorless gas is formed by incomplete combustion of coal, gasoline, and other carbon-containing fuels. By far the largest emission source is from transportation. Every 1,000 gallons of gasoline burned by internal combustion engines dispenses 3,000 pounds of CO through exhaust.[6]

The principal toxic property of CO is associated with its greater affinity than oxygen to combine with hemoglobin. Consequently, there is a reduction in the oxygen-carrying capacity of the blood. This affinity of hemoglobin for CO is over 200 times that for oxygen.[7]

It is well known that high concentrations of CO will cause many physiological and pathological changes and may ultimately result in death. Concentration levels of 200 p.p.m. (Parts of contaminant per million parts of air. .01 percent by volume of gas in air equates to 100 ppm.) for an extended period of time have produced headaches and dizziness.

2. *Sulfur Oxides:* The second most commonly generated gases are the oxides of sulfur (SO_x). Sulfur dioxide is formed when sulfur reacts with oxygen, and is primarily the result of burning sulfur-bearing fuels, such as coal and oil. Other emissions come from the refining of petroleum, the smelting of sulfur-containing ores, and

the manufacturing of sulfuric acid. Paper making, refuse burning, the burning of coal refuse banks, and miscellaneous industrial operations contribute to the total also.[9]

Sulfur dioxide can react with moisture in the lungs to form sulfuric acid, thus damaging lung tissue. Sulfur oxides cause crop and property damage, and were the chief cause of the pollution crises in London and Donora. Sulfur dioxide, through conversion to sulfuric acid in humid weather, contributes to visibility reduction. Test results, published by NAPCA in February, 1969, stated that adverse health effects were noted in areas tested when the annual mean level of sulfur dioxide exceeded 80 ug/m^3, and vegetation damage was observed at an annual mean of 60 ug/m^3.[9]

3. *Hydrocarbons:* The gas-phase hydrocarbons, particularly non-methane, are considered serious air pollutants because of their indirect role in the photochemical production of secondary pollutants (SMOG). Hydrocarbon emissions are generated primarily from the inefficient combustion of gasoline and from the use of hydrocarbons as process raw materials.[5] For every 1,000 gallons of gasoline burned, approximately 300 pounds of these vapors escape through the exhaust pipe.[6] It is no wonder, then, that 52 percent of the pollutants come from transportation sources, with additional amounts from organic solvent evaporation, industrial processes, solid waste disposal, and stationary combustion sources.[8]

4. *Oxides of Nitrogen:* The oxides of nitrogen (NO_x) are produced by high temperature combustion processes, during which part of the oxygen combines with atmospheric nitrogen. Oxides of nitrogen are also produced through naturally-occuring biological reactions, which normally take place over wide areas and therefore result in low concentrations. Technologically (man)-generated (NO_x), however, is emitted in concentrated emission source areas and, therefore, levels can reach high concentrations.

Leading man-made emission sources include transportation sources, power plants, industry and miscellaneous sources.[10]

Technologically-generated NO_x usually takes the form of nitric oxide (NO). This, in turn, is oxidized in the atmosphere, resulting in a more toxic nitrogen dioxide (NO_2). Photochemical processes involving hydrocarbons quicken this oxidation process. (See Interaction Products.) Nitrogen oxides have been shown to cause adverse effects on health and vegetation, and since they also aid in the formation of secondary air pollutants, they are classified among the serious air pollutants.

5. *Interaction Products:* Many contaminants released to the atmosphere are both chemically and physically unstable. Reactions with these contaminants may cause secondary pollutants, the cause of many serious smog conditions facing urban areas. Los Angeles, for example, is particularly affected because it is susceptible to stagnant air and temperature inversions. Because of the complexity of these reactions, some unexplainable, they will be developed in a simplified fashion.

These reactions are influenced by many factors, such as: contaminant concentrations, solar radiation, and meteorological conditions. The photochemical action of sunlight (ultraviolet radiation) on nitrogen dioxide (NO_2) causes the formation of atomic oxygen (O) and nitrogen oxide (NO). This, in turn, reacts with air oxygen (O_2) to form ozone. A small portion of the ozone and oxygen atoms reacts with certain hydrocarbons, resulting in the formation of oxidized compounds and free radicals, and subsequently reacts with the initially-produced nitric oxide to form more NO_2. Results of these reactions are the rapid oxidation of NO to NO_2, the increased concentration of ozone, and the formation of peroxyacetyl nitrate (PAN) and other radicals.[11]

Photochemical oxidants have been shown to adversely affect human health and have caused vegetation damage.

AIR POLLUTION EFFECTS

A. EFFECTS ON MAN: it has already been seen how extreme air pollution incidents caused the deaths of a large number of people in London and Donora. Damaging effects, however, are not always as obvious. Certain effects often appear too late for remedial action to correct the damage. Asbestosis, caused by the inhalation of asbestos fibers, is one example of this type of damage. The human body masks this problem by producing a protective body around the asbestos fibers, keeping their effect dormant. At some date, years later, this protection is destroyed, allowing the asbestos fibers to seriously impair the lungs.

1. *Respiratory System:* The human body's filtering and operating systems are not designed to function efficiently under the size, type, and concentration of pollutants found in many of our urban and working atmospheres. The efficiency of various stages of particulate removal in the respiratory system is mainly dependent on the particulate size. The tiny hairs and liquid mucus in the nasal passages can effectively screen out particles larger than 5 microns in size. Contaminants that escape this barrier, however, can enter the respiratory tract, which is equipped with millions of tiny hairs called cilia. Cilia tend to move the contaminant toward the mouth. A fluid mucus in this area also helps trap and remove particles. The cleaning mechanism of the cilia and fluid mucus adds a barrier, removing particles 2 microns and larger in size. Particles that escape these barriers can enter the alveolar, where tiny sacs in the lung exchange oxygen and carbon dioxide. When contaminated, these sacs become afflicted and the individual must breathe harder to get enough oxygen into his blood stream.

If the cilia are damaged, the collection efficiency in the respiratory tract may change markedly. This may happen when highly-polluted air or cigarette smoke comes in contact with cilia. Such a reaction could cause temporary paralysis and, thus, stop the cleaning action. If the condition exists for a long period of time, the cilia may be destroyed, allowing larger particles to enter the lung area.[1][2]

The synergistic effect of sulfur dioxide (SO_2), absorbed in particulate matter, can cause serious damage to the lungs. Sulfur dioxide, acting alone, would be largely absorbed in the trachea and only a small portion would reach the air sacs. Sulfur dioxide, absorbed in particulate matter, however, is carried into the lungs and acts on tissue that the gas alone would not otherwise reach.[14,15] This synergistic effect is so serious that the product of particulate matter and sulfur dioxide's ambient air concentrations is used in determining when emergency action, such as requiring the cutback of industrial production, should be taken, thus lessening pollution generation, to protect against endangerment to health.

Lung diseases that may be caused or aggravated by exposure to polluted air are chronic bronchitis, emphysema, asthma, and lung cancer. It should be noted that some of the same carcinogens (cancer-producing agents) found in cigarette smoke are found in polluted air.

In a recent study associating air pollution and chronic respiratory disease, the mortality of men aged 50 - 69 years was 64% greater in the highest air pollution concentration area than in the lowest.[16]

2. *Visibility:* Pollution can cause poor visibility which, in turn, can prevent recognition of danger when driving, thereby increasing accident potential. Another effect is the mental depression which can develop when pollution blocks out the sun and transforms a clean day into a hazy one. Reduced visibility can also affect man, by making a community unattractive, consequently depressing land values. Employers may find that they must pay a higher wage rate to attract talented workers. New business was attracted to Pittsburgh after that city made a concerted effort to attack its serious pollution problems.

B. ECONOMIC EFFECTS: There have been many estimates on what our polluted atmosphere is annually costing the individual. A report by Michelson (1967), which investigated the high cost of living in 17 counties in the New York/New Jersey area, concluded that an average family lost $620.00 per year because of heavy concentrations of pollution. Any effect on fabrics, metals, building products, paint, rubber, vegetation, or animals will be reflected in higher costs to the consumer. In order to more clearly understand additional costs caused by pollution, it is necessary to specifically examine the elements affected.

1. *Fabrics:* Fabrics can easily become soiled. Thus they require frequent cleaning and subsequently reduce their functional life. Contaminants also attack the fabric structure, decreasing fabric life.

2. *Metal:* Metal corrodes because of the electrolytic action started (set up) when pollutants, such as sulfur dioxide, are deposited on the metal surface in the presence of moisture. It has been shown that there is a correlation between the air pollution concentration and the level of metal corrosion.[18]

3. *Building stone:* Pollutant action on stone is principally that of an acid dissolving carbonate. Sulfur dioxide is by far the most damaging pollutant. By oxidizing to sulfur trioxide and combining with moisture, sulfuric acid is formed. The soot and grime on a building exterior provides a good surface for sulfur dioxide absorption. Cleopatra's Needle, for example, after being moved from Alexandria to London, corroded more in 80 years than it had in the 3,000 years it resided in Egypt.

4. *Rubber:* Ozone attacks the unsaturated carbon-to-carbon bond, causing rubber to crack. Communication and power transmission lines with rubber coverings are highly susceptible to damage from ozone.

5. *Vegetation:* Plants build complex organic compounds, necessary for cell structure and growth, from simple inorganic compounds in the air. The building mechanism is located mainly in the leaf, and if any serious damage is done to this structure, the plant's growth is seriously hampered. Air pollution interferes with plant and tree growth in several ways. Particulate matter settles on foliage and interferes with the plant's ability to take in these inorganic compounds and to absorb the sun's rays. Sulfur dioxide and nitrogen dioxide react with dew, forming nitric acid and sulfuric

acid; this, in turn, damages leaves on which the gas is deposited. Ozone, fluorides, and photochemical smog have also been found to have a very damaging effect on vegetation.

6. *Animals:* Animals are afflicted by pollutants through the normal ingestion of the contaminant on the food they eat. Pollutants can contaminate forage in the initial form, or can react in the plant to form a toxic substance.

Polk County, Florida, is an area that produces 1/3 of the world's phosphate fertilizer. Fluorides can cause severe fluorosis, leading to deterioration of the bones. Before correction measures were taken, 4 1/2 tons of fluoride were emitted daily into the surrounding atmosphere. From 1953 to 1960, the cattle population of Polk County, Florida, decreased by 30,000 head. (19)

C. DETECTION, MEASUREMENT AND ANALYSIS: The effect pollution is having on man's environment and health is causing the adoption of increasingly more stringent pollution control regulations. These regulations, however, are only as effective as the methods used for enforcement and the accuracy of the tests made to determine compliance.

Proper procedure in adopting air pollution regulations dictates that ambient air criteria and ambient air quality standards be developed to define the consequences of high ambient air concentrations on health and welfare. Ambient air criteria are scientifically sound statements about air pollutant exposures and the associated effects at a concentration level for a stated amount of time. The federal government has published Air Quality Criteria for several pollutants. Ambient air quality standards refer to a limit on the amount of a given pollutant that will be permitted in the atmosphere, and are based upon the air quality criteria.

Assuming present air quality is worse than the air quality standards, then (as dictated by the 1970 amendments to the "Clean Air Act") states had to adopt strategies, means of reducing pollution emissions.

These strategies were then applied to sources of pollutant emission in order to lower the concentration of pollutants in the ambient air down to the approved standards. The strategies that were selected had to be capable of implementation and had to be enforceable.

Attacking the problem at the source is the most effective strategy for emission control. This strategy could involve changing the source (low sulfur fuels), eliminating the source, or capturing and cleaning the emissions generated by the source. Emission standards restrict emission from sources beyond a specified amount of pollutant concentration or percent of process weight. Typical emission standards restrict particulate matter to .05 grains per standard cubic foot of stack gas. As a guideline for emission standards selection, states can adopt the best available technology to meet the ambient air quality standards. Transportation control and land use planning are also strategies that could be used to lower pollutant emissions. If the ambient air quality standards are violated, or could be violated by the addition of a new source, states can prevent or stop construction.

The U.S. Environmental Protection Agency passed additional regulations in accordance with the "Clean Air Act" mandate. National emission standards have been established for new and modified industries that have a major impact on ambient air quality. Hazardous pollutant standards are still being developed, under the Occupational Safety and Health Administration. Initial "target areas" included asbestos, carbon monoxide, silica, lead and cotton dust. Standards also govern exposure levels for beryllium, mercury, trichloroethylene, inorganic lead, sulfur dioxide, nitrogen oxides, chromic acid, anhydrous ammonia, and toluene diisocyanate.

AMBIENT AIR QUALITY SAMPLING

No one instrument is suited for testing all types of contaminants. The selection of the means of sampling should be determined by prior knowledge of the relative nature and concentration of the contaminant. Sufficient sample size should be obtained so that analysis can be conducted with precision and accuracy. Measurement instruments range from simple dustfall jars to sophisticated automatic gas analyzers.

Dustfall jars are positioned around a specific area to measure the particulate matter that settles out of the atmosphere. High-volume samplers pass a measured amount of air through a filter, capturing the particulate contaminant. Gases can be collected by drawing known amounts of air through reagent solutions. In most cases, samples are analyzed at a laboratory.

STACK SAMPLING

Testing at the source eliminates many variables that tend to cloud areas of responsibility, especially when a number of sources are associated with a given area. The total amount of gas emitted, the amount of gas sample withdrawn, and the amount of contaminant collected in a given time will be an index of the concentration of contaminant emitted to the atmosphere. Source testing will not only determine compliance to Emission Standards but, when correlated with meteorological conditions, will show the effect the emission has on neighboring areas.

Along with determining the concentration and the nature of the contaminant, a particle size analysis should be run on the sample collected. This should be a prerequisite in selecting a collection device.

Sampling methods and equipment are constantly changing because of tighter regulations. Enforcement requires an accurate determination of the amount of pollutant emitted. The following references are suggested for in-depth study of sampling methods and equipment.[22,23,24,25]

POLLUTION CONTROL

Once tests have been conducted that determine the nature of the contaminant and its concentration, a pollution control system can be developed. First consideration should be given to control at the emission source, by eliminating or treating the source. Elimination of the pollution's source can be accomplished by stopping or changing the process. Examples of changing the process would be the use of low sulfur fuel, resulting in lower SO_2 emission, or changing metal melting techniques to reduce particulate emissions. Stopping the source is only done when the company responsible for the emission feels the addition of control would not be too costly to warrant installation. The federal government is attempting to offset this financial burden by developing tax incentives. Source treatment by dilution for working area pollutant concentration control or removal of the contaminant directly from the point of generation are the next strategies to be considered.

The control of pollution in enclosed working areas can be accomplished through general ventilation. Contaminants that escape into the working area from the process source are prevented from reaching hazardous concentration levels by constantly adding clean (diluted) air. The amount of dilution air required is based on the toxicity and concentration of the pollutant. Air quality standards for enclosed working areas dictate the ventilation design concentration limits for each pollutant. Most dilution techniques are used for gaseous contaminants because of the ease with which they will diffuse throughout an area. Dilution used with particulate matter has only a minimal effect. For toxic pollutants, the quantity of dilution air required for effective control can be very large. Operations that are heated and/or air-conditioned normally cannot afford this control method because of high air volumes required. Removal at the source becomes the most practical solution, therefore.

A complete pollution removal system collects the emission at its source, transports it in ducts, removes contaminants with an air pollution control device, then discharges clean air into the atmosphere. This system should also incorporate a means of transporting the collected contaminants back into the operation for further processing, if of some value, or to disposal areas, if of no value.

Removal of the pollutant directly from the source to a tall stack is sometimes used to disperse the pollutant, thus (with proper meteorological conditions) diluting its overall ground-level concentration. In most cases, emission standards dictate the need for pollution control equipment before the pollutants are discharged into the atmosphere. As a result, the tall stack dispersion method, used as an only means of pollution control, is becoming less popular; however, it can be an additional-method safeguard against accidental emissions.

Applying the contaminant removal system as close as possible to the source can accomplish control with a minimum volume and the least number of variables to affect control (for example, wind in outdoor collection application). Effective capture is attained with a well-designed hood.

1. *Hoods* should be designed to enclose the source area as completely as possible. If total enclosure is not practical, the hood should be located as close to the generation area as possible. Careful consideration should be given to shape of the hood around the source area, and the subsequent ease of maintenance. If the hood causes difficult operational and maintenance problems, it will soon be eliminated and effective control destroyed.

   The capture velocity at the hood face should be high enough to reverse the direction of the emission before it reaches the outside air, but not so high as to adversely affect product flow or operation conditions.

   The more completely the emission source is enclosed, the less costly will be the fans, motors, ducts, cleaning equipment, and cost of maintaining and operating the system. This is because the capture velocity required at the hood face decreases as the source distance from the hood decreases, thus decreasing the volume (Q) of air needed. Lowering the volume of air required reduces the size of the collection system. The product of the total area of enclosure opening (A), and the selected capture velocity (V) for the type emission involved, gives the minimum volume required for control. $Q = VA$

2. *Ductwork* must be properly sized for a carrying velocity high enough to eliminate any contaminant fallout. Inlets, elbows, and transition sections should be designed for minimum flow resistance at these velocities so that undue wear will not occur. If the systems resistance is kept nominal, then fan and motor requirements can be maintained within realistic limits.

3. *Pollution Control Equipment:* When the type of pollutant, chemical characteristics, particle size, concentration, temperature, collection rate, and the method of disposal are known, the type of cleaning best suited for the application can then be determined. No one method of cleaning can meet all air pollution control applications.

   Having determined the nature of the problem, pollution code requirements should be studied to ascertain the degree of cleaning needed for compliance. Also, it may be valuable to conduct an in-depth study of the techniques other companies with similar equipment and processes are using to solve their problems. Many associations, such as those of the Steel and Foundry Industry, are providing information on past and new control systems to their members.

4. *Mechanical Separators:* Mechanical collectors are grouped into two types: gravity and inertial cyclone. Gravity collectors are very seldom used because of their inefficiency. The cyclone separator is the most widely used mechanical collector.

The simple cyclone collector consists of a cylindrical upper section, centrally placed exhaust pipe penetrating below the tangential inlet which admits the particulate loaded gas into the body, and a conical lower section, connected to the body. The conical lower section is attached to a dust hopper that stores the material after it is separated from the gas stream.

When the gas enters the tangential inlet, a whirling motion is imparted. Dust particles are subjected to a centrifugal force, and migrate outward to accumulate at the wall of the conical section. A combination of the spiraling motion and gravity drives the concentrated dust and gas downward along the wall toward the cone bottom. Near the bottom of the cone, the clean gases turn upward and form an inner spiral which departs through the clean gas outlet.

Cyclone collection devices offer high efficiency cleaning on relatively large high-density particles. Maximum removal occurs on particles over 10 microns in size. Pressure drops can range from 2 inches W.C. (Differential pressure — resistance to air flow — between two points is measured in inches of water column, W.C.) for low efficiency cyclones to 7 inches W.C. for high efficiency cyclones.

Today, most cyclone collectors are used as a pre-cleaner, prior to the final cleaning equipment. This reduces the dust load on the more efficient final collector unit. Cyclones are well suited for this application because of their low initial and operating costs and their ability to collect the contaminant dry. When used with scrubbers, a cyclone also serves the purpose of reducing the slurry solids concentration, and helps reduce the load on the settling pond. Several modifications of cyclones are used. A multi-cone collector uses several small diameter cyclones in parallel to gain efficiency. Cyclone walls can be wetted with water sprays to increase efficiency.

5. *Filter Bag Collectors:* Filter bag collectors utilize cloth-filtering media to capture contaminants from a carrier gas stream. The carrier gas is designed to flow through bags that are normally of an envelope or tubular shape, and the filtering media used for collection can be woven or felted fabric. These fabrics are available in various fibers, yarn construction, finishes, and weaves. Important considerations involved in selection of media are the method of cleaning, gas temperatures, and physical and chemical characteristics of the gas stream.

Fabric collectors operate in the 99% plus range of collection efficiency. This high degree of efficiency is attained by utilizing the dust to form an initial filter cake, filling the fabric voids for added filtration. On clean cloth, bleeding of contaminants will occur until sufficient particles are collected on the fibers forming the filter cake. As the contaminant is collected on the fabric supporting structure, the resistance to air flow caused by the dust layer increases. Since the collection system's prime mover, the fan, is designed to operate at some maximum resistance, the filter bags must be cleaned periodically.

Three types of cleaning systems widely used by industry are the shaker, collapse cleaning, and air pulse cleaning systems. Shaker cleaning uses mechanical or pneumatic shakers to periodically clean the collected material from the bags. The efficiency of this type cleaning dictates that the velocity of carrier gas flowing through the bags normally be below 3.5 ft./min. This helps eliminate a quick material build-up and also prevents the pollutant from imbedding itself too deeply in the fabric structure. If this occurs, removal will be difficult and may possibly cause blinding of the cloth. Both of these factors, unless controlled through cleaning, increase resistance to air flow above the design point.

Collapse cleaning is accomplished by passing a small volume of air in a reverse direction through the bags. This, in turn, causes the bags to collapse, thus breaking the contaminant cake. Reverse air cleaning is very popular where bag wear presents a serious problem. This is especially true for applications requiring the use of glass bags, because of their fragile properties. Because of this gentle bag action, filament yarn bags (impregnated with silicone and graphite for lubrication) are normally used, making cake release easier. The efficiency of this type cleaning dictates that the velocity of carrier gas flowing through the bag normally be below 2.5 ft./min.

Air pulse cleaning, a third technique, is accomplished by subjecting the individual filter bags to jets of compressed air. This method develops so much cleaning force that much of the initial filter cake will disappear after cleaning. Therefore, felted fabrics with built-in filtering ability are used. As a result of this efficient cleaning action, the velocity of air through the bag can be as high as 15 ft./min. Since the velocity is determined by dividing the total volume of air by the square feet of total filter bag area (normally referred to as air-to-cloth ratio), this method requires fewer filter bags than low air-to-cloth ratio collectors.

Several advantages of fabric collectors include this excellent filtering ability with moderate power requirements (resistance normally 3-4 inches W.C.) and their ability to collect the contaminants dry. Main disadvantages are the fabric's limited temperature range, which could require pre-cooling of gases entering the unit, and the fact that the system is highly affected by moisture, gas and fatty acids.

6. *Scrubbers:* Wet scrubbers use a liquid, usually water, to remove the entrained contaminants from the carrier gas stream. Scrubbers are used as particulate collectors, as well as gas absorption collectors. Contacting power is a concept developed by Lapple, Karmack and Semran for correlating scrubbing efficiency with the energy required to move the gas stream through a scrubbing stage (zone). The resistance to this flow (pressure drop) occurs when the contaminated gas comes in contact with the scrubbing liquid, thus creating a turbulent scrubbing action. (32) Because of this ability to relate energy requirements to scrubber efficiency, scrubbers can be grouped by their energy requirement: low (under 8 inches W.C.); medium (8 to 25 inches W.C.); and high (25 inches W.C. and above). Low and medium energy scrubbers

are applicable for contaminants in the larger micron ranges, while sub-micron particulates require high energy units.

There are many means of developing the energy levels required for different scrubbing applications. The type, size, and loading of the particulate in the gas stream are important variables to consider when selecting what scrubbing mechanisms or how many scrubbing stages (zones) are required to accomplish collection at the desired efficiency and within reasonable operating and maintenance costs. The first stage scrubbing mechanism could be designed to eliminate the major portion of the contaminant, thereby removing much of the load from a more restrictive higher energy section needed for sub-micron particulate collection. Contaminant build up at the wet-dry junction, where the carrier gas first comes in contact with the scrubbing liquid, may cause plugging, a situation requiring a special cleaning method.

Because of multiple scrubber mechanisms used in various scrubbers, only two common types will be discussed: the low to medium energy inertial scrubber; and the high energy venturi scrubber.

Inertial scrubbers obtain particle-liquid contact by impingement. Impingement can be accomplished in two ways: the carrier gas stream can impinge on baffles that are wet from constantly being flushed with liquid, or liquid droplets can be entrained in the carrier gas stream, then impinged on the baffle plate. Performance depends on the flow rate of carrier gas, with the normal pressure differential ranging from 2 to 12 inches W.C.

Venturi scrubbers consist of a venturi-type constriction, through which the contaminant gas flows. A curtain of liquid is injected near the constricted area where the gas impinges upon the liquid stream. The flowing gas atomizes the liquid into many fine droplets. As the energy input is increased, the turbulence created by the interaction of the droplets and contaminants leads to increasingly high collection efficiencies.

Several advantages of scrubbers are: their ability to handle extremely hot gases without pre-cooling, the reduced volume requirement of the fan located on the discharge side of the scrubber due to the scrubbing action cooling the gas, the fact that scrubber operation is not affected by the carrier gas moisture content, and the lower initial equipment and installation costs when standard construction materials are used.

Scrubbers also have disadvantages. Compliance with emission regulations usually requires the use of higher energy scrubbers, resulting in extensive operating costs; corrosion effects are usually increased by the cool wet gas generated in the scrubber and the slurry of collected contaminants; water treatment systems are needed when water pollution is a factor; water requirements are high; and scrubbing often results in a steam plume, causing undesirable complaints.

7. *Absorption Equipment:* Absorption is a diffusion process, involving the transfer of gas molecules into a liquid phase. All scrubbers have the capability of absorption, but differ widely in their efficiency of removal.

    Scrubbing time and turbulent scrubbing action are important criteria in design of absorption scrubbers. The packed tower offers long reaction time and good scrubbing action. In the packed tower scrubber, the scrubbing liquid is usually introduced at the top and is fed by gravity through various layers of packing. Packing may be fixed or in a floating bed of low density spheres. The carrier gas and scrubbing liquid are usually counter current in flow. Packed towers are used to remove particulate matter as well as gaseous contaminants.

    It should be noted that venturi scrubbers are also becoming popular for gaseous removal. They overcome a deficiency in reaction time by imparting more scrubbing action through an increase in power.

8. *Electric Precipitators:* Electrostatic precipitators rely upon the force imparted to the particulate by an electrostatic charge to remove the contaminant from the carrier gas stream. The carrier gases pass between parallel plates where high voltage wires are hung. The voltage developed between the wires and the grounded positive plate ionizes the air and creates an electron flow from the wire to the plate. Particles passing through this field are negatively charged and, thus, attracted toward and collected on the parallel positive plates.

The contamination layer which builds up on collecting plates decreases the collection efficiency of the precipitator, making it necessary to periodically clean the plates. Electrode cleaning is normally done while gas flow continues. This is accomplished through the action of rappers or hammers. The dislodged material falls into hoppers below and is removed for disposal.

The ability of a particulate to adhere to the collecting plates is greatly affected by the electrical resistivity of the contaminant. Two contrasting effects of resistivity are the ability of some contaminants to easily give up their negative charge and the inability of others to give up the charge. Carbon easily gives up its negative charge, thereby losing the attracting force and allowing re-suspension of the particles. Sulfur has a high electrical resistivity and does not give up its charge when it strikes the collection plate, which aids collection efficiency.

The advantages of precipitators are low pressure drop (normally around .5 inches W.C.), low maintenance and operation costs, collection of acid and tar mists (difficult to collect by any other means), and their suitability for use in high-temperature applications.

The disadvantages of precipitators are initial high cost (especially when an efficiency of 99% is required), unchangeable operating conditions, necessity for use of a precleaner to reduce dust load to the precipitator, and large space requirements.

9. *Adsorption Equipment:* Adsorption utilizes a phenomenon in which molecules of a gaseous contaminant contact and adhere to the surfaces of a solid. Two adhering forces that can be involved in this phenomenon are: the chemical combination of the gas with the free valences of atoms on the adsorbent surface, and liquefaction of the gas and its retention by capillary action.[30]

Since adsorption occurs only on surfaces, the more surface area that is exposed per unit volume, the greater the collection efficiency.

There are many materials that exhibit significant adsorptive properties. In applications involving the control of pollutants, activated carbon is by far the most commonly used. An important feature of activated carbon is its ability to allow the recovery of the adsorbed contaminant by regeneration.

When beds of activated carbon can no longer adsorb the gases at an efficient level, regeneration of the carbon is required. For continuous operation, a minimum of two units is needed, so that one unit can continue adsorbing while the other unit is being stripped of collected contaminant. Regeneration, removal of the gaseous contaminant, is normally accomplished by passing saturated steam through the carbon bed.

10. *Afterburners:* Afterburners are frequently used for the control of combustible particulate (dusts, mists, fumes) and gaseous emissions. Afterburners are available in two types, direct-fired and catalytic.

    With direct-fired afterburners, the contaminated gases enter a throat section, where they are mixed with combustion gases and flames. Completion of combustion reactions generated in this area occurs in the main chamber. The cross-sectional area of the main chamber is sized so that the velocity of the gases is reduced, to guarantee complete reaction. Well-designed systems discharge mostly carbon dioxide and water vapor to the atmosphere. Operation temperatures can vary from 850°F to 1,500°F.

    In catalytic afterburners, the contaminant first passes through a pre-heat zone, where the gas temperature required for catalytic combustion is reached. This temperature normally varies from 650°F to 1,000°F. The preheated gases then flow through the catalyst bed, where the remaining combustible contaminants are burned by the catalyst.

11. *Fans:* The proper size and type fan required for any given application cannot be determined until all the previously mentioned subjects have been examined. Properly analyzing the pollution problem at its source and defining the quantity of air evacuation needed for control determine the quantity (Q) of required air that the fan must move. Pick-up hood

configuration, transmission velocity, duct size and length of runs, and type of collection system determine the resistance the fan must be capable of overcoming. The type of contaminant may necessitate special construction in order to minimize maintenance.

Fans are normally placed between the cleaning equipment and the stack so they can operate on the cleanest possible gas. If the contaminant is not corrosive in nature and does not cause undue maintenance problems, the fan can be placed on the input side of the collector. This often simplifies ductwork, and, in cases where tempering air is required to lower gas temperature, helps in mixing the gases to prevent stratification.

When the above requirements have been determined, fan selection can proceed. Fan manufacturers can aid in selecting the fan best-suited for proper system performance.

12. *Contaminant Disposal:* Should be an important consideration when designing any of the previously-discussed collection systems. Disposition of collected material, whether for sale or as a waste product, has a vital bearing on the success of the control system. Air pollution problems caused in handling the material, solid waste pollution disposal problems, or water pollution caused by the contaminant are all important problem areas to be considered.

13. *Dust Suppression:* Dust suppression systems are used to condition coal, ore, and stone in material handling systems with moisture, to eliminate dust generation at crushing, screening, conveyor transfers, truck loading and unloading, and stock piling operations. Water is normally used, along with a specially-formulated wetting agent. The wetting agent lowers the surface tension of the water from 72 dynes/cm to as low as 27 dynes/cm. The mixed solution, if properly applied at good spray application points, can effectively control dust generation with the addition of as low as ½% moisture to the material.[33]

The purpose of dust suppression is to condition the material and cause dust particles, after wetting, to agglomerate. It is, therefore, primarily a dust control system, with the object of preventing dust from becoming airborne rather than of removing the dust from the air after it has become airborne.

A complete suppression system incorporates: a proportioning system that mixes the wetting agent and water in the designed ratio, a pump to transport this mixed solution at the proper pressure and flow rate to the spray application points, spray headers properly positioned for maximum material conditioning, and automatic controls to turn the sprays on and off.

14. *Maintenance of Control Systems:* The old saying, "An ounce of prevention is worth a pound of cure," should be taken seriously when discussing maintenance on pollution control systems. These systems, by their nature, require preventive maintenance or they may destroy themselves. A good example of self-destruction is given in the following discussion of a collection system that includes the pick-up hood, ductwork, bag filter collector and fan.

The bag filter collector has been sized with enough cloth area to keep the air-to-cloth ratio (velocity through the bag) within the limits required for effective filtration and cleaning. While in operation, one of the cleaning devices fails — taking one fifth of the cloth area off line. Because of the increased air-to-cloth ratio, the cleaning mechanism no longer is capable of cleaning the bags as effectively, and a build-up of material on the bags occurs. The material on the bags causes undue stress on the support mechanism, forcing it out of alignment and weakening the bags' supporting fabric structure, thus shortening bag life. The fan, not having been designed to overcome the added pressure drop (resistance), no longer operates efficiently and air volume decreases. The duct-work is sized for a certain transport velocity at a rated volume; since the volume decreases, the velocity decreases, causing pollutant fall-out in the duct. The fallout clogs the ductwork. Some time later, the operator notices a distinct change in evacuation efficiency and reports the condition to maintenance. This is normally too late. One problem caused many other problems, increased the required maintenance, added to collector operating costs, and resulted in process down-time. Periodic monitoring of static pressure readings across the compartments of the collector would have determined how efficiently the cleaning system was operating.

All control systems will have maintenance problems. To a large extent, how great these problems become depends upon the type of preventive maintenance program developed.

SUMMARY

The 1970 Amendments to the "Clean Air Act" took into consideration aspects of air pollution management that have been discussed in this chapter. These amendments speak of the effects of health and welfare defined by the air pollution criteria documents, require the setting of air quality standards based on these effects, define acceptable methods to measure the ambient air concentrations, and suggest control guidelines to help states select the best available control technology when the ambient air quality dictates the need for the adoption of strict regulations. One added consideration required by the "Clean Air Act" that can have a major impact on whether air quality goals will be reached at some future date is the potential pollution and industrial growth in areas being studied. This factor must be seriously considered when developing management strategies for the control of any environmental problems.

As the country grows, we must study indicators to allow us to determine how much emphasis should be placed on solving the many complex problems associated with protection of the quality of the environment. Population density is certainly a useful tool in indicating a possible pollution problem. An important consideration when considering the magnitude of this population effect is to determine at what level of affluence (life style) members of the population live, which will dictate what products and services he will demand. The need for development of these products and services will, in turn, require increased energy usage.

The generation of energy used in producing the items and services to which man has become accustomed is a major cause of pollution. Two cars, air-conditioning, and well-packaged food items are among the items and services that are becoming commonplace. The potential of pollution problems increases as man's affluence increases. If the public insists that such items are a necessity, "for a better life", and continues to require an increase of the per capita energy demand to meet increased affluence demands, at the same rate as in the past, major environmental problems lie ahead.

By the year 2000, if per capita consumption increases at the present rate, the energy demand could total nearly three times what it is today.[3,5]

These relationships help define, on a broad scope, what future problems could be anticipated in air pollution (more pollutants emitted per unit area), water pollution (added thermal pollution caused by increased electrical power generation, increased sewage and industrial process wastes), and solid waste disposal (more garbage generated per person and more industrial and municipal wastes collected by increasingly efficient collection systems). The challenge to stop the degradation of the environment will grow as our country grows.

Complications such as the 1973 worldwide fuel shortages can affect energy availability and compel alternative fuel selection which could offset gains in the area of pollution control. Researchers must then work overtime to develop counter measures, if progress in this discipline is to maintain momentum.

This chapter has dealt only with air pollution, but it is hoped that the problems and possible solutions presented have pointed to the many factors that must be considered when evaluating *any* pollution problems. Purifying the air through scrubbing may mean contaminating the water. Reducing our waste by incineration may mean polluting our air. Pollution solutions must be considered in a broad context. Sound pollution control and disposal techniques must be developed and utilized, if we are to meet increased environmental demands imposed by an expanded economy and population. These environmental demands cannot be met without great effort and expense. Unless man makes a serious beginning today, he may never catch up with the demands of tomorrow.

BIBLIOGRAPHY

1. "Air Pollution". New York, N.Y.: World Health Organization, Columbia University Press, 1961.

2. Schrenck, H.H. et al., "Air Pollution in Donora, Pennsylvania", Public Health Service Bulletin 306, 1949.

3. Selden, M. G., Jr., "Dust Hazards of Processing Minerals", *Minerals Processing*, May, 1969.

4. "Air Quality Criteria for Particulate Matter". Washington, D.C.: U.S. DHEW, PHS, EHS, National Air Pollution Control Administration, Publication Number AP-49, 1969.

5. *National Air Pollution Control Administration Reference Book of Nationwide Emissions.* Durham, N.C.: U.S. DHEW, PHS, CPEHS, NAPCA.

6. "Controlling Air Pollution from Motor Vehicles", Report by Motor Vehicle Committee to New Jersey Air Pollution Control Commission, September 21, 1964.

7. "Air Quality Criteria for Carbon Monoxide". Washington, D.C.: U.S. DHEW, PHS, EHS, National Air Pollution Control Administration, Publication Number AP-62, March, 1970.

8. "Air Quality Criteria for Hydrocarbons". Washington, D.C.: U.S. DHEW, PHS, EHS, National Air Pollution Control Administration, Publication Number AP-64, March, 1970.

9. "Air Quality Criteria for Sulfur Oxides". Washington, D.C.: U.S. DHEW, PHS, EHS, National Air Pollution Control Administration, Publication Number AP-50, 2nd printing, April, 1970.

10. "Air Quality Criteria for Nitrogen Oxides". Washington D.C.: U.S. DHEW, PHS, EHS, National Air Pollution Control Administration, Publication Number AP-84, January, 1971.

11. "Air Quality Criteria for Photochemical Oxidants". Washington, D.C.: U.S. DHEW, PHS, EHS, National Air Pollution Control Administration, Publication Number AP-63, March, 1970.

12. Kilburn K.H., M.D., "Cilia and Mucus Transport as Determinants of the Response of Lung to Air Pollutants", *Arch. Environ. Health*, Vol. 14, p. 77, January, 1967.

13. Dr. H.C. Wohlers — Private Communication.

14. Muskie, P.S., "Air Quality Criteria — Staff Report for the Subcommittee on Air and Water Pollution, Committee on Public Works, United States Senate", Washington, D.C.: July, 1968.

15. Goldsmith, J.R., "Effects of Air Pollution on Human Health" in *Air Pollution*. New York, N.Y.: Academic Press, 1968.

16. Winkelstein, et. al., "Suspended Particulates", *Arch. Environ. Health*, Vol. 14, p. 164, January, 1967.

17. Michelson, I., "The Costs of Living in Polluted Air Versus the Costs of Controlling Air Pollution." A Report to the U.S. Public Health Service Conference on Air Pollution Abatement in the New York — New Jersey Area, January 11, 1967.

18. Yocum, J.E., and McCaldin, R.O., in "Air Pollution". New York, N.Y.: Academic Press, 1968.

19. Special Subcommittee Hearings on Air and Water Pollution (1964) Committee on Public Works, United States Senate, Eighty-eight Congress, Second Session, Washington, D.C.

20. Stern, A. C. (Ed.), "Air Pollution and Its Effects", 2nd ed., 1 Vol., New York, N.Y.: Academic Press, 1968.

21. The Clean Air Act, as Amended 1970 — Section III.

22. Stern, A.C. (Ed.), "Analysis, Monitoring and Surveying", 2nd ed., 2 Vol., New York, N.Y.: Academic Press, 1968.

23. ASME Power Test Code 27, "Determining Dust Concentrations in a Gas Stream," *American Soc. of Mechanical Engrg.*, 1957.

24. Western Precipitation Corp., Bulletin WP-50, "Methods for Determination of Velocity, Volume, Dust and Mist Content of Gases," 7th ed., Los Angeles, 1968.

25. Dwyer, J.L., "Contamination Analysis and Control". New York, N.Y.: Reinhold, 1966.

26. Stern, A.C., (Ed.), "Air Pollution", 2nd ed., 3 vol. New York, N.Y.: Academic Press, 1968.

27. Hemeon, W.C.L., "Plant and Process Ventilation". New York, N.Y.: Industrial Press, 1963.

28. Sargent, G.D., "Dust Collection Equipment", *Chemical Engineering*, January 27, 1969.

29. "Industrial Ventilation", *American Conference of Government Industrial Hygienists*, 11th Edition.

30. *Air Pollution Engineering Manual*, U.S. DHEW, PHS, NCAPC, Public Health Service Publication No. 999-AP-40, 1967.

31. *Air Pollution Manual*, Part II, Control Equipment, American Industrial Hygiene Association.

32. Semrau, K.T., "1960 Correlation of Dust Scrubber Efficiency". JAPCA 10: 200-07 (June).

33. "Wet Dust Suppression", talk delivered to the National Crushed Stone Association, Dust Control Technology and The Crushed Stone Producer by The Johnson — March Corporation.

34. Sussman, Victor H., "New Priorities in Air Pollution Control", *Journal of the Air Pollution Control Association*, Vol. 21, No. 4, April, 1971, p. 202.

35. Anthrop, D.F., Environmental Side Effects of Energy Production", *Bulletin of the Atomic Scientists*, Oct., 1970, p. 41.

XIV. Engineering Controls

Safety begins with corporate commitment and is backed up by engineering controls. Company policy should stipulate that safety be designed and built into a project before it is executed.

In order to accomplish this, controls that result in safer equipment should be established before the design phase of engineering, and must be concerned with the design, construction, operation and maintenance of the facility or equipment. These concerns are interrelated; good design will help reduce operating hazards and will optimize maintenance experience. A well managed maintenance program will continue the original design level of safe operation and will continue to produce at the design rate.[1,3]

The prime objective, therefore, of any engineering group is to design equipment and process, and to plan job procedures so that personnel exposures to hazards and injuries are eliminated or controlled.

IDENTIFY THE PROBLEM

Safety texts and references serve a valuable function in aiding the identification of *common* hazards or exposures. However, no matter what the industry or government agency, the hazards *peculiar to that industry* must be identified prior to the formulation of a workable plan for engineering safety controls. A good place to start is to review all lost-time injury records, first aid cases, doctor referrals and major property damage. Probably the lost time injuries have been formally reviewed and, where recommendations have been made (perhaps to make modifications, or install guards or interlocks), they probably were implemented. Note these after-the-fact, added-on safety features so that they will not be overlooked in new facility design.

When a new plant facility is in the planning stage, it is important to evaluate comparable machines and equipment that are currently operating, in terms of safety and reliability. Conversation with operating and maintenance people, plus on-site observations, will provide the design engineer invaluable insight into operational stresses. The value of this on-site evaluation of the man-machine-materials-environment interface cannot be overstressed.

To clearly understand the importance of establishing a safety policy for engineering design, examine the model policy, which follows. (See Figure 1.)

MODEL POLICY
SAFETY POLICY FOR ENGINEERING DESIGN

DATE ISSUED_____ DATE EFFECTIVE _____

SECTION _____

ISSUED BY _____

JURISDICTIONS AFFECTED _____

AUTHORITY _____

Good business practice dictates that management provide a safe and healthy work place. Safety is more satisfactorily accomplished when it is given a proper priority in engineering design. Costs are lower when safety provisions are designed and incorporated before-the-fact, rather than added at a latter date. Under the Occupational Safety and Health Act of 1970, the need for an effective safety and health policy is even more evident. Many of the consensus standards once offered as recommended practices are now mandatory. It is imperative that all applicable standards be consulted in the design engineering phase.

A standing safety committee will audit the project in terms of health and safety at various stages. The staff safety engineer will coordinate the activity. Engineering drawings and requests for appropriations will be considered complete only when signed by the staff engineer. Audits will be documented and will become a part of the permanent project folder.

Figure 1.

EQUIPMENT SAFETY SPECIFICATIONS

1. PURPOSE

Safety specifications are provided to influence the design, purchase, installation and operation of equipment. It is intended through conformance, that a safe process and working environment will result and that maximum protection will be afforded to individuals directly or casually involved. Since safety is more satisfactorily accomplished by removing the hazard than by guarding it, it is urged that ample consideration be given to process and equipment design.

Special attention shall be given in the design of safeguards to the following hazards that might arise out of use or maintenance of this equipment:

    a. Direct contact with moving parts of the equipment.

    b. Work in process—ejection, conveying, discharge, etc.

    c. Mechanical failure.

    d. Human failure—curiosity, distraction, fatigue, chance-taking, fear, etc.

2. RESPONSIBILITY

Except where design is stipulated by purchaser's drawings or other written agreement, the responsibility for providing safeguards shall lie with the equipment vendor and shall, as a minimum, conform to the standards of the American National Standards Institute, applicable State Safety and Health Regulations, and the Safety Regulations of the Walsh-Healey Public Contracts Act Part 50-204. Where differences occur, the regulation adjudged by the manufacturer to provide the greater safety shall apply. Appropriate notification of such judgments shall be in writing to the purchaser.

Figure 2.

One effective technique in evaluating equipment reliability is to examine the production efficiency and determine why it is less than 100%. Another effective yardstick is to review the maintenance work requisition system with a critical eye to parts and components that are being adjusted, serviced, replaced, overhauled or repaired. It is important to be concerned with equipment efficiency rates because when a machine is down (not running), it is in an unnatural condition. During these times, a series of new exposures to injuries is introduced.

Process difficulties and equipment breakdowns may place undue pressure on the supervisor and his people to get the product out at budgeted rates. In his effort to maintain production rates, the employee may take short cuts, improvise or modify standard practice.

In evaluating an existing piece of equipment with an eye to determining some of the hazards, a "people watch" is effective. It is important to look closely at what the people are doing.

One word of caution, however ... the observer must be conscious of his natural tendency to watch what the machine is producing and consequently miss the hazard exposures.

What is the operator's interface with the production equipment? Is there rapid movement while executing duties? Is the operator particularly attentive while working at a nip point, or does he carelessly reach in to make an adjustment? Nip points are particularly hazardous in manufacturing where the product or material must be rolled or calendered. (A nip point exists when two or more shafts or rolls rotate parallel to one another in opposite directions. They may be in close contact or some distance apart. There is little or no danger from a nip point where shafts rotate in the same direction.) Rubber, steel, and paper companies are especially susceptible to this exposure.

Once the hazards are identified, the next step is to develop check sheets and flow diagrams to serve as aids for the design engineer.

One control mechanism that enables the designer to highlight exposures is a flow diagram, as shown in Figure 3. This particular diagram illustrates the elevation of a fourdrinier paper machine. As can be seen, nip points common to this type of equipment are numbered. It is the designer's responsibility to indicate on the engineering drawings the type of guard, interlock, procedure or control mechanism that will protect the operator from this known hazard.

As part of the engineering commitment to design out hazards, there must be a recognition of the function and the ability of the

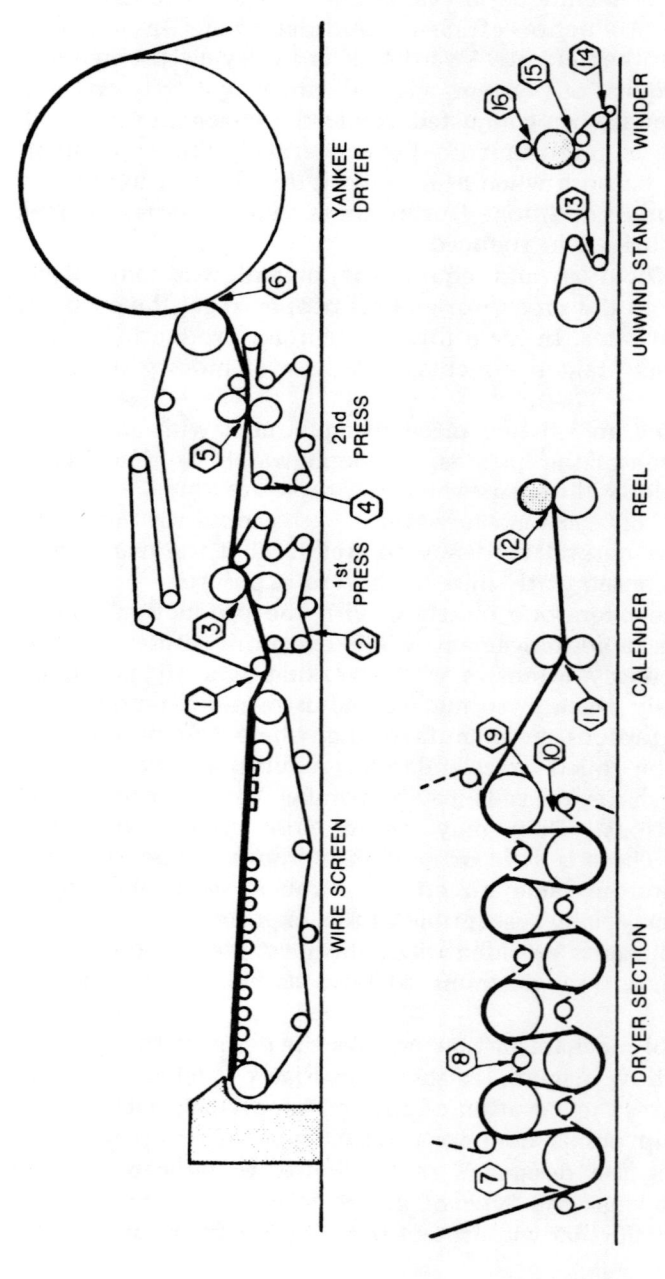

Figure 3.

persons using the equipment and facility, as well as others who will work in the area. Concern must be given to the behavior of man; repetitive tasks should be evaluated in a different manner than once-in-a-while activities. It should be noted that a routine task may lull the operator into a false sense of security that may result in an accident. Equipment must be designed with Human Factors Engineering in mind to take this exposure into account.

Numerous professional organizations publish literature in the area of safety, hazard control and loss prevention; however, much of this information is sidetracked to the engineering library or the safety manager's office and is not readily accessible as a before-the-fact resource.

In the design phase, inputs relative to hazards of former projects should be evaluated to avoid repetition of the same hazards or safety "negatives". Many potential or subtle hazards are overlooked because safety per se is not approached in the same formalized manner as production rates, maintainability, or return on investment. These criteria are vigorously investigated in the feasibility phase of engineering design, but safety is too often added-on rather than built-in.

From a practical standpoint, it is easy to see why the project engineer may give safety a low priority. Generally the project engineer's own performance is measured in terms of whether or not the project was installed on schedule (or earlier), at a pre-determined cost (or lower), and whether it satisfies or exceeds design criteria.

Understandably, the engineer aligns his priorities with those of his manager. The pressure of the Occupational Safety and Health Act of 1970 will operate as a force to bring before-the-fact safety engineering into its proper perspective, because this legislation contains provisions for citing violations and penalties. Economics now favor making safety an essential factor in the project design.

A WORKABLE PLAN

Now that the objective of reducing hazards has been established and the hazards identified, a plan is required to accomplish the goal.

1. Consult the standards promulgated under the Occupational Safety and Health Act. Initial standards were derived from two sources: previously-existing established federal standards and national consensus standards. New and/or modified

OSHA standards become effective upon publication in the *Federal Register*.

2. Examine the loss experience and accident record of the industry. What were the individual accidents that served as inputs for the statistics? How does the industry, as a whole, compare to the individual company with the best safety record? Why does X company have the best safety record? Perhaps the corporation has less inherent exposures, a master all-accident prevention approach is used, or less manual labor (therefore, less exposure) . . . or is it "luck?"

3. Determine what checklist, sketch, procedure, specification, memory jogger or communication vehicle could have been used to design out previous hazards that resulted in personal injury or property damage.

4. Conduct a meeting with the production and maintenance supervisors for the purpose of discussing safety and reliability features. Part of this consideration is the concern for safeguards and handling devices that will be utilized when the equipment is shut down for repair, overhaul or turnaround.

5. Enlist the aid of insurance underwriters who are vitally concerned with construction material, storage, use and occupancy of the facility.

6. Consult the local and state codes relative to employee health and safety. Both state and local health and safety requirements are considered minimums by most safety professionals. Therefore, the goal should be maximum performance, not minimum conformance.[1]

7. Develop check sheets and guides that cite legal requirements for the various components that comprise the project.

8. Develop check sheets and flow diagrams that highlight the greatest exposures found in the above categories, to enable the designer and design engineer to be aware of potential hazards that must be eliminated, enclosed or reduced in some other manner.

9. Establish an order of priority and a timetable for the accomplishment of the objectives.

10. Break the task down into definable segments and assign the responsibility for accomplishment to a specific individual.

DESIGN GUIDELINES

The elimination or control of potential hazards at the blueprint stage of new plant facilities is the most effective engineering control. Corrections or modifications at this stage of the project often reflect the experience gained by a company over the years. As a result, this same company can profit from the safety audits and evaluation of earlier projects. Safety oversights and omissions that came to light after the facility was in operation, as well as other experience factors, serve as excellent resources for the development of check lists. Figure 4 is representative of a check list that evolved from experience and safety awareness into a meaningful tool because of a commitment to safety.

The following check list (Figure 4) is designed to review common hazards. It does not preclude that other hazards may be present. The intent is to focus attention on the potential risks to employees who will operate the equipment or work in the project area. Engineering provisions will be made in the design to eliminate or reduce to a minimum any dangers to the health and safety of employees.

The standards, data sheets, and other safety references are considered a minimum requirement and should not limit ingenuity or imagination or prevent action that will provide a greater safety factor.

The interface of production management, maintenance management, and engineering management is required early in the design stage and should be frequent throughout the engineering phase. By interfacing, the design engineer will profit by the experience of operating and maintenance people, and some of the subtle hazards will be engineered out as a result of this relationship. Reference to historical data and reports on similar equipment can serve as an effective tool to improved design. A review of operating performance vs. the projected or designed capacity, as well as an evaluation of maintenance costs and downtime will aid in the identification of weak links in earlier design. Evaluation, feedback and correction are required to close the loop for engineering control.

An "audit" control mechanism should be applied at various stages of a project to be effective and non-restraining. Clarence H. Schatz, Corporate Facilities Engineer, Scott Paper Company,

ENGINEERING SAFETY CHECK LIST

1. **ADJUSTMENTS** YES NO
   a. Can adjustments be made while the person is in a normal stance or is the person in an unnatural position? ☐ ☐
   b. Can adjustments be made while equipment is running without exposing the employee to a hazard? ☐ ☐

2. **CHEMICALS**
   a. Will the project involve personal exposure to, or actual handling of chemicals? ☐ ☐
   b. Are the chemicals on the approved list of the Chemical Control Committee? ☐ ☐
   c. Does the project include recommended safeguards for handling and storage of chemicals? ☐ ☐
   d. Does corrosion present a safety hazard? ☐ ☐
   e. Are safety showers and eye wash fountains provided? ☐ ☐
   f. What personal protective equipment or facilities are required? _____
   g. What is the threshold limit value for the chemical as stated in The American Conference Governmental Industrial Hygienists Guide? _____

3. **CLEANING FACILITIES**
   a. Does the project provide for adequate means of maintaining good millkeeping (water, air, steam, drains)? ☐ ☐
   b. Is the project area accessible for good millkeeping? ☐ ☐
   c. If air pressure is required, is it regulated to 30 PSI? ☐ ☐

4. **COLOR CODING**
   a. Does the project provide for color coding of piping, conduit, aisleways, hazard points, etc.? ☐ ☐
   b. Are direction of flow and labeling provided for? ☐ ☐

5. **STANDARDS**
   Have the appropriate ANSI, NFPA and OSHA standards been consulted? ☐ ☐

6. **CONTROLS**
   a. Are controls fail-safe? ☐ ☐
   b. Are panels clearly identified, standardized and designed to avoid operating mistakes? ☐ ☐
   c. Are controls located so that they allow safe thread-up, start-up, etc.? ☐ ☐
   d. Is the operation of each control device compatible with any corresponding display and with common human response tendencies? ☐ ☐
   e. Are the operational requirements of force, speed, precision, etc. within limits of all persons who use the system? ☐ ☐
   f. Are the control devices arranged conveniently and for optimum use? ☐ ☐
   g. Is adequate clearance provided between control panels, lever switches, etc and hazard points? ☐ ☐

Figure 4.

developed a chart (Figure 5) that reflects the phase, time, details, responsibility and method of audit. In several instances, check lists are recommended in the audit review procedure, and may be developed around broad hazard areas such as electrical, lighting and noise, chemical and environmental, equipment-personnel interfaces, personnel physical requirements, walks — ladders — steps, and others.

Properly instituted, an Engineering Safety Audit can serve several purposes:

1. Create a greater awareness of safety.

2. Serve as a filter to catch many of the hazards naturally related to the project.

3. Force people into learning applicable codes and practices. The most effective and least costly approach is to touch on these areas at the time the project is in the design phase. This is when the process is being defined in broad terms as to equipment styles and types, raw materials and the like. When awareness is present in the design phase, hazards can be eliminated before-the-fact.

4. Provide a formal feedback from the field as an assist to future projects.

Review procedures differ from company to company, mainly because of varying degrees of management emphasis relative to safety. The military has found that the success of design reviews is directly proportional to the formality with which reviews are conducted and to the extent of interest evidenced by management. As in any managed activity, the review at the various levels requires planning, some type of formal procedure, a mechanism for resolving differences, and the recording of results.

To formulate a firm set of rules for a successful design review program would be impractical, principally because every agency, company, project and task has its own requirements. General guidelines, however, have been formulated by analysis of successful design programs. The following two basic characteristics appear to be common to most of them:

    a.   Plans for the project include a schedule for design reviews at significant points in every phase of the design and development project.

ENGINEERING PROJECT AUDIT SAFETY AND HEALTH

PHASE	TIME	DETAILS	PEOPLE	METHOD
Initial	Before engineering is started, in conceptual stage.	1. Review basic concept of process or operation, personnel hazards, etc. 2. Establish experience and relevant safety codes.	Project Eng. Project Mgr. Operating Dept. Head Safety Eng.	1. Meeting 2. Checklist 3. Summary outline
Final Design	1. Upon completion of major design on smaller projects 2. Periodically on large projects. 3. Before going out for bids.	1. Identify safety considerations. 2. Insure code conformance. 3. Review satisfaction of anticipated hazards.	Project Eng. Project Mgr. Apptd. Safety Committee Member Safety Eng.	1. Checklist 2. Review of preliminary hazard list. 3. Safety codes
Pre-Startup	During final phases of installation.	1. Safety compliance of total system. 2. Inspect workmanship for hazards. 3. Uncover hazards not detected in prints.	Project Eng. Plt. Safety Mgr. Operating Dept. Mgr.	1. Checklist 2. Comparison with local & plt. codes 3. Inspection tour
Operational	1. One to three months after startup.	1. Hazards that have developed out of change.	Project Eng. or Mgr. Safety Mgr. or Plt. Safety Mgr. Operating Dept. Mgr.	1. Tour 2. Review of safety record 3. Report with recommendations

Figure 5.

b. Documents are published to enumerate review policy and to formally establish review procedures for the project; in addition to clarifying requirements, this puts teeth into the program by emphasizing management's interest in it.[4]

PLANT LAYOUT

The use of process and operation flow sheets is an excellent device for process and plant planning, by increasing the detail of the flow sheet to show hazard points and the built-in provisions for hazard control. A drawing depicting the entire plant layout should be prepared to show the overall relationship of product flow between buildings and facilities, railroad trackage, designated roadways, traffic pattern, utility lines (compressed air, potable water, city gas, electric, etc.), fire protection lines and the location of bulk storage of hazardous substances. (Figure 6 shows major processing sections.)

Individual layouts of the specific buildings, showing equipment layout and space utilization, supplement the overall plant layout. This detail can be further refined by check sheets and flow diagrams to highlight the hazards inherent in the respective machine, equipment or process. (Figure 7 illustrates a flow diagram of a chemical process.)

Figure 3 showed a general flow diagram of material through a fourdrinier paper machine. Nip hazards (identified by encircled numbers) are the greatest exposure in papermaking. Utilizing such a flow diagram, the chances of a designer overlooking the need for a guard at a hazardous point are reduced significantly.

The type of building that houses the manufacturing process is of paramount importance. The building and layout must complement the task efficiently. The plant layout and process configuration relationship is closely allied with health and accident hazards. Therefore, the designer must be thoroughly familiar with accident prevention principles. (Figure 8, an elevation of a building, shows the various safety features incorporated in this design.)

It is important that the safety professional be able to assess the strength of his engineering department and that he provide the safety awareness that will cause the designer to go to readily-available resources when he recognizes that he lacks depth in a given exposure area. *The Accident Prevention Manual* of the National Safety Council has a chapter devoted to "Sources of Help for the Safety Man", for such reference.[1]

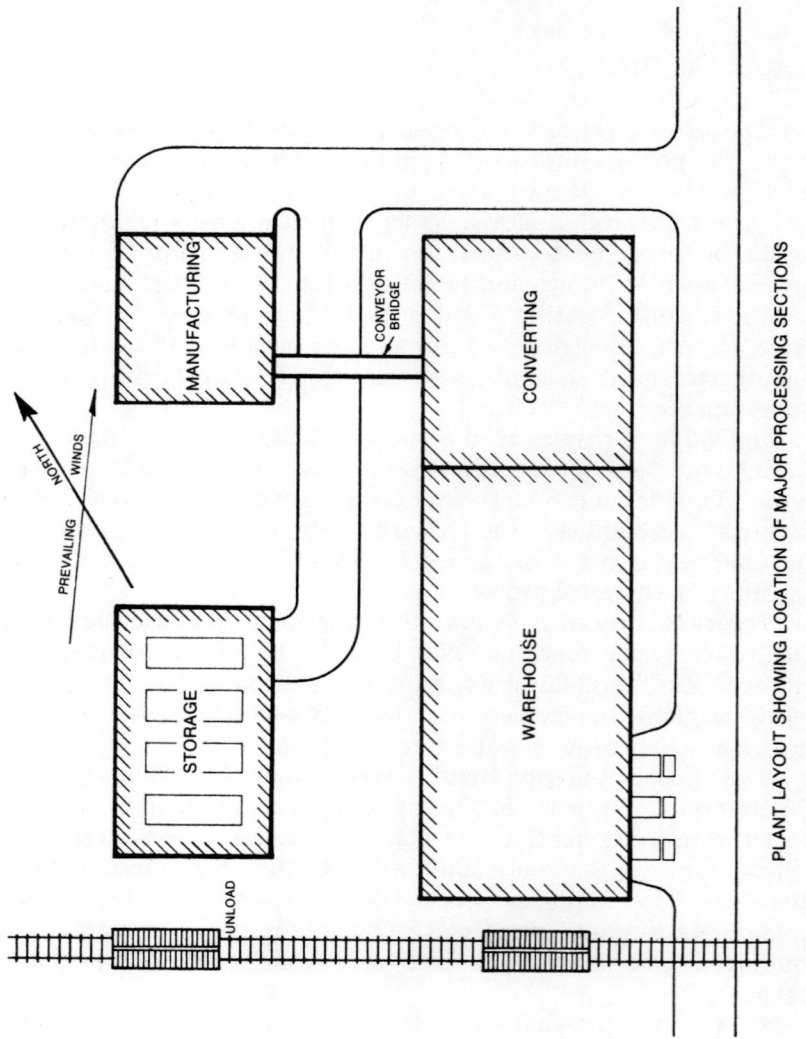

Figure 6.

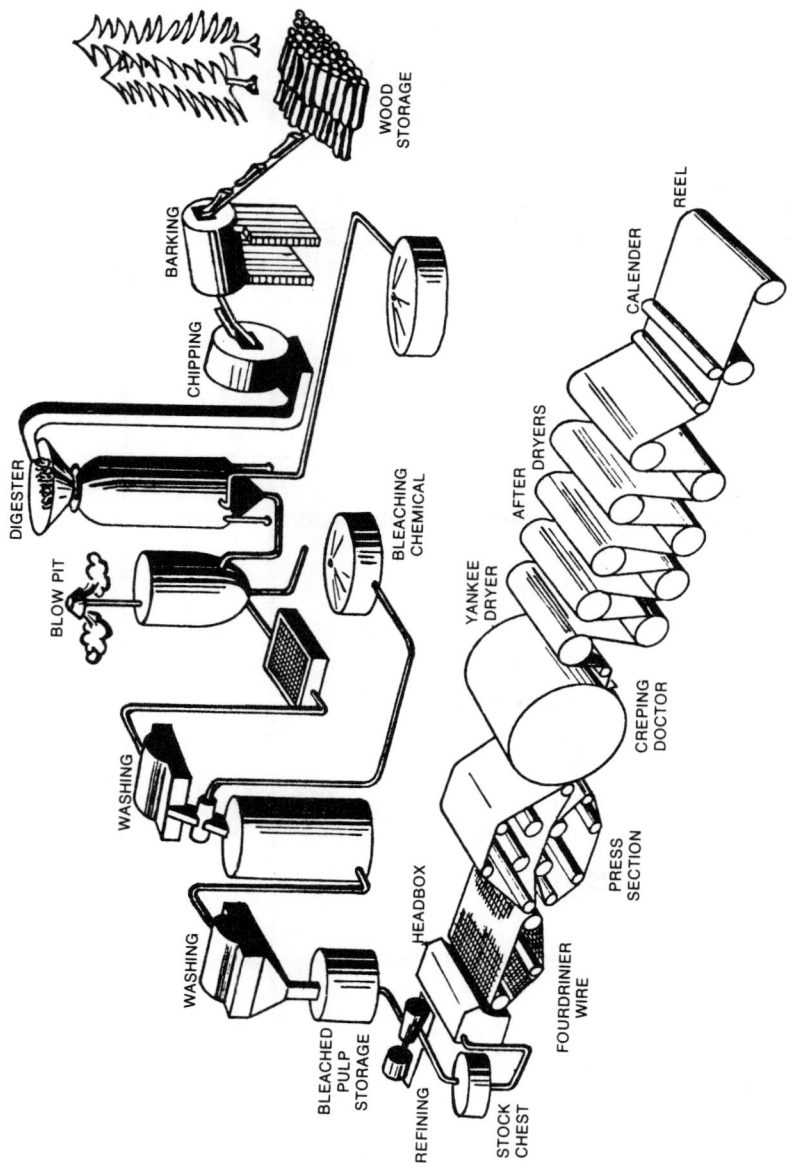

Figure 7.

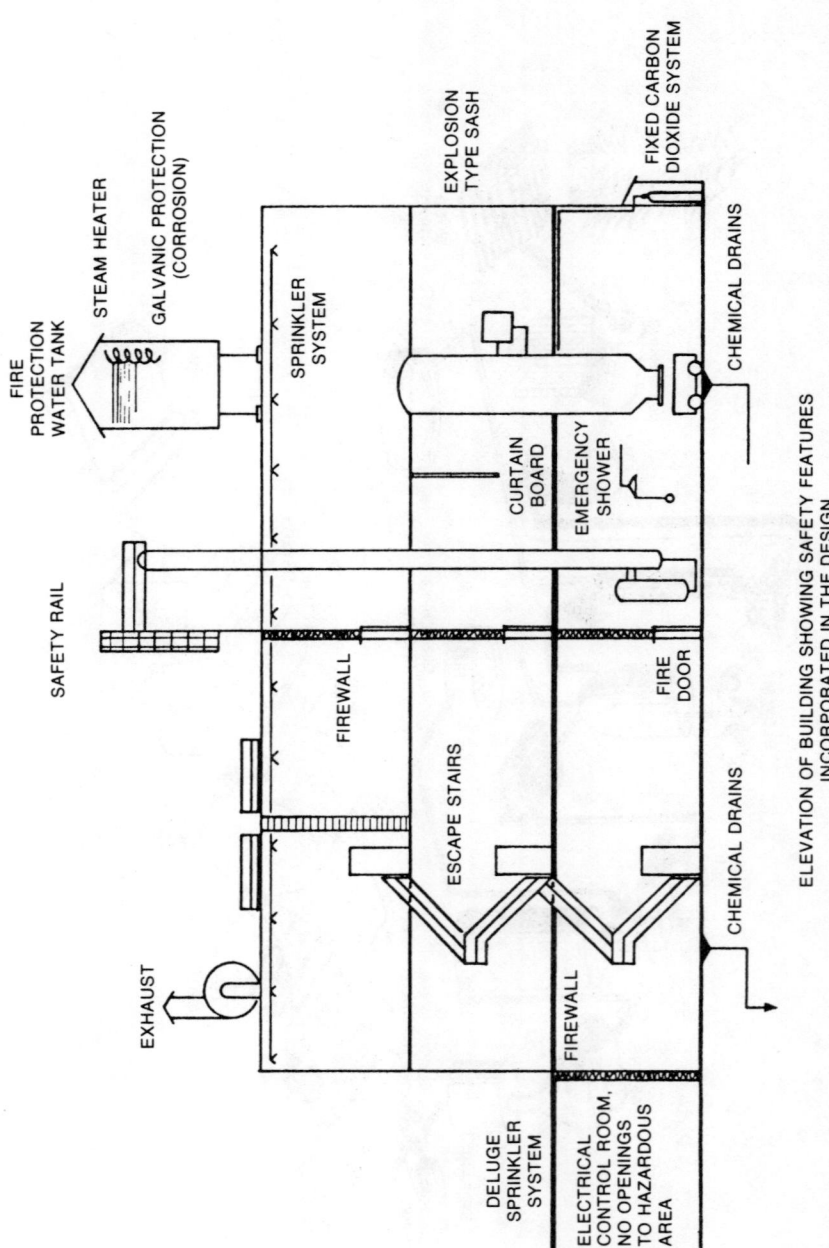

Figure 8.

SYSTEM SAFETY ANALYSIS[1]

Although the engineering controls previously outlined are formalized and effective, they were not determined through any exacting sophisticated analytical methods or new theoretical concepts.

The demands of extremely high reliability, safety specifications and cost in aerospace projects resulted in the development of various methods to analyze a system. A system, in our context, is a predetermined task or function; components of a system can cover a wide spectrum and include men, machines, materials and environment. When any part of a system malfunctions or fails, it can affect the performance of the task.

It was the recognition that the failure of one component could degrade the system's effectiveness that led to the four principal methods of analysis: failure mode and effect, fault tree, THERP, and cost-effectiveness. Each method has variations, and more than one may be combined in a single analysis.

1. *Fault Tree Method* — An undesired event is selected, and all possible happenings that can contribute to the event are diagrammed in the form of a tree. The branches of the tree are continued until independent events are reached. Probabilities are determined for the independent events. The tree is then simplified and the chain of events leading to the undesired event and the probability of occurrence can be computed. To effectively utilize this method, a computer and an in-depth knowledge of mathematics are prerequisites.

2. *Failure Mode and Effect Method* — Each component is considered independently as to a failure or malfunction, and the effects of the failure are traced through the system and the ultimate effect on the task performance is evaluated. This method has limitations, in that it considers only one failure at a time; thus some possibilities may be overlooked.

3. *THERP, Technique for Human Error Prediction* — This method provides a means for quantitatively evaluating the contribution of human error to the degradation of production quality. It can be applied to human components in systems and can be combined with the fault tree or failure mode and effect methods. The choice of method would be dependent on computer and mathematics depth.

4. *Cost Effectiveness Method* — System performance is weighed against dollars of cost. The cost of a desirable system modification or system is ruled out if the gain in performance computes to be too small to justify the cost.

J. L. Recht, writing in the *National Safety News*, indicated there is ample reason to believe that these successful techniques can be equally useful in the non-military and/or space situations.

HUMAN FACTORS

Human factors engineering takes into consideration the abilities and limitations of man. Technology in the past fifty years has exceeded all other previous history before World War I, yet the physical capabilities of man have remained relatively constant. A man's individual limitations determine the level of his human performance. Therefore, the designers of systems must take into consideration the operator's training, acquired skills, type of tools and the environment in which he works. Without this consideration, the human component in a man-machine system can become the variable that prevents optimum operation of the system.[4]

According to "Murphy's Law": "If there is a wrong way to do something, sooner or later someone will do it that way." There is a natural curiosity in people that urges them to explore, test and modify existing conditions. The "wet paint" sign is an example of this curiosity; how many times have we observed a person touch a newly-painted surface to see if the paint is indeed wet? Therefore, it is necessary that we understand the behavior of man when a system is designed, so that we recognize the "wet paint" signs in the system.

An analysis of accident reports will bear out "Murphy's Law." When a man's activity is machine-paced or controlled, he will find ways of circumventing this control. Upper limit electrical stops exist on most overhead cranes to prevent the load cable from being wound about the drum to the point where the cable snaps. It is not uncommon for an operator to depress the control button until the limit switch cuts off the electric power, stopping the ascent of the hook, thus using a safety device in a manner for which it was not intended. Warning signs on production equipment will caution: "Don't remove guards or reach into nips when the machine is running"; yet, how many workers are injured making adjustments while the equipment is running? Too many. It is the responsibility of the design engineer to take this behavior into consideration. In aircraft design, human factors engineers have considered it essential that "go right or no go" and "work right or no work" concepts be adopted uncompromisingly in the design of aircraft components and controls.[14]

Some call this approach to equipment design "idiot-proofing"; actually, this approach takes into consideration the persons who will use the equipment and designs it accordingly. The human factors approach to safety will reduce the number of accidents that an investigation places in the UNSAFE ACT category.

PURCHASING

The purchasing function must be included in the loss prevention effort. A cooperative arrangement whereby the safety director reviews all specifications for the purchase of all plant equipment and materials is helpful in keeping safety in the forefront. In many organizations, all specifications originate in the engineering department. The safety engineer or safety manager provides the impetus for the safety effort; it is his responsibility to oversee the controls that will insure that all items are purchased to specifications and applied to situations within design criteria. This approach will reduce the misapplication of equipment, switchgear and chemicals.

An effective method of aiding the purchasing function in its role requires the engineering and technical departments to initial all purchase orders and/or specifications for chemicals and equipment. A good source of direction and recommendation for a chemical specification can be the chemical vendor.

Quotations on new equipment and purchase requisitions should include a statement that the equipment, machinery component or chemical must meet all local, state and federal requirements. When specifying companies are aware of a particularly high exposure from a hazard or legal aspect, they should indicate this at the time of negotiation.

Although construction of new facilities is usually by outside contractors rather than in-plant labor, an important construction consideration is worthy of emphasis in this chapter. At the time of bidding, regulations relative to sanitation, health, safety, housekeeping and fire prevention should be formally set down and the company and contractor should reach agreement as to which party is responsible for the various facets of the joint loss prevention effort.

SUMMARY

Designing a reliable system that has hazards eliminated or well protected is the goal of engineering. Once the facility is operating, it is the objective of the maintenance department to keep it maintained to original equipment specifications.

BIBLIOGRAPHY

1. *Accident Prevention Manual*, Chicago, Illinois: National Safety Council.
2. Bureau of Labor Standards, *General Safety and Health Standards for Federal Supply Contracts* under Walsh-Healey, Title 41, Part 50-204, Section 1, 1966.
3. Gerse, J. and Holler, W. E. (Eds.), *Maintainability Engineering*, Martin-Marietta Corporation.
4. *Industrial Noise Manual*, Second Edition. Detroit, Michigan: American Industrial Hygiene Association.
5. Manufacturing Chemists Association, *Guidelines for Risk Evaluation and Loss Prevention in Chemical Plants*, 1970.
6. McCormick, E. J., *Human Factors Engineering*. New York, New York: McGraw-Hill, 1964.
7. National Safety Council, *Fundamentals of Industrial Hygiene*, edited by Olishifski, Julian B. and McElroy, Frank E., 1971.
8. National Safety Council, *Industrial Safety Data Sheets*.
9. *Nip Hazards on Paper Machines*, Second Edition. Chicago, Illinois: American Mutual Insurance Alliance, 1968.
10. *Occupational Safety and Health Act of 1970*. Public Law 91-596.
11. Recht, J. L. "Systems Safety Analysis: An Introduction," *National Safety News*, December, 1965.
12. Recht, J. L., "System Safety Analysis: Error Rates and Costs", *National Safety News*, June, 1966.
13. Simons, R. H. and Grimaldi, J. V., *Safety Management*, Revised Edition. Homewood, Illinois: Richard D. Irwin, Inc., 1963.
14. U.S. Army Materiel Command, *Engineering Design Handbook*, 1966.
15. United States Department of Labor, *Inspection Survey Guide*, Bulletin 326, 1970.

Motor Fleet Safety XV.

FLEET EFFORTS AND DAMAGE REDUCTION

People ordinarily associate loss reduction with accident reduction. Unquestionably, the elimination of accidents should be a major goal in our society generally and in fleets specifically, but real progress in loss reduction requires a carefully planned mix of a number of mutually-beneficial approaches.[30] It is essential to have (1) pre-accident involvement, meaning pre-event evasionary countermeasures that reduce the frequency (or at least the likelihood) of destructive occurrences, or (2) activities that lessen the harmfulness of the collision itself, and (3) procedures that reduce the end losses through post-crash emergency, salvage and repair services.[29,30]

These matters are by no means the sole responsibility of the federal government, vehicle manufacturers, highway engineers, or community hospital services. Fleet owners, union leaders and fleet supervisors have decision-making responsibility and authority for actions which have much bearing on crash counter-measures. Their convictions, preferences and actions can be effective in highway, vehicle, trailer and body design; the use of driver restraint systems; the availability of fire-fighting and rescue equipment; and procedures for cargo loading and stabilizing.

The relevancy of fleet efforts in crash and damage reduction can be demonstrated by citing the destructiveness of accidents, by briefly examining the changes at work in the transportation industry, and by suggesting the impact of government regulations. Following this overview will come four basic and essential topics for fleet operation and safety: selection, training, maintenance and supervision. Meaningful and important areas which cannot be covered within the scope of this chapter include accident investigation, public relations, recordkeeping, work appraisal, communications and the special problems of various kinds of fleets (i.e. ambulance, police, transit, motorcycle and sales). Neither can this chapter cover earth-moving, heavy construction or materials-handling equipment.

Recent developments are highly relevant. The number of cars, buses and trucks registered in this country is increasing twice as fast as the population. In 1960, there were 73.9 million motor vehicles for a population of 180 million. At the end of 1969, there were 107 million vehicles for 107.5 million drivers.

Rising trends in motor vehicle registrations, in the number of drivers, in vehicle utilization and in total mileage, passengers and freight tonnage, might seem to add up to a happy future for fleets. The truck and bus industries, however, entered this decade with most of their old problems unsolved and some serious new ones.

Among the newer concerns are the burgeoning fuel crisis; a growing interest in the welfare of the consumer; public awareness of pollution and the need to control it; union activity and the impact of new, national labor agreements; tremendous population growth; advances in transportation technology and a growing trend toward professionalism; and governmental regulatory action. These newer problems are being added to the old and constant dangers of death, injury and property damage from vehicle crashes.

All the wars which this nation has fought since 1775 have killed far less than half as many as have died on our highways since 1900.

In any case, accidents seldom "just happen"; they are caused. Poor vehicle maintenance, although it ranks a distant second to driver failure, probably plays a more important role in accidents than is generally recognized. During 1969, the U.S. Department of Transportation's Bureau of Motor Carrier Safety checked a large number of trucks and buses and found safety violations to be so serious that between 12 and 23 per cent were ordered out of service; 10,828 of 46,731 trucks inspected and 47 of 397 buses[55] were removed from service. In fact, of 3,516 chartered and scheduled buses inspected by the government during one month, more than half had at least one safety violation. The violations which were discovered most frequently in charter buses included faulty brake hoses, cracked windshields, defective speedometers, broken lights and empty fire extinguishers.[10]

Recently, an intensive analysis of 1,029 truck accidents revealed that, for the sample, the most frequent type of tractor-trailer accident was a collision with a passenger car; in second place came single-vehicle accidents, followed by collisions with another tractor-trailer. Significant accident factors included drinking, vehicle defects and speed.[19] In truck-passenger car collisions, for every truck driver killed, 38 automobile drivers died.[40] The frequency of injury to truck drivers, however, is nearly twice as high as for automobile drivers.[73]

One result of such statistics is an increase in governmental regulation. By middle 1971, there had been sixteen standards issued under the Highway Safety Act of 1966 (P.L. 89-564) and nearly every one of them affects fleet operation.

1. Periodic M. V. inspection
2. M. V. registration
3. Motorcycle safety
4. Driver education
5. Driver licensing
6. Codes and laws
7. Traffic courts
8. Alcohol in relation to highway safety
9. Identification and surveillance of accident locations
10. Traffic records
11. Emergency medical services
12. Highway design, construction and maintenance
13. Traffic control devices
14. Pedestrian safety
15. Police traffic services
16. Debris hazard control and clean-up

The Highway Act is not a direct regulator with specified sanctions, but calls instead for a cooperative effort, whereby the national government gives grants-in-aid to the states to accomplish agreed-upon safety goals, with the state utilizing its own constitutional powers toward those ends.

Far more publicized and of more immediate importance are the vehicle and equipment programs and their related standards under the National Traffic and Motor Vehicle Safety Act of 1966 (P.L. 89-563). A large number of minimum standards of performance have been issued by the Department of Transportation's National Highway Safety Bureau for passenger cars, for vehicles, including buses and trucks (in interstate and/or foreign commerce), and for equipment to protect the public against the risk of accidents, death and injury.

Consequently, many observers think federal rules are remaking the face of the motor fleet business. The Bureau of Motor Carrier Safety has rules which apply to trucks, their operations and drivers; the National Highway Safety Bureau has adopted rules on how new vehicles are to be built; the U.S. Department of Health has ruled on engine emissions in pursuance of its anti-pollution and health effort; the U.S. Food and Drug Administration has moved to curtail the use of amphetamines, or pep pills, in part as a result of their misuse by truck drivers. Seat belt installation (required by NHSB) and use (required by BMCS) are now necessary for commercial operators of buses and for drivers and co-drivers of trucks and tractor-trailers. Furthermore, sleeper berths must be equipped with restraint systems. Other standards are probably on the way, based on recommendations from research groups.

It has been suggested that passenger car drivers are often irritated and sometimes terrified by the noise, air turbulence, spray and even rocks thrown by passing trucks.[8] As a result, a new government standard on truck and bus tires has been adopted; wheel guards and wind flap standards may also be enacted. Noise pollution is an additional problem; since 1968, the U.S. Department of Labor has had the authority to establish limits of industrial noise to which a worker can be exposed; there are also ordinances in some cities limiting vehicular and equipment noise.

Fleets will not only have to avoid polluting streams with drainage from flushed tanks or waste from cleaning or washing vehicles; they will have to consider, along with growth, development, profits and taxes, the building of terminals and facilities that do not detract from the landscape. External pressures on industrial fleets are great, indeed, and only the carefully managed businesses can expect to survive.

The common carrier principle, whereby regulated transportation companies must carry people and/or goods at published fares on established schedules, requires the servicing of unprofitable as well as profitable markets. As a result of circumvention of the common carrier principle by shippers and carriers alike, some suggest that deregulation or a superforwarder system (special companies acting as travel agents for the industry) may be the only solutions.[6,9]

DRIVER SELECTION

Sophisticated driver selection and training are the first steps not only in accident prevention but in the operation of a profitable business. Although one cannot establish successful driver screening and instructional methods overnight, there are available sound guidelines and scientific methods to follow.

Specifically, there are (1) tough screening rules and regulations mandated by the U.S. Department of Transportation, (2) research findings in driver task analysis and in identifying the traits of safe drivers, (3) more effective application blanks, (4) a considerable battery of psychological and skill tests, (5) standard driving history checking methods, (6) a better knowledge of what to look for in physical examinations, (7) more refined interviewing methods, and (8) a storehouse of training materials and methods.

Thus, a sizable body of knowledge has become available containing research findings, success and failure experiences, statistical techniques, workable tools, wisdom and judgment which can be utilized in fleet safety management.

Bureau of Motor Carrier Safety Rules and Regulations, Parts 391 and 392

Under the Motor Carrier Act of 1935, the Interstate Commerce Commission (which had been created by act of Congress in 1887) was given an entire new bureau with both broad economic and safety powers over motor carrier operations. Rules were established governing (among other matters) driver qualifications, accident reporting, hours of service for drivers, inspection, maintenance, and the transportation and handling of explosives and dangerous articles. In 1966, the Interstate Commerce Commission's safety authority and responsibility were transferred to the U.S. Department of Transportation.

The Department of Transportation rules, parts 391 and 392, pertain to the driving of commercial motor vehicles operating in interstate or foreign commerce.

RESEARCH FINDINGS IN DRIVING TASK ANALYSIS

In recent years, driver educators and administrators have tended to reach the somewhat embarrassing conclusion that many driver education, licensing, training and rehabilitation programs do not really prepare or improve drivers significantly. Since these programs must be tailored to the attitudes, knowledge, skills and habits required for safe and effective driving, one must know in operational and performance language just what *is* required in accident-free driving. This requires that the driving task itself be far better understood.

Thanks to studies sponsored some years ago by the Accident Prevention Division of the Public Health Service and to research funded by the National Highway Safety Bureau, there is some agreement as to the parameters of the driving task.[44] Work is proceeding toward computer-sorted, expert judgment lists of proven safe driving practices, characteristics of good driver performance and driver capability.[53] This amounts to a reduction of drivers' tasks into component behaviors and a criticality evaluation, whereby each aspect of behavior is assessed in terms of its importance to safe and efficient driving. The goal is a scientific and systematic analysis of how one moves into, within and out of traffic, doing so safely and with minimum disturbance to the flow of traffic and to other drivers.

In summarizing what is known about the human factors required in driving, logical approaches to basic fleet management problems will become apparent, especially in the selection and training areas.

The driver is a critical part of a man/machine/environment system where there are Information-Receiving, Decision-Making, and Driver-Action factors — see, think and do.

1. *Information-Receiving.* One must think in terms of the variety of stimuli which reach the human organism of all the events external to the driver to which he must attend. The visual, auditory and kinesthetic senses are the more important ones in driving.

Driver vision includes five areas of concern, in order of importance: the active search for clear fields ahead; a check for adequate maneuvering room on the sides; an awareness of being overtaken from the rear (except when backing, where the "field ahead" is again functional); observance of visual displays within the vehicle; and signs, signals and other communications from the road and other drivers.

The process of perception stands somewhere between sensing and thinking; it employs both the sensations aroused by stimuli and the experiences of the past. It is, therefore, influenced by needs, values, personality and intelligence, as well as by the nature of the stimulus, its movement, color form, tone, loudness, acceleration, location, etc. And perception involves some degree of interpretation, integration, and organization which is immediately prior to, if not actually a part of, decision-making.

2. *Decision-Making.* Information processing (or thinking) and driver judgment are the mental processes that are involved here. This area is somewhat arbitrarily named; it could just as well be called problem-solving, which begins with a stimulus situation involving a definition of goals.

Decision-making by a driver can be considered in terms of: (a) the likelihood that certain events will occur so that one has a choice under conditions of certainty, risk and uncertainty, (b) the recognition of the number and kinds of choices open and the elimination of all but one, (c) the rate of speed at which it is possible to make successive decisions, (d) the methods by which the driver evaluates his predispositions toward certain classes of decisions (e.g., cautious or risky), and (e) the adoption of a plan.

3. *Driver-Action.* Driver behavior is sometimes almost entirely the result of decisions made, while at other times driver actions are immediate reactions to stimuli and are almost simultaneous with perception. Whether automatic, judgmental or somewhere in between, the driver's control actions primarily have to do with (a) steering or changing course (lane changes or turning), (b) speed regulation (accelerating, maintaining speed or slowing), (c) stopping (panic, quick or regulated), (d) emergency and recovery control (not necessarily having to do with other actions listed here), and (e) signaling (by lights, arm and hand, or horn). Furthermore, these actions may be carried out smoothly or erratically, correctly or incorrectly, singly or in combination, and with success or failure — all are dimensions which can be scaled.

Dilemma: Thinking Versus Habit. When a roadway is relatively clear and familiar, the driver is experienced, and the distraction factors are minimal, then the driving task itself is well within man's capabilities. But if traffic density, weather conditions, roadway complexity and signal control devices are all "bad," then the motorist is wisely urged to "concentrate" on his driving.

One might usefully compare the driving task to walking or handwriting. In learning to walk and to write, the child proceeds through a series of difficult, sometimes painful steps. Once mastered, these abilities become, for the most part, automatic. Only when the walker is confronted with another walker, neither of whom is sure what the other is going to do, do difficulties arise.

Locomotion around obstacles or in the midst of other walkers presents the unexpected — so sometimes the walker runs into other walkers and other objects. After handwriting has been "overlearned," the writer becomes perplexed about problems incidental to writing itself — spelling, capitalization and punctuation.

In driving, too, the novice has considerable difficulty in the tasks which later become largely automatic and smoothly coordinated. Only when he is suddenly confronted with problems outside the usual situation does he have difficulty and sometimes fail. Driving along a road is automatic for the most part, except for emergency conditions or unexpected events.[32]

It is useful to view the driving task as composed of three basic subtasks and their fundamental parts.[44] The most basic of these is the general driving task, *guidance*. This considers the driver's lateral control; keeping the car on the road in proper lane position at a specified velocity (longitudinal control), and the correct speed for conditions. The task by itself requires the driver's awareness of only a limited environment; it is necessary for him to scan visually only a median-to-shoulder, bumper-to-horizon area. However, he must monitor this environment continuously for it is constantly changing.

The second group of subtasks are *situational* — they involve the driver's reacting, through his vehicle, to conditions imposed by traffic, roadway characteristics and the environment, that is, to factors beyond the driver's control. Open-road driving, following, passing, merging and overtaking are more than a matter of lateral and longitudinal control on one's own vehicle.

The final subtask may be labeled *subsidiary*, for it is concerned with such factors as route-following, sign-reading and the operation of the vehicle's accessories. A driver must control his vehicle according to roadway limitations and acceptable rules of operation to arrive safely at his destination.

To summarize, safe driving requires maintaining a vital "safety cushion," or space between one's own vehicle and other objects in the environment. To do this, the driver must be able to *sense* and *identify* the objects and/or events which are important. As he detects and identifies those things which are important to driving, he must *interpret* this information and *make decisions* for the control of his vehicle. And once he has decided what control action is necessary, he must make the required *driver response* and do so with full awareness of his vehicle's interaction with other vehicles. As one initiates *control action*, the vehicle responds to command, according to the kind and condition of the vehicle and the nature of the roadway. A trailer-truck combination will certainly not behave like a parcel truck or handle like a passenger car. And a family car is different from a sports car.

Elementary as these points may seem,[26] they are highly relevant to fleet driver training. Driver selection procedures, too, should be built realistically around the actual driving task and tailored to the type of vehicle and to the kind of driving.

Research Findings Identifying the Traits of Safe Drivers

The great majority of drivers of passenger cars and commercial vehicles seldom become research subjects. Driver research on human variables and characteristics has focused on, or used as study models, drivers who have attracted attention in one way or another. They include at one extreme such groups as accident and/or violation repeaters, drivers involved in fatal accidents and drinking drivers. At the other extreme, a few professional truck or bus drivers who have won awards (and even, on occasion, racing drivers) have been studied. The adjective which best describes these groups is "exceptional." This is true in the sense that none belongs to the typical driver population — they have all been studied because of some particular driving behavior(s).

These studies provide only hints, rather than firm evidence, because they suffer from one methodological shortcoming or another and are seldom really comparable to one another or additive in nature. Furthermore, in the final analysis, the small number of drivers carefully studied cannot be considered a random sample from the driver categories they purportedly represent.

However, if treated with caution, certain randomly chosen summations are at least suggestive in nature. They tend to fit *accident* and/or *violation-involved drivers* more closely than drivers selected randomly from the population or those with better records. The categories include drivers who have previously failed driver skill tests, had driver's licenses revoked, have been cited for negligent/reckless/drunken driving, or admit to making driver errors and to driving while tired/speeding/being preoccupied/passing on curves/driving too fast for conditions/driving cars requiring repair/following too closely/disregarding stoplights and signs.

They may also have certain personal limitations. They were often in a stress or crisis situation; they were excitable, unstable, impulsive, adventuresome, fatalistic, eccentric or resistant to authority; they were more aggressive, had less control of their hostility and exhibited low tension tolerance.

On the other hand here are a few of the characteristics of exceptionally *safe* drivers who were studied because they had such unusual, *atypical* records.

In *childhood*, safe drivers had fewer fears, less temper and greater shyness, were more apt to be followers rather than leaders, and were less interested in being the center of attention.

In *adolescence*, they seemed to have been less aggressive, had fewer fights, and had more group interests and less truancy than other teenagers.

Family history? There was a smaller percentage of divorce among their parents and they had better relations with their own parents.

Employment record? Safe drivers had longer periods of employment with fewer employers and showed indications of greater dependability, greater job satisfaction, higher workmanship standards, and a preference for outdoor jobs.

Marital adjustment? They had fewer fantasy satisfactions and were more cheerful and faithful individuals.

Inter-personal relationships? They were more conservative, had greater tolerance for, and were more at ease with people, were more receptive to authority and more sensitive to the needs of other people.

In their *hobbies* and avocational pursuits, they were less oriented toward social activity, had fewer satisfactions derived through people, were more inclined to gardening and other home-based hobbies, and tended to be churchgoers.

Health? These individuals seemed to have more complaints and illnesses which, when diagnosed, were probably functional or psychosomatic rather than related to failing health or resulting from an actual physical deficiency. They were more aware of their physical limitations and seemed to make greater use of compensations.

Driving habits? As one might expect, these individuals seemed to be somewhat compulsive about safety. They were less easily distracted, more courteous, more concerned with the condition of their vehicle, less likely to believe that accidents are due to luck, and more realistically aware of the extent to which driver, vehicle and highway contribute to accidents; and they probably would maneuver their vehicles more skillfully.

In summary, very safe drivers may be less interested in betting or in risk-taking, less sensitive to criticism, more concerned over their future, and better adjusted to routine and to discipline.

They are more even-tempered people who accept responsibility, are more tolerant of tension and are more in control of their own hostilities.

However, these apparently differential traits of safe drivers are at best hypotheses to be tested in application blanks, interviews and on-the-job performance rather than accepted as guidelines or recommendations for selection. The research which produced these data on exceptional drivers did not control for mileage and exposure to traffic hazards. These measures would have to be equalized for "good" and "bad" drivers in order for biographical characteristics to be predictive of collision involvement.

MORE SOPHISTICATED APPLICATION BLANKS

One can only judge the future by the past, and the best predictor of driving success is a person's demonstrated capacity in a similar job in the past. If one establishes a research program to determine which biographical characteristics objectively relate to job success in specific driving situations and then devises a scoring system, he will have a weighted application blank with validity for driver selection in his own operation. Ordinarily, he would need a consultant or professor of behavioral science to help in this research.

Even without a statistical study to determine the relevance of selected aspects of personal history to a driving job, a carefully designed application blank can be of value in selection, especially when used in combination with an interview.

The easiest way to construct one's own application blank is to follow the lead of associations established to assist specialized fleets, for example, the American Trucking Associations (ATA), the National Association of Motor Bus Owners (NAMBO), the National Association of Fleet Administrators (NAFA), the National Tank Truck Carriers (NTTC), or the American Transit Association (ATA). Another method is to examine application blanks used by other companies and include items that seem logical and appropriate based on one's own driver job description(s).

An application blank can be shortened considerably by dropping items which do not meet these criteria: Is the item necessary to identify the applicant? Is this the proper medium for gathering this information and will it really be used in selection at all? Is the information in keeping with state and federal laws? Is it probable that the applicant's answers will be honest and reliable?

It is a safe bet that no more than half of the application blanks in use really fit the needs of the organization depending upon them, and only half of these are creatively used by those persons responsible for hiring. A local college or university department of psychology or business administration could help considerably in customizing one's selection tools.

Unless there are good reasons for gathering materials that reflect the applicant's grammar, vocabulary and sentence structure (or even his handwriting), it is best to have a well-organized, relatively-simple, short application blank that either provides spaces for short answers or is set up as an autobiographical questionnaire with multiple choice answers. This structured or patterned format will lend itself equally well to statistical analysis and weighting, overlay keys for objective analysis or as background data that can be quickly examined for use in the selection interview.

The Motor Carrier Regulations specify the information required in applications for employment with interstate commercial carriers. This includes the usual identifying data, previous employers, the issuing state, number and expiration date of each unexpired motor vehicle operator's license, a list of all motor vehicle accidents and violations during the previous three years, and a "statement setting forth in detail the facts and circumstances of any denial, revocation or suspension of any license." The following is also required:

> The nature and extent of the applicant's experience in the operation of motor vehicles, including the type of equipment (such as buses, trucks, truck-tractors, semi-trailers, full trailers, and pole trailers) which he has operated (391.21).

PSYCHOLOGICAL, PSYCHOPHYSICAL, KNOWLEDGE AND SKILL TESTS

Testing is probably the least understood and most abused area in selection. Too many people treat psychological tests either as magic, mysterious devices for accomplishing miracles, or as something to be damned and condemned out of hand. The truth lies between, and a selection system (including tests) which is relevant to the driving needs of one's own organization (or at least pertinent to that situation) will pay off. Paper and pencil tests that have been validated for a specific setting are inexpensive tools indeed.[64] However, driving jobs, conditions and labor pools differ enough from place to place that serious errors can be made in transferring techniques that work in one situation to another setting for which they were not developed.

It is, of course, possible that one may find oneself in a situation where tests are not used and where the selection process is deemed successful as judged by small labor turnover, business growth and profits, organizational climate and employee job satisfaction. But if such criteria of successful selection (and effective management) are not present, one should seriously consider the installation of a testing program. Two factors, however, should be present:[2,5] a situation in which there are more applicants than vacancies, and a clearly defined criterion — a carefully written and realistic specification as to what the job duties are. Then, one can try out personality inventories and aptitude tests that have proved useful in the past in similar settings.[2,4] If the situation permits, through the help of a consultant and the application of sophisticated statistical methods, one can determine which psychological tests work and how best to apply their results.

Driving a motor vehicle is a highly complex task, involving as it does many facets of behavior, including learning ability, attention, emotional control, psychophysical judgment, sensory motor coordination and reaction time. It is obvious why researchers and personnel selection and safety supervisors have tried to make use of psychophysical, knowledge and skill tests in the selection of drivers. If one takes the general research literature on psychomotor skills during only a three-year period, there were at least 45 studies dealing with aiming, control, coordination, dexterity, reaction time, speed and steadiness.[5,6] In the more specific driving area itself, there were at least 31 skill studies in the space of ten years.[3] This work has the potential of helping develop an understanding of what constitutes good driving performance, and how to develop and maintain it.

Unfortunately, however, for the immediate needs of fleet people, sophisticated simulators,[3,6] instrumented cars,[2,8] dynamic vision-testing devices,[1,2] and precise data gathering methods[1,8] are far too expensive to be practical in employee selection.

The new Motor Carrier Safety Regulations require a written examination of at least 30 questions selected from a list of 100 true-false items provided to test the applicant's knowledge of parts 390-397 of the BMC Safety Regulations. It is up to the carrier to construct additional questions relating to its own driving jobs, the equipment used, routes followed, passengers or goods carried, required maintenance knowledge and other important duties (customer relations, loading requirements, roadway conditions, etc.).

The most important skill test is the one given behind-the-wheel, which is necessary to enable the job applicant to demonstrate and be evaluated on his ability to handle a vehicle or combination under actual traffic conditions.

> The road test must be of sufficient duration to enable the person who gives it to evaluate the skill of the person who takes it at handling the motor vehicle and associated equipment, that the motor carrier intends to assign to him (391.31)

These Motor Carrier Rules also specify the minimum operations required of the commercial driver-applicant, including pre-trip inspection, coupling and uncoupling (if appropriate), use of vehicle controls (including brakes), turning, backing, parking, slowing by means other than braking, and operating in traffic.

STANDARD HISTORY CHECKING METHODS

Even the honest applicant will tend to present himself in the most favorable light; even though he may make no conscious attempt to falsify, he omits facts in his background having to do with failures, legal problems or negative character traits.

However, it is possible to confirm or deny much of the data gathered through application blanks and interviews by follow-up telephone or mail contacts with previous employers, associates, schools and banks. When handled by organizations which specialize in such work, these follow-ups often provide much broader information than financial condition, and although the service takes time and may seem costly, it is often worth the investment.

Minimally, as a matter of common sense, one should contact recent employers and/or examine school transcripts, military discharge papers and "to whom it may concern" letters. The U.S. Department of Transportation regulation requires a number of investigations and inquiries. These include:

> An inquiry into the driver's employment record during the preceding 3 years to the appropriate agency of every state in which the driver held a motor vehicle operator's license . . . (and) investigation of the driver's

employment record during the preceding 3 years within 30 days of the date his employment begins (391.23)

A copy of the response by each state agency and a written record with respect to each past employer who was contacted must be retained in the motor carrier's driver qualification file.

A state-by-state directory for fleet management entitled "How and Where to Check Driving Records," detailing whom to contact and how to obtain a record of violations, accidents and licensing information, is available from the Department of Safety of the American Trucking Associations (2).

WHAT TO LOOK FOR IN PHYSICAL EXAMINATIONS

Physical examinations are a standard procedure in most businesses and industries, since they not only provide a legal safeguard but also determine the applicant's ability to meet physical job requirements.

The new Motor Carrier Safety Regulations are highly detailed concerning the physical qualifications for drivers of commercial vehicles operating in interstate or foreign commerce. They specifically define what is acceptable and provide a model of the medical form to be used by the examining physician (391.41 through 391.49).

If a fleet is not engaged in interstate or foreign commerce, the carrier is not legally bound by these rules, but they are useful guidelines for hiring drivers. For example, the rules on limb loss and physical impairments should be reason for thought; the remarks about blood pressure, diabetes, hearing loss, vascular diseases and neurological signs should be of serious concern; and the test on visual acuity, drugs and alcohol should constitute minimal requirements even when one is not legally required to conform. Snellen (value) visual acuity of at least 20/40 in each eye, with or without corrective lenses (contact lenses are permitted only if there is medical proof the driver tolerates them and he has a spare set of lenses); field of vision of at least 70 degrees; and the ability to recognize the colors of traffic signals should be required unless a fleet or the state(s) in which it operates has higher standards. The above are static vision test standards; when dynamic vision testing becomes practical, minimum visual standards for recognition of objects in motion (which are much more relevant to driving) should certainly be adopted.[1,3]

There should be immediate disqualification of a driver applicant who uses amphetamines, narcotics or other habit-forming drugs or who has a current clinical diagnosis of alcoholism.

MORE REFINED INTERVIEWING METHODS

In proper hands, the interview can be one of the most powerful screening and selection tools, but it can also provide the opportunity for the injection of bias, prejudice, ignorance and poor judgment. A properly handled interview can also begin to condition the applicant for a lifetime of respect for and loyalty to the organization that hires him.

Interviews can be used to gather information, to give information and to modify behavior. The major goal of interviews in the selection process is to gather relevant information which cannot be more expeditiously obtained in some other way, and to digest and evaluate all information available. Information from application blanks, previous employers, medical examinations, psychological, knowledge and skill tests and the interview itself must be analyzed to make a selection. The interviewer should have a plan, knowing in advance what information he wants to get, how to go about getting it and how to control the interview so he learns what he needs in a limited time.

The best results seem to come when the interviewer studies as much information as possible about the applicant prior to the interview,[41] makes use of a patterned interview format which provides a standardized situation for all applicants,[20] and permits the applicant sufficient freedom to select and volunteer information he feels is related to the job requirements and his own experience. When the interviewer follows an established procedure with each applicant, he can not only record more meaningful information but can compare the applicants more objectively. What the interviewer must do is reach a rational decision to hire or reject[71] based on valid applicant data related to job requirements. In contrast, however, it is only fair to mention that some experts believe the interview is useful only for an initial screening-out of obvious misfits. They would limit themselves to these appraisal techniques: personal history data (via application blanks, references and contact reports), medical examinations and objective tests (knowledge, skill and personality measures).

TRAINING

Experienced fleet specialists would readily agree that the position of driver trainer is a highly specialized, very important, full-time job with a tremendous amount of responsibility. They would further acknowledge that a trainer should be not only a masterful driver but a knowledgeable person concerning essential motivational, educational and communication skills, who must be given wide authority to back up his major responsibilities.

Fleet Training is Rare

Few fleets have a full-time driver training program because most managements do not believe they can justify the expense. It is important to realize, however, that accident reduction is only one of the objectives of a good training program; others include the prevention of painful and expensive injuries, the reduction of vehicle costs per mile, improved customer and public relations, and an increase in profits.

Surveys have shown that fewer than half the fleets have training programs for new employees and fewer than one-fourth have remedial, refresher and rehabilitative training. There are a number of professional driver training schools (some of high quality, such as the one at North Carolina State College), and there are frequent training courses and institutes sponsored throughout the country by the National Committee for Motor Fleet Supervisor Training, headquartered at Pennsylvania State University, which can benefit small fleets. In large truck and bus fleets and big transit systems, there are some excellent, showcase programs, such as those of Ryder, Greyhound, McLean Trucking and a number of other truck fleets; Montreal, Toronto, Chicago and the maintenance training academy in New York City are just a few of the cities with good programs in the transit area.[35,65]

Essentially, there are four ways to set up a program to orient and train one's newly-hired drivers: (1) selecting one's best driver and sending him to school for an intensive, short course in professional driving and instructor training (or recruiting such a man), so he may teach what he has learned, (2) contracting with a consultant organization, which comes in with a package or custom-built program, (3) developing and staffing a corporate driver training school (perhaps through a combination of methods 1 and 2 above), and (4) using on-the-job training and an extensive student trip under the supervision of one of the employer's best drivers during the probationary period.

DRIVER REHABILITATION

All of the above methods are also applicable to training for new, different and specialized equipment, for refresher training (to update) and for remedial, rehabilitative or retraining (to salvage drivers in trouble). This driver improvement training may be combined with other techniques including disciplinary interviews, clinics, case studies, group discussion meetings and role-playing.

Driver rehabilitation works best if it combines three phases:

1. *Assessment and diagnosis* — determining driver patterns, habits and limitations which might be related to accident and violation behavior.

2. *Remedial programs* tailored to the specific needs of drivers — giving skill training on-the-road to those who need it. Giving others direct council which acquaints them with their strengths and weaknesses and teaches compensations for psychophysical limitations. Providing re-education to some in the form of information about updated federal, state and company rules and regulations, "defensive driving" and functional traffic knowledge. Counselling drivers with attitude problems through individual or group work, which leaves each driver with new insights into his own behavior and more positive feelings toward law enforcement, traffic laws and safety procedures.

3. *Evaluation and research* — collecting, studying and utilizing driver performance data in a way that will help determine what works and how well. One can modify procedures and systems for more effective rehabilitation.[33]

Assessment and diagnosis should include interviews, driver records analysis, and psychophysical testing to enable the driver trainer or supervisor to pinpoint, so far as is possible, the reasons for the driver's accidents, violations, vehicle abuse, rule violation, etc. Often a pattern will be discovered which points to the reasons for the driver's failures and permits successful "treatment."

Psychophysical testing is primarily useful in driver diagnosis, education and training. If a comprehensive inventory of a driver's relevant personal characteristics is to be made, one or more tests are essential for determining visual acuity, depth judgment and color recognition.

A well-planned battery of tests can be quite useful in making a driver aware of his strengths and weaknesses so that remedial training, new habits, compensations and correction can be implemented and practiced. Recent research has suggested that some aspects of vision relate to accidents and violations, while others do not. Dynamic visual acuity shows a consistent relationship[12],[13] followed by static acuity, glare recovery and visual field. Glare threshold and phoria (eye muscle balance) do not show a consistent relationship to accidents, though they may turn out to be related to specific accident types.[13] Visual field shows deterioration over short periods of time, suggesting a need for frequent screening.[14]

DRIVER IMPROVEMENT SYSTEMS

Driver improvement can be built around defensive driving materials (such as those offered by the National Safety Council) which include a driving code on avoiding preventable accidents: observing all traffic rules, alertness to driving errors of others, adjusting driving to changing conditions and hazards, yielding right-of-way to avoid accidents, and driving with an attitude of confidence.

Retraining can also make excellent use of the Smith System, which includes five seeing rules: aim high in steering; get the big picture; keep your eyes moving; make sure they see you, and leave yourself an "out."

Strategic driving[38],[66] can also be the focus of a rehabilitation program which includes the PACE of driving (Preparation, Anticipation, Communication and Emergency).

Strategic driving begins with planning. *Preparation* means selection of the best route; determining in advance where to eat, sleep or "take a break"; obtaining marked maps and learning the location and extent of road repairs which might require detours. The strategic driver also prepares himself and his vehicle to insure maximum driving efficiency. His plans include time allowances for interruption and delay, and he has an alternative plan for every operation.

Anticipation is the broadest aspect of strategic driving. It concerns the many actions and expectations upon which the motorist makes judgments and decisions as he drives; preventive and defensive behavior to prevent an accident situation from arising and to avoid the situation if it does arise.

Communication includes the essential non-verbal signs and signals used by a driver to "tell" others that he will turn left or turn right or stop, the road and traffic signals to show drivers what to expect, and the vehicle noises and sounds that alert the motorist to his car condition, road adhesion and responsiveness.

Emergency (or sudden danger) represents the fourth element of strategic driving. Sudden dangers require three kinds of action knowledge: (1) learning to *leave an "out"* to help avoid a serious accident situation if it suddenly develops; (2) learning *what to do if* suddenly confronted by the threat of an accident; and (3) *practicing* the various evasive behaviors that can help avert an accident.

It is extremely important for a motorist to practice imagining various kinds of sudden-danger situations, and to think of what could be done to avoid an accident or minimize damage if such a situation should occur. Then, if it does occur, his responses are not determined wholly by what confronts him but also by the time he has spent preparing himself to act in that particular situation.

Considerable space has been devoted to driver improvement activities, since it often is difficult to obtain a training budget, at which time some energies are devoted to helping the driver and salvaging the employee. The sad reality is that treatment (rather than prevention) is too often emphasized in fleet policies.

HOW TO ESTABLISH TRAINING PROGRAMS

Some advice is necessary for those fleets large and/or foresighted enough to have the funds for initial training. Training is no longer an art, but rather is a technology[4,5] where it is possible to determine clearly WHAT we want to achieve, WHY we want to do it, WHO will do it, HOW to write meaningful objectives, and WHERE and WHEN to implement these objectives via appropriate lessons and materials.[4] It is also possible to evaluate how well the objectives are achieved and how to determine when there is a new training need.[1,6]

What? The answer is highlighted by an acronym, *A.S.K.* (*A*ttitude, *S*kill and *K*nowledge). *Attitude* is the driver's interest in being alert, his motivation for decency and courtesy, his application to the driving task in control of his vehicle. *Skill* includes his perceptual talents, his coordinating capacities, his visual search patterns and other functional driving habits. *Knowledge* refers to relevant insights concerning one's own state of physical and mental health and to the information a driver possesses, state driver regulations, federal and company rules. All these the driver must translate into meaningful, dutiful behavior behind the wheel.

Each of these basic parts of the *A.S.K.* must be made more functional for training by writing objectives for the course and by carefully planning the performance or end result that trainees are expected to achieve. The statement of objectives is the key document, the blueprint, the description of the final goals to be reached.[46] These objectives can be written more readily and are more likely to be achieved if they are built upon behaviors required in the driving task.

Topics which should be covered in the classroom include orientation (vehicle problems, traffic regulations, company policies), human factors (causes of accidents and violations), safety rules and procedures, vehicle inspection, freight or passenger handling, accident and breakdown reporting and other necessary paperwork. Field experience and practical training exercises should include terminal operations, fire-fighting and vehicle maintenance. Behind-the-wheel training should cover all driving requirements, including basic maneuvers, turning, backing, stopping and negotiating in traffic.

Why? In all truck and bus operations there are schedules to meet. Time is precious. Therefore, to conserve time and make for greater efficiency, "selective training"[67] which concentrates on exactly what is needed by each driver can be given to meet those specific requirements.

Who? Training should be carried out by an experienced, professional driver with a spotless driving record,[42] who is thoroughly familiar with the company's equipment, routes and policies. He should be empathetic, get along well with other drivers and be able to teach in the classroom as well as demonstrate, observe, correct and train behind the wheel. When this superman cannot be found, the employer should recruit or upgrade the best man available, pay his expense to a top-flight driving school (or at least an instructor's course), and give him up-to-date facilities, full authority and complete cooperation. If it is necessary to settle for less, make sure the trainer has a well-studied copy of the American Trucking Associations' *Truck Driver Training.*[17]

How? It is not necessary that the latest programming techniques or teaching machines[43] be used, to have effective classroom learning, or that simulators be used prior to entrusting personnel with expensive equipment. However, it is essential that realistic course objectives be prepared in advance, as suggested under "WHAT?" above, and that techniques, methods and training aids be appropriate to the educational goals. Also, one must use techniques that are functional and relevant to the required behavior. If a driver is expected to be able to trouble-shoot and find a blown fuse, instruct him so he can perform, not so he can write an essay on "the steps required to locate a burned fuse."

How long? It is not possible to determine in advance how long training should last, although some excellent guidelines are available which can be used as starting points. High school driver education for teenagers varies, from 30 to 60 hours of classroom and four to six hours of simulation, besides 12 hours of behind-the-wheel training. Commercial school training (e.g., Link Driving Centers) calls for four hours of classroom, six hours of simulation and six hours (minimum) of behind-the-wheel training. Professional truck driver training varies from four to 60 hours of full-time training. Bus companies give licensed drivers three to ten days of classroom, followed by five to 27 days of on-the-road training. The actual number of training hours will be determined by the level of knowledge, skill and experience of the personnel hired, the efficiency of the training program itself and the complexity of the driving job in the particular company. The very nature of the job is also important. Trucks haul cargo; buses carry people.

Where? The most effective training is probably that given by fleets with specially-designed training classrooms, custom-made mock-ups or simulators, off-street driving ranges and carefully laid-out routes through traffic for advanced behind-the-wheel training. However, no fleet has all of these, and it is necessary to compensate for the absence of or a limitation within any of these facilities by additional coverage in the area for which there is adequate equipment and training materials.

When? Driver training should begin when the man is hired, continue in some form throughout his period of employment and end only when he is promoted to a non-driving job or leaves the company. If a company does not have initial or continuous training, or automatically-initiated remedial training programs, then it should begin training whenever there is evidence of a training need.

The analysis and determination of a company's overall training needs can be an involved and difficult task. However, a fleet's specific need for driver training is much easier to determine if a standard of performance for the job has been agreed upon. When a statement has been written describing the conditions which must exist when the job is properly carried out, it is possible to compare what *is* going on with what *should* go on. If there is a gap between the two, there is a training need, and there should also be clues as to the kind and amount of training required.

Comparisons of a company's drivers' performance with those of comparable fleets is another way of deciding when there is a training need. To oversimplify, if one's accident rate is 6.00 per 500,000 vehicle-miles, while one's competitors have an average rate of 4.00, one presumably has a need for training. However, one would not initially know whether the biggest need was for supervisor training, driver skill, attitude or knowledge training, or maintenance training.

This decision would require further evaluative efforts, including analysis of behavior, critical incidents, equipment failures, organizational structure and climate, performance records, etc. and the use of some of the following techniques: checklists, interviews, questionnaires, task forces, tests, workshops and careful observation.[16] Logical analyses and careful thought will suggest what additional knowledge, understanding, attitude or behavior changes a group or individual may need to be more productive, thus bringing the matter full-circle to the initial *"WHAT?"* of training.

VEHICLE INSPECTION AND MAINTENANCE

Here is a typical advertisement for a "fleet manager, maintenance." While it is no substitute for the development of a job description by each specific organization, it a representative (though brief) example of what is required in this position.

Fleet Manager — Maintenance

Manager to supervise personnel, administer multi-million-dollar budget, and upgrade the maintenance and service operation of a large heavy-duty vehicle fleet.

Candidate interested in this challenging job should have experience in maintenance of mobile equipment or aircraft and a familiarity with union negotiations. Should be able to develop and implement new policies. A combination of managerial experience and plant maintenance experience would be ideal.

Vehicle inspection is basic to adequate maintenance and safe driving. It is obvious that a driver should visually inspect his vehicle before he takes it out. Not so evident is the fact that a driver's attitude and emotional state of safety awareness begins with proper preparation for a trip, and manifests itself in an intelligent examination of critical systems: tires, lights, brakes, mirrors, steering, wipers and warning devices/emergency equipment, for example. Methodical pre-departure checks (both visual and operational) can reveal unsatisfactory conditions which must be remedied before the vehicle departs. The bureau of Motor Carriers has worked with the American Trucking Associations and the Nation Association of Motor Bus Operators on the preparation of driver pre-trip vehicle inspection manuals.

If the garage or maintenance department has the major responsibility for inspection, the driver should receive a copy of the checklist for his own additions during the trip as part of his daily report. En route inspections (especially operational) should also be made by the driver, and thorough and conscientious vehicle defect reports made to his superior or to the terminal manager.

Inspection and maintenance recordkeeping is an essential part of any program. Minimally necessary is information identifying the vehicle (including specifications), driver reports on vehicle condition, inspection dates and details, lubrication events, and repair details with dates.

When a vehicle is assigned to a driver and comparative records of efficiency, abuse, economy of operation and safety are kept, there is likely to be an increase in driver responsibility and satisfaction, as well as a reduction in fleet operating costs.

Some of the very best advice appears in the trade journals. It pays to subscribe to and regularly skim newspapers like *Automotive News* and *Fleet Management News*, and magazines like *Fleet Owner* and *Commercial Car Journal*. For example, an article on diesel fuel injection pumps pointed out that a fleet that is using diesels for pick-up and delivery operations as well as over-the-road business might be damaging one of its fleets if it is using only one grade of fuel in both. Furthermore, extra care should be taken to make sure that the equipment is precisely geared for the type of operation and the terrain on which it will be used. Driver training geared to the particular vehicles being driven is important, since the operation of diesel engines puts a premium upon the maintenance of engine speed within rather narrow limits for optimum performance and durability.

A 50-item checklist for terminal managers[5] demonstrates ways of eliminating damage to trailer landing gears, roll-up doors and "hoses." Many of the hints given would not be ordinarily thought of as maintenance matters, but as guides for loading, tarpaulin storage, uncoupling trailers, distributing and securing freight, use of ether in cold weather, snow removal, proper methods of getting into and out of cabs, proper towing of disabled tractors, etc.

The development of a well-rounded maintenance program is essential, since accidents and downtime due to vehicle failure represent management at its worst. A balanced maintenance program adapts itself to a particular operation and to changing conditions. The essential elements include adequate facilities, scientific diagnostic methods, proper personnel and equipment, standard operating procedures and comprehensive recordkeeping.

The maintenance practices of four groups of drivers should be of interest. These data were collected in a large questionnaire study completed in early 1970 for the Department of Transportation's National Highway Safety Bureau.[3,4]

Four-page questionnaires were mailed to a systematic random sample of 58,344, drawn from the polulation of 47,065,425 owners of American passenger cars registered in the 50 states and manufactured within the years 1958-1968. Of the 17,024 returns, 14,503 were complete enough to be used in the data reduction and analysis and were demonstrated to be representative of the 58,344 sample.

Eighty-three percent of the respondents were male (12,027 of 14,503 cases) and they reported their feelings about governmental compulsory vehicle inspection and regular maintenance servicing.

Specially-designed questionnaires were also mailed to a random sample of 2500 truck owners and 2500 motorcyclists. For truck drivers, 175 usable questionnaires were returned; of these, 13.14 percent came from truck owners and 79.14 percent from truck driver/owners. About 62 percent were replying concerning pick-ups; 13 percent, single-stake bodies; five percent, vans; eight percent, single-closed-body, and three percent, tractor-trailers.

A 24-page questionnaire was distributed to 440 members of the National Association of Fleet Administrators. One hundred and nine of the 124 returns from "fleet managers" or "fleet administrators" were usable. These men managed fleets ranging from 46 to 8,600 vehicles and employed drivers whose average age was 34 and who had spent 8.6 years on the job. Half of these drivers were salesmen driving passenger cars. It is interesting to note that passenger vehicles were normally used for 28 months prior to disposal, while trucks were used for an average of 55 months.

It was found that 58 percent of the fleet managers provided for periodic preventive maintenance, while 30 percent left this to the driver's discretion. In almost all these fleets, monthly maintenance reports were submitted to the fleet administrator or field supervisor. In about 45 percent of the fleets studied, no reports were used. In fact, 28 percent of all the respondents felt many engine tune-ups were unnecessary and 15 percent felt tire rotation was superfluous; however, 19 percent felt there were no driver maintenance practices that should be considered unnecessary. Some 66 percent felt their vehicles were better maintained than those of the general public, and 61 percent felt their vehicles were as well maintained as those in other fleets.

Approximately 25 percent of the fleet managers stated that oil changes, lubrication and fuel changes were most frequently neglected; 23 percent named tire condition; and 19 percent said it was proper tire pressure that was most neglected by drivers. They felt drivers neglected these items through inattention, forgetfulness or ignorance (24%), lack of time (22%) or carelessness (17%).

The large majority of the fleets polled did not operate their own maintenance facilities. Most of their maintenance work was performed at independent garages and service stations, except for warranty repairs which were accomplished by franchised dealers.

Unless the value of a program can be measured, it cannot be improved, and the heart of maintenance evaluation lies in accurate recordkeeping. Just keeping track of the use of oils and fluids for each vehicle can be a difficult chore, but a valuable procedure. Fuel, transmission and differential lubrication, oil, packing in wheel bearings, battery water, antifreeze and hydraulic fluid for vehicles, and chemicals in fire extinguishers are just a few of the fluid items whose use plays an important role in maintenance.

This logic can be carried a step further in suggesting that basic maintenance be categorized as either lubrication or mechanical, and within these as either minor or major. Thus a code system, A (lubrication/minor), B (lubrication/major), C (mechanical/minor), and D (mechanical/major) could be easily translated as: A — lubrication, B — oil and filter change, C — engine tune-up, and D — wheel bearing, steering linkage and brake check. The required frequency of these inspections can be determined only by an examination of other interdependent factors and an answer to the questions of how serious is a part failure and/or breakdown, and how important is dependability.[7,2]

There is not only the consideration of maintaining and servicing the fleet to increase the likelihood of safe vehicles, but there is also the need to develop a maintenance and repair operation that itself is safe. The heavier the vehicle, the more serious the potential injury in its care and repair. Therefore, adequate mechanical aids such as hoists and jacks must be available for lifting parts, to save time, make work easier and prevent accidents. For instance, to insure shop safety, the serviceman, supervisor and driver should know that truck and bus tires must be lifted and moved only with mechanical devices and inflated only in a cage or under conditions where the hazard from a blow-out or blown locking ring will be minimal.

Finally, although one may have skilled drivers and excellent mechanics, emergencies do occur. It is advisable to maintain a current and complete list of vendors who can make emergency repairs to equipment anywhere on a seven-day-week, twenty-four-hour-a-day basis.

SUPERVISION

In a small fleet, the owner is likely to be the fleet manager, the safety director, the terminal manager and the dispatcher. The larger the fleet, the more likely there will be a man in each of these positions. In large fleets, there will, of course, be a number of terminal managers, dispatchers and possibly separate fleet managers of, for example, long-haul and short-haul truck operations, passenger cars and buses.

The Supervisor is a Key Figure

The supervisor in direct contact with employees is a key man who must master the science of getting things done well *through people*. Although a representative of management, he must also be employee-oriented and fully aware of the workers' needs, problems and satisfactions. Fleet administration is different from most other kinds because the driver (employee) is more or less on his own most of the time.

All Men Have Needs

The supervisor should be aware that man is goal-seeking from birth onward, and, as a result of internal drives, takes action to reach his goals. Thus, man is motivated as a result of internally-generated needs or drives. Men need to meet their physiological and survival needs. They want security, adequate livelihood and safety. They also need opportunity for self-expression in their work and some degree of independence. They want to make effective use of their muscular and intellectual equipment. They need fair and reasonably-consistent treatment from others, and they want status and respect (or at least recognition as individuals). Finally, men are social creatures who need satisfying relationships with others.

It is helpful to categorize and rank these human drives into a conceptual hierarchy, beginning with the most primitive and basic drives, and ranging upward to the apex of the hierarchy . . . which is self-actualization. Thus, following Maslow,[48] is a Hierarchy of Needs:

Self-actualization

Esteem

"Belongingness" and Love

Safety

Physiological

The graphic model of steps from the bottom upward underscores the fundamental point that until one's basic need is fulfilled, a man's behavior cannot be motivated by the next higher need on the next level. By the same logic, once a need is satisfied, it no longer motivates. These needs appear in sequence as one grows from infancy to adulthood, and the pattern is repeated many times as man encounters new experiences throughout his adult life. Thus an individual can be characterized as being at an observable level under a given set of circumstances, which may be higher or lower than it was at a given time under other conditions.

Needs are Frustrated in Daily Living

In life and work, and at each need level, there are multiple obstacles to need satisfaction: discomfort, poor health, unemployment; accidents and fears of injury; social class, racial, ethnic, educational or caste barriers to employment; task repetitiveness, low skill requirements and the predominance of tools and techniques . . . all potent obstacles to one's need for self-expression.

Some of these can be categorized as "frustrations" — *real* obstacles to need satisfaction — and others as "anxieties" — *imagined* obstacles to meeting needs. Both produce barriers to the achievement of human needs and organizational goals. Profits, productivity, customer satisfaction and high employee morale are likely under good leadership and impossible under poor management and supervision.

A good manager has a balanced concern for production problems and employee problems. The ideal manager assumes that workers want to be productive and that this can be brought about through the creation of an organizational climate that encourages creativity and participation.[7]

Clues to Bad Management

Evidences of inadequate supervision are everywhere, in fleets of all kinds and sizes and in all geographical locations. Broadly speaking, symptoms of poor management at any and all levels run from low productivity, red-ink profit-and-loss statements and union-management troubles to high turnover and high insurance premiums. There are also a multitude of specific events that alert top management to bad middle management: poor customer relations; lost merchandise; combinations sidelined in a roadblock for equipment reasons; broken-down buses with passengers standing by the side of the road; sloppy dock handling; and dirty terminals and garages.

There are also, on occasion, clues to poor top management. Poor supervision, driver and maintenance problems can only be solved if top management supports safety programs. Its attitudes are critical and its policies have to be positive and consistent. It must be fair but firm in enforcing federal rules and contract provisions[3,7] and it must itself know where it stands on driver discipline.[9]

There are overwhelming reasons why management should want safety, and all of these have to do with money ... not altruism. All safety procedures in the long run are economically sound. There are excessive costs in accidents: public liability and property damage insurance premiums; fire, theft and equipment damage losses; workmen's compensation payments; cargo or personal injury claims; court costs; and the expense of replacing injured employees with inefficient substitutes and training these replacements. However, sometimes these facts have to be sold to top management by a combination of recordkeeping and public relations which includes (1) the generation of a policy statement, (2) the definition of goals, (3) listing the functions required to satisfy the policy and goals, (4) asking management to correct and approve the package, (5) preparing evaluation procedures for measuring achievement, and (6) preparing reports of successes and failures for presentation to, and involving, management.[2,2]

Need for Middle-Management Training

Very few fleets have formal training programs for supervisors, or hire outside training organizations to provide such services. Nor do many fleets have structures upgrading policies, much less management development programs. However, this neglect is slowly changing, and fleets that survive the crunch of the 1970's must have behavioral science inputs,[60] enlightened management[7] and regular supervisory training.

Much can be said for promotion from within, providing seniority and popularity are de-emphasized and some objective and workable employee-rating system is in operation (57). When a pool of experienced men is available, and methods of evaluating job knowledge, leadership, intelligence, motivation and loyalty exist, it is possible to select supervisors wisely and boost morale too.

Of course, outside recruiting may be essential to find the needed combination of knowledge and talent, and to prevent "inbreeding." Experienced men from other companies, freshly-graduated college men and armed services veterans with some previous relevant experience have all worked out well in many companies. Carriers may have to go entirely outside the industry, for there are signs of a declining management personnel pool within fleets.[6]

The most important ingredient is supervisory or management development training for men as they move *up*, whether they have been promoted from within or recruited from outside.

SUMMARY

This chapter has attempted to demonstrate the importance of selection, training, maintenance and supervision in fleet operations. Rather than provide a blueprint for the construction and management of a fleet loss control program, it has tried to involve the reader in research findings, logical principles and practical issues relevant to fleet safety. While simple solutions are seldom offered, the basic issues and appropriate source materials are cited.

After the first reading, one may not automatically reduce accidents, manage more effectively, select better employees or train them more professionally, but one should be more insightful, flexible and knowledgeable ... and more critical of pat answers and panaceas, and more willing to try out and test sophisticated solutions.

Dig out the recommended sources; study and apply the research findings, and do not be afraid to seek out the expert or the professional when necessary, in universities, insurance companies and consulting or training firms.

BIBLIOGRAPHY

1. *Accident Facts, 1973 Edition.* Chicago, Illinois: The National Safety Council, 1973.

2. American Trucking Associations, 1616 P Street, N.W., Washington, D. C.

3. Arthur D. Little, *The State Of The Art of Traffic Safety.* Cambridge, Mass.: prepared for the Automobile Manufacturers Association, 1966.

4. Automotive Safety Foundation. *A Resource Curriculum in Driver and Traffic Safety Education.* Washington, D. C.: ASF, 1970.

5. Bald, J., "Maintenance Checks for Terminal Managers," *Fleet Owner,* 1970, 65 (9), pp. 57-59.

6. Bauman, M. H., as quoted in New Jersey Motor Truck Association *Bulletin,* November 1, 1970, 7 (21), 15.

7. Blake, R. R., and Mouton, Jane S., *The Managerial Grid.* Houston, Texas: Gulf Publishing Company, 1964.

8. Bloomberg, R., "Douglas Toms' Fresh Outlook on Highway Safety," *Fleet Owner,* 1970, 65 (2), pp. 70-71.

9. Buck, D., "How Tough is Your Driver Discipline?" *Commercial Car Journal,* 1967, 113 (5), pp. 93-98.

10. Bureau of Motor Carrier Safety. *Reports* Number 67-2 through 67-14 for 1967; 68-2 through 68-12 for 1968; 69-1 through 69-4 for 1969. Washington, D. C.: U. S. Department of Transportation, Federal Highway Administration.

11. Bureau of Motor Carrier Safety Regulations. *Revised Part 391,* "*Qualifications of Drivers.*" Effective January 1, 1971. MCSR Amendment No. 10, DOT Federal Highway Administration, *Federal Register,* April 22, 1970.

12. Burg, A., "An Investigation of Some Relationships Between Dynamic Visual Acuity, Static Visual Acuity and Driving Record." Report No. 64-18, Department of Engineering, University of California at Los Angeles. April, 1964.

13. Burg, A., "The Relationship Between Vision Test Scores and Driving Record: General Findings." Report No. 67-24, Institute of Transportation and Traffic Engineering in cooperation with Department of Motor Vehicles, USPHS No. AC-00015, State of California, June, 1967.

14. Burg, A., "Vision Test Scores and Driving Record Additional Findings." Report No. 68-27. Institute of Transportation and Traffic Engineering in cooperation with Department of Motor Vehicles, USPHS Grant No. AC-0015, State of California, December, 1968.

15. Conger, J. J., Miller, W. C. and Rainep, R. V., "Effects of Driver Education: The Role of Motivation, Intelligence, Social Class and Exposure." *Traffic Safety Research Review,* 1966, 10, (3), pp. 67-71.

16. Craig, R. L., and Bittel, L. R. (Eds.), *Training and Development Handbook*. (Sponsored by the American Society for Training and Development.) New York: McGraw-Hill, 1967.

17. Darmstadter, N. (Ed.), *Truck Driver Training — A Manual For Driver Trainers*. Washington, D. C.: American Trucking Associations, Inc., 1968.

18. Edwards, Dorothy S., Hahn, C. P., and Fleishman, E. A., *Evaluation of Laboratory Methods for the Study of Driver Behavior: The Relation Between Simulator and Street Performance*. Public Health Service Grant No. 8 R01 0I00695. Washington, D. C.: American Institute for Research, May, 1969.

19. Ernst and Ernst, *Truck Accident Study. Report of Procedures and Findings*. Cleveland, Ohio: Prepared for Automobile Manufacturers Association, Inc., August, 1968.

20. Fear, R. A. and Jordan, B., *Employee Evaluation Manual for Interviewers*. New York: The Psychological Corporation, 1943.

21. Fellmeth, R., *The Ralph Nader Study Group Report on The ICC and Transportation*. New York: Grossman, 1970.

22. Givens, W. E., "How to Interest Top Management in Safety," *Traffic Safety*, September 1970, 70 (9), pp. 20, 21, 36, 37.

23. Goeller, B., "Modeling the Traffic Safety System," *Accident Analysis and Prevention*, 1969, 1 (2), pp. 167-204.

24. Goldstein, L. G., *Research on Human Variables in Safe Motor Vehicle Operation: A Correlation Summary of Predictor Variables and Criterion Measures*. Driver Behavior Research Project, The George Washington University, Washington, D. C., June, 1961.

25. Goldstein, L. G., *Driver Selection — The Lure, Logic and Logistics*. Presented at National Safety Congress, Chicago, October, 1963.

26. Goldstein, L. G., "The 'Case' Against Driver Education," *Journal of Safety Research*, December, 1969, (1), 4, pp. 149-164.

27. Gordon, B., "Jackknife!" *Trucking Business*, 1969, 63 (3), pp. 14-15.

28. Greenshields, B. D., and Platt, F. N., "Development of a Method of Predicting High-Accident and High-Violation Drivers," *Journal of Applied Psychology*, 1967, 51, pp. 205-210.

29. Haddon, W., Jr., "The Changing Approach to the Epidemiology, Prevention, and Amelioration of Trauma: The Transition to Approaches Etiologically Rather Than Descriptively Based," *American Journal of Public Health*, 1968, 58, pp. 1431-1438.

30. Haddon, W., "Why the Issue Is Loss Reduction Rather Than Only Crash Prevention," presented before the Society of Automotive Engineers, Detroit, January, 1970.

31. Henderson, H. L., and Kole, T., "The Contributions of New Jersey Accident Prevention Clinics Through Accident Reduction and Knowledge of Motorist Characteristics," *Aspects Techniques de la Securite Routiere*, December, 1963, 16, 5. 1-5.17.

32. Henderson, H. L., "Designing a Safe Car for the Average Driver," *Traffic Digest and Review*, October, 1967, 15 (10), pp. 9-14.

33. Henderson, H. L., and Kole, T., "New Jersey Driver Improvement Clinics: An Evaluation Study," *Traffic Safety Research Review*, December, 1967, 11 (4), pp. 98-105

34. Henderson, H. L., Dunstone, J. J., Hettich, P. I., Hartsough, W. R., Carter, N. A., and Wanschura, R. G., *Motor Vehicle Owner Maintenance Practices.* Contract FH-11-6939, U.S. Department of Transportation, N.H.S.B., April, 1970, INTEXT, Transportation Research Division, Scranton, Pennsylvania.

35. Henderson, H. L., Carter, N. A., Dalle, G. V., et al., *Training Needs Study.* For the American Transit Association and U.M.T.A., in 1971.

36. Hulbert, S. and Wojcik, C., "Driving Simulator Research," *Driving Simulators and Application of Electronics to Highways.* Highway Research Board, Bulletin 261, 1960.

37. Kerrigan, J. D., "Are You Creating Problem Drivers?" *Commercial Car Journal*, October, 1965, III (10), pp. 90-93.

38. Kole, T. and Henderson, H. L., "Strategic Driving," *Safety Education*, April, 1962, 41 (8), pp. 3-6.

39. Kole, T. and Henderson, H. L., "Cartoon Reaction Scale With Special Reference to Driving Behavior," *Journal of Applied Psychology*, 1966, 50, pp. 311-316.

40. Leavitt, H., *Superhighway—Superhoax.* New York: Doubleday & Co., 1970.

41. Lopez, F. M., Jr., *Personnel Interviewing*, New York: McGraw-Hill, 1965.

42 Lorenzen, E. F. N., *A Study of the Driver Records of California Public Secondary School Driver Instruction Teachers.* Doctoral dissertation (abridged). Automotive Safety Foundation, Washington, D.C., A82200, 1969.

43. Lumsdaine, A. A., and Glaser, R. (Eds.), *Teaching Machines and Programmed Learning.* Washington, D.C.: National Education Association, 1960.

44. Lybrand, W. A., Carlson, G. H., Cleary, P. A., and Bauer, B. H., *A Study on Evaluation of Driver Education.* Development Education and Training Research Institute, The American University, Washington, D.C., Contract FH-11-6594, July, 1968.

45. Mager, R. F., *Preparing Instructional Objectives.* Palo Alto, California: Fearon Publishers, 1962.

46. Mager, R. F., and Beach, K. M., Jr., *Developing Vocational Instruction.* Palo Alto, California: Fearon Publishers, 1967.

47. Malfetti, J. L., and Fine, J. L., "Characteristics of Safe Drivers: A Pilot Study," *Traffic Safety Research Review*, 1962, 6 (3), pp. 3-9.

48. Maslow, A. H., *Motivation and Personality*. New York: Harper and Row, Inc., 1954.

49. Mazelsky, B., *Development of Standards for a Heavy Vehicle Underride Guard*. Aerospace Research Association, California. Contract FH-11-6878, December, 1968.

50. McFarland, R. A., and Moseley, A. L., *Human Factors in Highway Transport Safety*. Boston: Harvard School of Public Health, 1954.

51. McFarland, R., "Psychological and Behavioral Aspects of Automobile Accidents," *Traffic Safety Research Review*, 1968, 12 (3), pp. 71-80.

52. McGuire, F. L., *A Brief Outline of Techniques in Driver Selection*. Beverly Hills, California: Western Psychological Services, 1962.

53. McKnight, A. J., "System Analysis Pinpoints Driver Task," *Traffic Safety*, December, 1970, 70 (12), pp. 16-17, 35-36.

54. Moynihan, D. P., "An Opinion About Traffic Accident Statistics," *Traffic Digest and Review*, 1965, 13 (10), pp. 2-7.

55. "News from Capitol Hill," *Traffic Safety*, October, 1970, 70 (10), p. 37.

56. Nobel, C. E., "The Learning of Psychomotor Skills," in Farnsworth, P.R., et al. (Eds.), *Annual Review of Psychology*. Palo Alto, California: Annual Reviews Inc., 1968.

57. Odiorne, G. S., *Management by Objectives. A System of Managerial Leadership*. New York: Pitman Publishing Corporation, 1965.

58. Preston, C. E., and Harris, S., "Psychology of Drivers in Traffic Accidents," *Journal of Applied Psychology*, 1965, 49, pp. 284-288.

59. Quenault, S. W., "Driver Behavior—Safe and Unsafe Drivers," Parts 1 and 2. *Australian Road Research Board Proceedings*, 1968, 4 (1), pp. 838-881.

60. Rush, H. M. F., *Behavioral Science Concepts and Management Applications*. New York: National Industrial Conference, Inc., 1969.

61. Schuster, D. H., and Guilford, J. P., "The Psychometric Prediction of Problem Drivers," *Traffic Safety Research Review*, 1962, 6 (4), pp. 16-20.

62. Schwenk, Lillian C., *Psychophysical Tests, Their Administration and Interpretation*. Ames, Iowa: Safety Education Laboratory, Iowa State University, Revised October 15, 1969.

63. Shaw, L., "The Practical Use of Projective Personality Tests As Accident Predictors," *Traffic Safety Research Review*, 1965, 9 (2), pp. 34-72.

64. Super, D. E., and Crites, J. O., *Appraising Vocational Fitness By Means Of Psychological Tests*. New York: Harper & Row, 1962.

65. "The 5-T Program: Total Training Today for Tomorrow's Transportation," *Fleet Owner*. A series of ten articles reprinted from the issues of July, 1961, through July, 1962.

66. The Strategy and Tactics Driver Safety Transparency Series. Driver Testing Equipment, INTEXT, 1968.

67. The White Motor Company. *Selection and Training of Drivers: A Manual For Truck Operators.* Cleveland, Ohio, 1964.

68. Tillmann, W. A., and Hobbs, G. E., "The Accident-Prone Automobile Driver: Study of Psychiatric and Social Background," *American Journal Psychiatry*, 1949, 106, pp. 321-331.

69. "Transportation Needs a Drastic Overhaul," *Business Week*, November 14, 1970, 2150, pp. 68-76.

70. Uhlaner, J. E., Goldstein, L. G., and Van Steenberg, N. J., "Road User Characteristics: Development of Criteria of Safe Motor Vehicle Operation," *Highway Research Board Bulletin 60*, National Academy of Sciences, National Research Council, 1952.

71. Webster, E. C., *Decision Making in The Employment Interview.* Montreal, Canada: Industrial Relations Center, McGill University, 1964.

72. Williams, W. E., "The Basics of Good Fleet Maintenance," *Traffic Safety*, 1970, 70 (3), p. 9.

73. Wolf, R. A., "Truck Safety is Being Studied, Too," *SAE Journal*, 1969, 77 (2), pp. 40-43.

XVI. System Safety

The continuing quest for more effective means of conserving our natural and material resources has focused attention on a spin-off of aerospace technology commonly referred to as *System Safety*.

Many experienced safety professionals claim, with some justification, that System Safety is nothing new ... just the basic concepts of good accident prevention wrapped in fancy terminology and aerospace gold plating. The fact remains that System Safety is gaining acceptance as a technical discipline associated with, but independent of, the more traditional fields of industrial safety, industrial hygiene, transportation safety, flight safety, etc.

This chapter describes the concepts, terminology, background and current application of the System Safety discipline. Major attention is given to the methodology and approaches to System Safety analytical techniques which can be adapted for use by safety professionals in a variety of activities.

THE SYSTEM SAFETY CONCEPT

What is it that makes System Safety different? The System Safety Concept envisions a pre-planned, organized effort which has as a primary goal the conservation of resources associated with a given system, product or operation by:

1. The pre-accident identification of potential hazards.

2. The timely incorporation of effective safety-related design and operational specifications, provisions and criteria.

3. The early evaluation of design and procedures for compliance with applicable safety requirements and criteria.

4. The continued surveillance over all safety aspects throughout the total life-span, including disposal.

This probably sounds much like what any experienced safety or loss control specialist would say applies to his function or field activity. What then makes System Safety different from any other safety-related activity?

First, the meaning of the term "system" must be understood by the reader in the same context as it is used herein. Figure 1 provides an illustration of the conceptualized system.

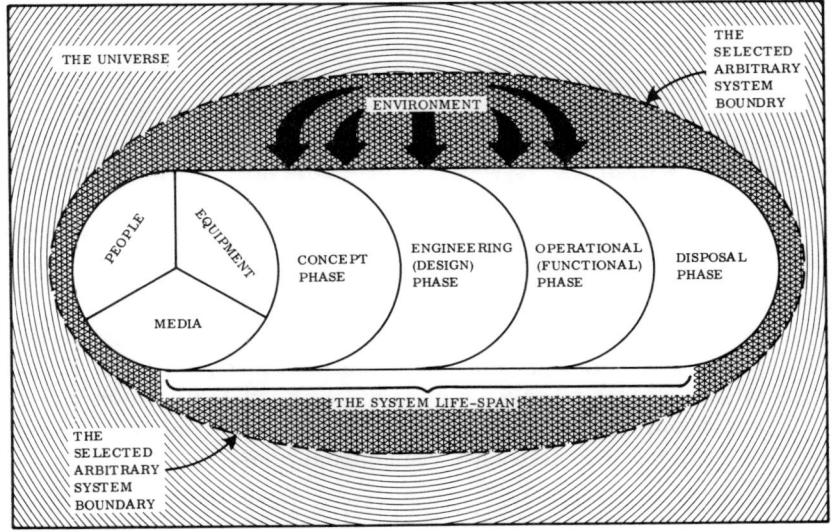

Figure 1. The System Concept

As can be seen in this view, a system refers to a given entity as well as to time.

Basically a system is the sum total of all elements working together within a given environment to achieve a given purpose or mission. Normally these elements include equipment, people and media (procedures, instruction tables, etc.). The dimensions of a system are dependent on the scope of the activity selected. For example, let's examine the brake system of an automobile. The auto itself, the entire U.S. highway system, and the world-wide automotive industry could each be considered.

Note that, in each of these examples, the boundaries at which the system interfaces become uncertain, and must be arbitrarily chosen. In the case of the individual automotive brake system, the mechanical portion of the system can be fairly easily identified when separated from the car, but not so easily after it is placed within an operating vehicle. Consider the driver. Should the condition of the tire treads be included, the road surface, and the weather? If the mission of the brake system is to stop the car within a given distance, then all these other (interface) factors should be considered part of the brake system. However, if the mission is defined only as applying a given amount of friction to the wheel drums, the interface factors can be ignored.

The other essential consideration in defining the specific dimensions of a system is time. Obviously such factors as driver alertness, tire tread, road surface and environment will vary significantly in time. However, these are still only time-factor considerations associated with the functional, or operational, phase of the brake system. Another aspect of time which must be considered within the system safety concept is that of system life-span phases.

In general, a typical system will pass through the following life-span phases:

1. The *conceptual* phase, when the basic purpose, mission and preliminary design/approaches are considered and formulated. Unfortunately, many existing systems, such as the automobile/road system, evolved without passing through any effective system-level conceptual phase.

2. The *engineering* phase, where the objectives, requirements and criteria developed during the conceptual phase are translated into actual equipment and procedures. This phase includes testing and analysis for evaluating compliance with stated criteria and goals.

3. The *operational* phase, which starts when the equipment, personnel and procedures function together in order to achieve the intended purpose or mission of the system.

4. The *disposal* phase, which begins when the equipment and personnel are no longer needed and must be effectively disposed of or placed into storage.

There are many variations in terminology and detail relative to the system life-span principle. In most weapon system programs initiated by the Department of Defense (DOD), the line between the beginning and end of each life-span phase is delineated by contractual provisions. However, in many non-military or non-aerospace systems, the demarcation between the various system life-span phases is often obscure or non-existent.

This should not, however, become an excuse for failing to apply the system safety concept in non-military/aerospace endeavors wherever possible. Hopefully, the systems concept and the foresighted aspects of the safety program associated with it will prove to be of such benefit that the system approach will be incorporated into all relevant fields of endeavor.

What then are the key elements of this concept, and how can they be applied to one's particular field of loss prevention?

KEY ELEMENTS

In the previous definition of the system safety concept, certain key words must be further explained. This is necessary for the broad-based understanding that will enable one to mold the concept into a usable approach in preventing accidents.

Pre-planned, organized effort. How many safety programs have been instituted as a semi-panic reaction to some major accident or catastrophe? In general, these reactive safety efforts are limited in scope to the specific accident causes previously experienced, and soon lose effective management support. They often continue, giving some false sense of control, only because bureaucratic and self-interests often inhibit the natural death of on-going, but non-effective safety programs.

The system safety concept, if properly applied, would tend to alter this "accident re-action" syndrome. It would place emphasis on organizing the safety program around the anticipated *hazard potential* rather than around the dictates of panic and political expediency. The term "pre-planned" implies that the type, scope and extent of the safety effort to be instituted in a given program is based on a rational evaluation of the potential hazards and risks involved during the conceptual phase. The formulation of the radiological safety program by the Atomic Energy Commission, at the beginning of the program to supply radioisotopes to the general public, is an example of the pre-planned approach to safety. Attempts to regulate automotive design safety and to draw up coal mine safety standards are examples, on the other hand, of reactive safety programs.

Conservation of resource. Too frequently, loss control programs are developed to deal with only one aspect of the total problem. For example, many plant safety programs are concerned only with injury to company employees, and therefore tend to ignore material damage, customer safety, public safety and environmental protection. The system safety concept, if properly applied, would transcend arbitrary boundaries of concern within the total environment of the system involved.

The term "conservation" is defined[1] as ". . . planned management of a natural resource to prevent exploitation, destruction or neglect." This is surely the ultimate goal of all safety professionals.

By adopting the broadest possible term, "conservation of resources," the system safety concept attempts to establish an unrestricted scope of concern and attention relative to accident prevention on a given system program. This concern encompasses human life and well-being, material damage and loss, contamination of the environment and the financial resources available to the program.

System, Product or Operation. Ideally, the single term "system" should be sufficient to describe the intent of this phrase. However, since many people think of system primarily in terms of a complex, organized set of equipment, such as a missile system, a submarine, or a wheat-harvesting combine, the additional terms "product" and "operation" are included for clarity. When properly understood and judiciously applied, key elements of the system safety concept can be effectively used on any size program of activity. Obviously, the scope and nature of the system safety effort planned for a nuclear missile system would be far different from that applied by a toy manufacturer planning a new type of plastic toy rocket. However, the same basic elements discussed below would apply to both the weapon and the toy.

Pre-accident Hazard Identification. Fortunately, for the sake of technical progress, creative man tends to follow a

---

[1] *Webster's Seventh New Collegiate Dictionary.* Springfield, Massachusetts: G. & C. Merriam.

"think-positive" attitude. It is doubtful if Orville and Wilbur Wright would have gotten off the drawing board if they had seriously evaluated all the potential accident modes prior to their first flight. However, while Kitty Hawk provided an environment which minimized the risk to the public, not all of our think-positive creators have been similarly foresighted in launching new ventures. The newspapers and product liability-related periodicals are continually reporting cases in which the courts are finding manufacturers negligent for failing to adequately foresee and overcome the hazardous aspects of their products.

The state of our polluted streams and environment attests to the lack of meaningful attention to the conservation of resources prior to the accident or contaminating incident.

The single most important and characteristic element of the system safety concept is a pre-accident hazard evaluation. In order for this element to be accomplished in practice, two conditions must co-exist:

1. There must be a management structure and attitude which promotes the asking of the question, "What are the potential hazards, and what are risks to the conservation of resources associated with this contemplated endeavor?"

2. There must co-exist the technical capability within the management structure to effectively analyze the problem to provide an acceptable, accurate answer to this management question, so that the necessary decisions as to how to proceed can be resolved.

It appears reasonable to assume that if useful answers are not soon forthcoming, the appropriate questions will no longer be asked. It is, therefore, as important for safety professionals to become adequately trained and equipped to answer such management questions as it is to work to create a management environment that will promote the asking of the proper questions.

The challenge is sometimes voiced, "Why the emphasis on pre-accident hazard evaluation? It seems as though we could be spending wasted money and effort looking for problems that don't exist. Why not channel what limited safety resources we have to those problem areas which are obviously causing accidents right now?"

The response to this valid question is relatively straight-forward:

1. Traditional safety activity is, and no doubt will continue to be, concentrated on those areas shown by accident statistics or publicity to be most dramatic. There is an old safety maxim that "there always seem to be plenty of funds available to investigate an accident, but hardly a nickel to prevent one."

2. There is no assurance that current statistics realistically show the relative importance of accident problems in terms of the loss of essential resources.

3. The cost of correcting an inherent design-related hazard in a system during its conceptual phase is inconsequential; during its engineering phase the cost is acceptable. But during the operation phase, when the accidents are most likely to occur, the cost is often prohibitive. Therefore, hazards (potential accidents) corrected prior to the accident are often never identified as being safety-significant, and hazards identified only because of real-life accidents may never properly be corrected due to the costs involved in redesigning existing equipment.

Timely incorporation of safety inputs to system requirements. Inherent to the aerospace systems engineering approach is the concept of "requirements before solutions." Translated into common language, this concept proposes that a thorough, comprehensive study be performed to identify and detail the various mission objectives and functional requirements and constraints of a given system before designing the equipment to be used. For illustrative purposes, assume that a system is to be developed to lift wrecked automobiles from a conveyor belt and over a hopper, into which the wreck is dropped for compressing and baling. In simplified form, the requirements established might read something like this:

1. *Functional.* A means is required to (a) connect to, (b) lift to a height of 50 feet, and (c) release upon command, a metallic object of irregular shape, weighing not more than 5000 lbs., and encompass it in an envelope of not more than 300 cubic feet.

2. *Mechanical.* The means will be designed with a mechanical safety factor of 6:1. All structural components shall meet the provisions of Section 12 of the State Industrial Safety Code.

3. *Electrical.* The means shall generate its own electrical power. All electrically actuated controls will remain in the existing position should a power failure be encountered. All electrical items will comply with the provisions of the National Electric Code.

4. *Electro-mechanical.* All load-connecting and releasing functions will be remotely controlled by the operator. Redundant systems will be provided so that no single malfunction or operator error can cause the load to inadvertently fall. The connecting and releasing components shall be designed to accept a free-fall of the object 12 feet without releasing it upon sudden constraint by the lifting cable prior to the object's striking the ground.

5. *Human factors.* The means shall be compatible for operation by a trained and experienced operator. The controls shall be designed to minimize inadvertent operator error.

6. *Safety.* The means shall be so designed that no single malfunction or error will cause release of load, damage or injury. A complete malfunction analysis and safety demonstration will be conducted prior to use, which must be observed and approved by the system safety engineer.

7. *Interfaces.* The means shall not require more than an area 20 ft. x 40 ft. in which to operate. It shall be designed so that it can be transported by rail or truck on standard roadways.

8. *Availability.* The means shall be capable of operating an average of at least 50 days between each day of down-time due to maintenance. It shall be capable of rapid field repair, with a maximum of 4 hours required for any maintenance operation.

As seen from this example, system requirements normally refrain from pre-selecting a design solution. Therefore the general term "means" is used instead of "crane" or "hoist," which would have indicated a specific design solution to the system's functional requirement. Theoretically at least, the designer could satisfy the requirements by means of a modified helicopter, a simple "A" frame and pulley overhead crane or a boom hoist, among other conceivable approaches. In practice, a formal trade-off study is often performed to evaluate the relative advantages and disadvantages of each possible approach before arriving at the "selected" solution.

This "requirements before solution" approach provides the best possible opportunity for the system safety specialist to evaluate potential hazards in the contemplated system's mission, and to formulate needed safety—related constraints in the design solution/selection process. Safety considerations are normally interposed throughout the document, and become a part of each section. In the illustration, the system safety engineer had foreseen the potential hazards of inadvertently releasing the car while suspended above a possible working area and of the lifting mechanism's failing and allowing the load to free-fall for some distance. The hazards associated with electrical shock and problems encountered in transporting the device from the factory to the operating site were likewise cited. In addition to the specific safety requirements, the system safety engineer further stipulated that he must personally observe all safety-verification testing before the system was placed in operational use.

In complex military/aerospace systems, the requirements formulation activity becomes quite extensive and involved. It is not unusual to find hundreds of engineers engaged full-time in a requirement formulation activity for a single system. The findings of these requirement studies and the resulting documents are carefully reviewed by the governmental procurement agencies before authorizing funding for the design and development effort.

Under such a rigorous and formal approach to requirement definition, one can readily see why "timely incorporation of safety inputs to system requirements" becomes a key element of the military/aerospace system safety program. Hopefully, these considerations can find increasing application in commercial and public-type programs incorporating the system safety concept.[2]

Early evaluation of compliance with safety requirements and criteria. It is unrealistic to suggest (as some managers are prone to do) that the system safety engineer "quit and go home", after completing a good set of safety inputs to the system requirements specifications. The effort to find the optimum amount of safety for the specific risks involved is a complex function, often

---

[2] For a more thorough explanation of the DOD system Engineering Management concept and approach, the interested reader is directed to MIL-STD-499 (USAF), dated July 17, 1969.

continuing throughout the entire life-span of the system. Techniques (which will be detailed later) have been developed to assist the system safety engineer in performing these engineering phase evaluations.

In addition to design safety analyses, the system safety engineer must maintain close surveillance over the methods, scope, objectives and results of all safety-significant development and qualification-testing activities. This is necessary for two basic reasons:

1. To identify and assure protection against accident risks during the conduct of the testing procedure.

2. To evaluate whether adequate testing is being conducted on the prototype equipment to verify compliance with safety criteria and objectives.

The extent and nature of such safety testing considerations vary extensively from system to system. Therefore, each program must be formulated on its own particular characteristics. The individual responsible for formulating and conducting such safety test surveillance efforts is advised to review carefully each identified design safety requirement against the proposed test plan, and to insist on modification and additions where needed. It is often most helpful to establish a set of accept/reject criteria for safety before initiating the testing effort. This will minimize the problem of arbitrary, "on-the-spot" decisions made by test conductors.

Continued safety surveillance throughout system's life-span, including disposal. Like that of the proverbial housewife, the job of the system safety engineer is never done. Even upon the satisfactory demonstration that all applicable safety criteria and requirements have been translated into functional hardware, software and procedures, the system safety activity must continue into the operational or functional phase.

This is true for a variety of reasons, including:

1. *Nobody's perfect.* The best of engineered systems normally requires a shakedown period during the initial operation to iron out functional and interface problems that were unforeseen. If any of these "bugs" should be potential hazards, the system safety engineer should be alert to identify and correct such problems before they result in accidents. An accident occurring during initial operation can be highly embarrassing

and often quite disruptive to the orderly flow of check-out evaluation. It, therefore, behooves the manager of a system to carefully weigh the risks of cutting back on the system safety surveillance effort at this critical period in the system life-span.

2. *Changes occur.* For any number of valid or invalid reasons, systems change periodically throughout their operating life-span. These changes may be design, procedural or environmental interface-related. In each case, the potential exists for degradation or loss of some safety aspect of the system. Continual review of all proposed changes is needed to minimize the risk that a change intended to improve some functional aspect of the system could result in a serious hazard or accident.

3. *Disposal.* Inevitably, sooner or later the system, or some portion thereof, will be assigned to the scrap heap. Not infrequently, this final disposition becomes a safety problem. The nerve gas problem is the classic example, but many less dramatic problems continue to plague industries and government agencies. A few examples are pressurized spray cans that explode in fire; pyroforic metals, such as magnesium, which burn for weeks in buried fills; radioactive materials; gasoline left in tanks of cars being scrapped; and toxic chemicals of all types. The system safety concept should plan an ever-increasing role in the control of undesired pollution and hazards during the disposal phase, through the application of proper pre-planning.

THREE DISTINCT ASPECTS

For the sake of clarity, the author has adopted the following choice of terms to identify which of the three aspects of system safety practice is being discussed:

1. *System Safety Engineering* (SSE). This term is used exclusively within this chapter to denote the formal, contractual effort by a contractor's system safety organization to satisfy a special contractual obligation. This can be at any contractual level from prime to a subcontractor's supplies.

2. *Product Safety Engineering* (PSE). This term is used herein to denote the efforts of qualified personnel toward the applica-

tion of the system safety concept in commercial product development and marketing endeavors.

3. *Product/System Safety Engineering* (P/SSE). This term is used in cases where some combination of contractual safety obligation and company product liability risks co-exists on the same program. For example, a company may be under contract with a DOD agency to provide an advance medical monitoring instrument, under the provisions of government standards, but may determine that the use of commercial safety standards provides certain additional criteria which should be incorporated into the design because of liability considerations.

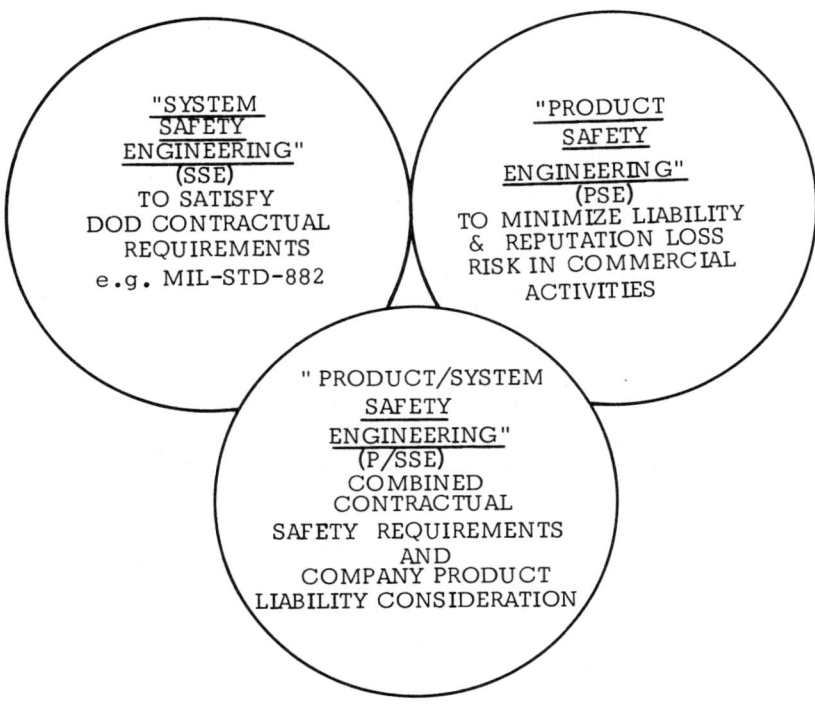

Figure 2. Three Aspects of the Practice of System Safety

SYSTEM SAFETY ENGINEERING (SSE)

The remaining emphasis of this chapter will be to provide an overview of the practice of system safety engineering, together with key references. It will be left to the reader to determine the optimum means for translating these SSE techniques and approaches into usable tools for his own needs and interests. The basic SSE terminology as defined by MIL-STD-882 is shown in Figures 3 and 4.

The methodology used within the system safety engineering discipline is to a large extent a synthesis of the system engineering management and analysis concepts associated with the aerospace industry, and classical accident prevention concepts associated with industrial loss control activities. SSE methodology is not yet fully matured; however, a rapidly growing literature exists within technical journals, symposium proceedings, and DOD and NASA documents.

Safety - Freedom from those conditions that can cause injury or death to personnel, damage to or loss of equipment or property.

System - A composite, at any level of complexity, of operational and support equipment, personnel, facilities, and software which are used together as an entity and capable of performing and/or supporting an operational role.

System Safety - The optimum degree of safety within the constraints of operational effectiveness, time and cost, attained through specific application of system safety management and engineering principles throughout all phases of a system's life cycle.

System Safety Management - An element of program management which insures the accomplishment of the system safety tasks including identification of the system safety requirements; planning, organizing, and controlling those efforts which are directed toward achieving the safety goals; coordinating with other (system) program elements; and analyzing, reviewing, and evaluating the program to insure effective and timely realization of the system safety objectives.

System Safety Engineering - An element of systems engineering involving the application of scientific and engineering principles for the timely identification of hazards and initiation of those actions necessary to prevent or control hazards within the system. It draws upon professional knowledge and specialized skills in the mathematical, physical, and related scientific disciplines, together with the principles and methods of engineering design and analysis to specify, predict, and evaluate the safety of the system.

Figure 3. System Safety Terminology (Per MIL-STD-882, 15 July 1969)

> Hazard - Any real or potential condition that can cause injury or death to personnel, or damage to or loss of equipment or property.
>
> Hazard Level - A qualitative measure of hazards stated in relative terms. For purposes of this standard the following hazard levels are defined and established: Conditions such that personnel error, environment, design characteristics, procedural deficiencies, or subsystem or component failure or malfunction:
>
>> Category I - Negligible
>>
>> ..... will not result in personnel injury or system damage.
>>
>> Category II - Marginal
>>
>> ..... can be counteracted or controlled without injury to personnel or major system damage.
>>
>> Category III - Critical
>>
>> ..... will cause personnel injury or major system damage, or will require immediate corrective action for personnel or system survival.
>>
>> Category IV - Catastrophic
>>
>> ..... will cause death or severe injury to personnel, or system loss.

Figure 4. Hazard Classifications (Per MIL-STD-882, 15 July 1969)

When reviewing SSE methodology, two basic and interrelated aspects must be considered: system safety management and system safety analyses. The effectiveness of the system safety concept is predicated to a large extent on the proper integration of these aspects.

SYSTEM SAFETY MANAGEMENT

A fundamental premise of the system safety concept is that the identification and assessment of hazardous risks must be developed and presented to responsible program management for appropriate action in a time frame which allows for making corrections with minimum cost impact. In aerospace activities, this normally means during the design phase, where changes can be accomplished without involving retrofit of existing hardware. The preferred order of choice in providing solutions to safety requirements during design is shown in Figure 5.

---

1. <u>Design For Minimum Hazard</u> - Select design approaches which provide optimum inherent freedom from hazard.

2. <u>Safety Devices</u> - Known hazards which cannot be eliminated through design selection shall be reduced to acceptable level through the use of appropriate safety devices.

3. <u>Warning Devices</u> - Where it is not possible to preclude existence or occurrence of a hazard by design, devices shall be employed for the timely detection of the condition and the generation of an adequate warning signal.

4. <u>Special Procedures</u> - Where it is not possible to reduce the magnitude of a hazard through design, safety and warning devices, special procedures will be established.

Figure 5. SSE Precedence for Satisfying Safety Requirements
(Per MIL-STD-882, 15 July 1969)

The implementation of system safety analysis procedures must be planned as a scheduled aspect of the total management program on a given systems project. This is normally accomplished by incorporating specified milestones for reviews and approvals at key points in the overall program schedule, compatible with system safety objectives, before final approval of a given design, and again prior to placement of a new system into operational use. This is normally documented in a System Safety Program Plan (SSPP) which is a requirement of MIL-STD-882 (See Figure 6 for the recommended outline).

Another key factor must be planned during the initial phases of a complex system program, where several contractors are providing sub-systems to a total system (e.g., in a missile project, the booster may be provided by Company A, the guidance set by Company B, payload by Company C, and the launch site by Company D). This is to assure suitable integration of the various subsystem hazard analyses into a composite analysis.

This is normally accomplished by the procuring agency's designation of one company as the integrating contractor for system safety. This contractor prepares the integrated system safety plan, which defines the specific nature and schedules for the system safety analyses to be submitted by each associate con-

1. General
   1.1 Introduction
   1.2 Scope and purpose
   1.3 Application and implementation
   1.4 Applicable documents
2. Safety organization, responsibilities, and authority
   2.1 Integrating contractor organization and responsibilities
   2.2 Associate contractor organization and responsibilities
   2.3 Subcontractors responsibilities
   2.4 System safety working groups
3. System safety program milestones
4. System safety criteria
   4.1 Definitions
   4.2 Hazard level categories
   4.3 Special contractual requirements
   4.5 Identification and dissemination
5. System safety analyses
   5.1 Identification of analysis techniques
   5.2 Qualitative and quantitative analyses
   5.3 Preliminary hazard analysis
   5.4 Subsystem hazard analysis
   5.5 System hazard analysis
   5.6 Operating hazard analyses
6. Safety activities
   6.1 Safety data
      6.1.1 Identification of data requirements - deliverable and non-deliverable data
      6.1.2 Acquisition and use of safety data
         6.1.2.1 Hazard data collection
         6.1.2.2 Document tree and data flow
         6.1.2.3 Documentation and files
         6.1.2.4 Format for reports and data submittal
         6.1.2.5 Accident prevention, investigation, and reporting
         6.1.2.6 Safety reports
   6.2 Training
      6.2.1 Crew qualification, training and certification
      6.2.2 Maintenance personnel training and qualification
7. Audit program
8. Ground handling, storage, servicing and transportation
9. Facilities and support requirements
10. Other system safety matters (not otherwise covered)

Figure 6. System Safety Program Plan Outline (Per MIL-STD-882, 15 July 1969)

tractor. These separate inputs are then integrated into a composite hazard analysis. The importance of this process in complex systems cannot be overstated, since the subtle interactions of unplanned events across the interfaces of subsystems often result in spectacular and costly accidents.

In summary, system safety management provides the framework wherein the findings and recommendations resulting from the application of system safety analysis techniques can be effectively reviewed and acted upon by responsible management. Without proper planning and implementation of the management integration aspects, the usefulness of a thorough system safety analytical effort may be erased.

It is likely that system safety management practices will, in the long run, provide more benefit to the loss control profession than the adoption of system safety analytical techniques to non-aerospace uses, a factor overlooked in articles devoted to system safety engineering.

SYSTEM SAFETY ANALYSIS

Although "hazard recognition" has long been a primary function of the safety professional, formalized methods for accomplishing this critical prerequisite to hazard control have been, to a large degree, lacking in the typical industrial safety situation. Experienced safety professionals are often reasonably successful by using an intuitive approach to hazard recognition and evaluation, but normally only if they are working in familiar territory. It has long been a common practice to evaluate reported statistics of accidents. However, this is at best merely an "after-the-fact" method of hazard recognition. The inherent limitations of the informal intuitive and "after-the-fact" approaches are readily apparent when hazard recognition is considered in the context of the system safety concept.

THE BASIC ELEMENTS

A basic tenet of the system safety concept is that safety considerations must be effectively integrated into the mainstream of program management, planning, engineering and operational activities, so that the maximum practicable inherent safety is planned, designed and built into the product or system with the lowest possible impact on cost, schedule and performance (i.e., optimum safety). As a primary means to help achieve this optimum safety objective, various system safety analytical techniques have been developed which are collectively oriented toward achieving a balance of the three essential elements of effective hazard recognition: identification, evaluation and communication.

Identification. The identification of a potential causative factor (i.e., hazard) is obviously needed initially to insure that proper safety provisions are incorporated for its control or elimination. (Mere identification of a hazard provides little assurance by itself that it will be properly controlled, unless it is adequately evaluated and the results are communicated to those with prime decision-making responsibilities.) Effective hazard identification requires a systematic, rigorous approach if the more subtle hazards are to be uncovered, particularly in complex systems.

Evaluation. It is well recognized that not all hazards are of equal importance. In some cases the anticipated consequences of the unsafe event may be minimal, while in others they are catastrophic. The extent of anticipated injury or damage which could result from the hazardous event's occurrence is referred to as the "hazard severity." A set of qualitative hazard categories has been established in MIL-STD-882 (See Figure 4), which serve as the most widely accepted standard for assessing hazard severity. A more sophisticated approach to evaluating relative severity potential would be to assign an estimated dollar cost factor to the anticipated loss, which is quite difficult in many cases, particularly if human life is involved.

The second aspect of hazard evaluation involves determining the likelihood of the hazardous event's actual occurrence. This may be reported in non-numeric (qualitative) terms such as possible, unlikely or improbable, or in numeric (quantitative) terms such as once in ten thousand flights, 0.001 per flight, or 1×10^{-4} / flight. Numeric estimates of the likelihood of a hazardous occurrence are referred to as "hazard probability." For complex systems, safety logic models (such as fault trees) must be developed to arrive at a demonstrable hazard probability estimate.

Communication. In many industrial safety situations, there may be no significant problem in transmitting the findings of a hazard evaluation into needed corrective action, because of the limited number of managers or foremen with whom the industrial safety engineer must interface to initiate corrective action. Communications in the area of hazard analysis data, however, become a major problem to most system safety engineers involved in complex programs. This is a result of various factors, including:
1. The number of components, interfacing subsystems, associate contractors and supporting activities within a large system

program, which may be impacted by design changes recommended on one part of the system, and which should contribute in some measure to the hazard analysis process and trade-offs involved.

2. The limitations which may be placed on effective safety communications across organizational jurisdictions, due to contractual or administrative considerations.

3. The need to effectively compete in management/engineering trade-offs involving less subjective and more immediate problems, such as cost, performance and schedule.

It must be emphasized that the basic objective of System Safety Analysis is not just to quantify hazards or to satisfy contractual obligations, but to provide the decision maker with the information needed to make effective choices relative to safety. In other words, the objective is to communicate. Of course, communications is a two-way street, and management must also be alert to the need for assuring that the necessary data (input) is made available to the system safety analyst as needed.

For good communications, system safety analysis data should be:

1. *Timely*. The earlier within a system's life cycle the valid need for a safety provision is identified, the less should be its impact on cost and schedule. Early detection also insures its effective implementation in the system. Therefore, the system safety approach is oriented toward conducting hazard analyses during the conceptual planning and design phases of a given project, rather than waiting until the hazards are physically present in an operating environment.

2. *Concise*. In order to provide management with the information needed to make effective decisions relative to safety, the data presented should be as concise and objective as possible, to minimize the possibility that critical safety information will be misinterpreted or ignored by busy managers.

3. *Factual*. Safety recommendations presented for management consideration must be as factual as possible, to assure good decisions and continued reception. The rapport developed between the system safety engineer and program management, which is a key ingredient of effective system safety,

must be continually reinforced by the factual nature of the safety-related information submitted. Every reasonable effort should be made to assure maximum practicable objectivity in the hazard analysis data submitted, which is especially important in the system safety approach because of the many competitive requirements (performance, cost, weight, etc.) with which the manager/designer/decision-maker is confronted.

A GENERAL DEFINITION

Now that the three basic elements of system safety analysis have been described (identification, evaluation and communication), the next step is to develop a working definition covering both the basic objectives and scope of system safety analyses, such as:

"System safety analyses provide before-the-loss identification, evaluation and communication of those factors and interactions within a given system which could cause inadvertent injury, death or material damage during any phase or activity associated with the given system's life-cycle."

THE ENCOMPASSING SCOPE

From this broad perspective, one can readily see why a variety of analytical techniques must be employed by the system safety analyst. Not only are there wide variations between the types of systems to be analyzed (e.g., an electric iron or an airplane); there are also many phases within each system's life cycle, each having unique levels of detail and needs relative to safety. Some of the more generally accepted reasons that system safety analyses are conducted are:

1. To obtain initial assessment of safety-significant aspects of a contemplated (or actual) product, activity, system or program.

2. To establish an objective basis for defining safety tasks, analyses, testing, training, etc., on a given project.

3. To identify potentially-hazardous equipment failure modes and improper usages.

4. To provide guidance for the proper selection of specific safety-related criteria, requirements or specifications.

5. To assist in the evaluation of safety considerations during design/procedural trade-studies.

6. To evaluate hazardous design considerations and establish relative corrective action priorities.

7. To organize baseline data for quantitative deductive (FTA) analyses.

8. To document subsystem level data for use in performing system level analysis.

9. To identify safety-significant problems/requirements across subsystem/environment interfaces.

10. To determine causative factors and interactions leading to specified unwanted/hazardous events.

11. To evaluate probability of specified unwanted/hazardous event occurence, and identify critical parts of causative factors.

12. To identify, describe and establish relative importance of potential hazardous conditions associated with contemplated (or actual) activities involving the use, testing, storage, handling, transportation, maintenance or disposal of an item of equipment, subsystem or system.

13. To establish objective basis for specifying precautions, personal protection, safety devices, emergency equipment/ procedures/training, and safety requirements for facilities, support equipment and environment.

14. To provide documented evidence of compliance with specified safety tasks, objectives and design requirements.

ANALYTICAL TECHNIQUES

A variety of analytical approaches and techniques have evolved within the system safety discipline in response to the differing nature of the various life-span phases, and types of products or operations involved. Although some notable exceptions exist, in general, each of the existing techniques can be grouped under one of the following three categories:

- Preliminary Hazard Analyses (PHA)
- Design Safety Analyses (DSA)
- Functional Safety Analyses (FSA)

PRELIMINARY HAZARD ANALYSES (PHA)

Why perform a PHA? Preliminary hazard analyses are performed to identify those conditions which, from previous related systems and experiences, can be expected to be potential hazards in the system being planned. For example, any contemplated system which may involve high-pressure compressed gases must be considered to have a potential inherent hazard which must be properly safeguarded by design and procedures. The findings and recommendations developed in performing a systematic preliminary hazard analysis (covering all basic components and operational considerations of the contemplated system) become the primary guide for safety-related planning within the program. This would include determination of those areas where safety requirements must be imposed on design, those safety characteristics which must be verified by development testing and design safety analyses, and those use-phase safety activities (such items as special safety equipment, training, testing, facilities, etc.) which must be planned and budgeted.

When to Perform a PHA. For maximum effectiveness, the preliminary hazard analysis effort should be initiated at the earliest practical time after the start of the program's conceptual phase. It should be treated as a continuing effort, requiring periodic updating and review for compliance implementation status, rather than a one-time-only task. PHAs can be beneficially performed as the initial system safety effort on any type of system, operation or product, regardless of the life-span phase involved. Therefore, the PHA should not be omitted just because the system safety program is not instituted during the conceptual phase.

How to Perform a PHA. To best assure a systematic, rigorous coverage, the PHA should be conducted in conformance with a pre-established matrix format and/or set of ground rules, tailored to best reflect the nature of the system to be analyzed, the extent of available information about the system, and the planned use of the analysis output data.

A typical safety analysis matrix format, both basic and preliminary, is shown in Figure 7. To conduct a PHA using such a format, the analyst should:

1. Survey the overall scope of the system to be analyzed and, on paper, subdivide the system into convenient (1) physical subsystems or elements (items), (2) operations or activities (functions), and (3) life-cycle phases, such as testing, repair, transportation, storage, use and disposal (modes).

2. Check to see that the entire system being analyzed is encompassed within the items, functions and modes identified in step 1, and that descriptive terminology is assigned. If certain areas are to be excluded from analysis, they should be specifically noted to avoid possible misunderstanding later.

SYSTEM: _____ SUBSYSTEM: _____		PRELIMINARY HAZARD ANALYSIS		PREPARED BY: _____ PG __ OF __ ISSUE DATE: _____ REV ____	
1	2	3	4	5	6
ITEM/ FUNCTION	MODE	HAZARDOUS ASPECT	HAZARD CATEGORY	NEEDED SAFETY PROVISIONS	CORRECTIVE ACTION PRIORITY

Figure 7. Typical Matrix Format for a Preliminary Hazard Analysis

3. Identify the hazardous aspects associated with each identified item or function for each mode. This is accomplished by first entering an "item" designation in column 1 of the PHA matrix form. Under the item description, enter the various general functions that the item is to perform. Next, consider each life-cycle mode and note any hazardous conditions associated with the specific item, function and mode being considered. The mode and identified hazardous condition (summarized) are entered in columns 2 and 3 of the matrix. This basic process is then continued until all items and functions have been systematically considered.

4. Evaluate the potential severity of each hazardous event identified, and assign a relative hazard category (per Figure 4) in column 4. For this step, the anlayst should consider that no special provisions to control the severity of the potential accident have been employed other than those inherent within the system as already defined. Relative likelihood of occurrence should not be a factor at this step.

5. Designate those safety provisions needed to control or eliminate the severity and probability of each hazardous event noted. Only the key words need be entered in column 5, with details provided in supplemental sheets or reports as appropriate, to assure proper attention and action. In determining the corrective safety provisions needed, the order of preference listed in Figure 5 should be followed.

6. Note in column 6 the relative safety priority for initiating/complying with the needed safety provisions described in column 5. The safety priority ranking is a subjective determination on the part of the analyst, used to highlight those areas within the analysis which should receive priority attention. Factors used to assess criticality ranking include hazard severity, likelihood of occurrence, and the impact on program to institute corrective action (i.e., cost schedule and performance).

A suggested criticality ranking guideline is as follows:

1. *Routine* ... should be adequately handled through routine channels for obtaining corrective action.

2. *Special* ... requires special follow-up action because of some unique aspect which might prove a problem when handled through routine channels.

3. *Critical.* .. requires special management attention because of the extent of program impact.

4. *Critical Priority* ... is the same as (3) except that a special time constraint exists which dictates immediate management attention.

The status of action taken in response to the findings contained in the analysis should be kept current by the person responsible for safety surveillance. At periodic intervals, the criticality rankings should be re-evaluated, adjusted as needed, and properly distributed to the affected persons.

DESIGN SAFETY ANALYSES (DSA)

Why Perform a DSA? Design safety analyses are of two basic types, inductive and deductive. The inductive approach involves the systematic evaluation of the effect each system component or element would have on the safety of the system, should it fail to operate in the prescribed manner and time; it is the "what happens if ..." approach. The deductive approach starts with a given identified hazardous event and systematically evaluates all predictable events, conditions and influences which could cause the event to occur — a "how could it happen" approach. Although each has its own unique advantages and limitations, which will be described later, both approaches are basically oriented to obtain the following set of objectives:

1. Verification that the design satisfies the established safety-related criteria.

2. Identification of specific design features which, if modified, would provide effective improvement in inherent safety, and the relative benefits in initiating such changes.

3. Documentation of objective baseline data which can be used in evaluating: (a) safety aspects of contemplated design changes, (b) interfacing safety consideration with other subsystems, and (c) compliance with legal or contractual obligations in the area of safety assurance.

When to Perform a DSA? In the idealized system development program, the basic design safety requirements and criteria will be established during the concept phase, based on data provided from the preliminary hazard analysis effort. These requirements (e.g., pneumatic pressure vessels and system components shall be designed with a 4 to 1 safety factor above maximum operating pressure; a pressure relief valve shall be installed so as to relieve any pressure within the system greater than 110% of maximum operating pressure; etc.), are factored into detailed equipment specifications and then into the design during the initial part of the engineering phase. Following testing and evaluation, the design is approved as meeting all specified requirements, and is released for production.

In such a system development approach, the optimum time to perform the DSA is prior to the final design approval and as soon as sufficient design detail is available to permit a thorough analysis. Acceptable findings by the DSA should be a prerequisite to final design approval.

In situations where the above conditions are not present, the DSA can be performed and utilized at any time a design safety assessment is wanted, provided sufficient design and operational details are available.

As previously indicated, two basic types of DSAs exist, inductive and deductive. In most situations, the inductive analysis should be conducted first. The data contained in the inductive analysis normally provide essential inputs and guidance for planning and performing the deductive analysis. The following discussion will illustrate the unique features of each DSA approach, and how they can be oriented to provide complementary functions.

Inductive Methods. As with the PHA, inductive design safety analyses are most effectively and systematically performed with the use of a pre-established matrix form. The primary difference between the PHA and DSA format and approach is that the DSA is considerably more detailed in terms of specific design features, failure modes, and design improvements needed. The DSA matrix is also oriented toward organizing certain data helpful in performing deductive analyses (logic modeling) on critical hazardous situations.

The basic advantages of the matrix format approach are:

1. It encourages the analyst to ask (and answer) certain key questions on each item under study.

2. By locating the data entered in prescribed location on the sheets, it makes review and recovery of wanted information easier.

3. It promotes a more systematic and comprehensive investigation.

Although the specific titles of the vertical columns used can be altered to best suit the situation pertaining to a given program or system, all matrix forms used for the inductive DSA on the same program should be similar.

A typical DSA matrix format example is shown in Figure 8. This matrix is oriented for a quantitative safety evaluation on either a total system or a subsystem (should the analyst's concern be limited to specific subsystems only, due to assigned responsibilities or contractual obligations). For situations where a qualitative safety analysis is to be performed, a supplemental sheet, attached to basic matrix and cross-referenced by the item number (column 1), should be used, as shown in Figure 9.

DEDUCTIVE METHODS

The deductive safety analysis techniques provide one of the most effective and versatile approaches to predictive safety analyses available to the safety professional. The basic concepts involved can be used to perform simple, qualitative evaluations (requiring no special training other than basic instructions and knowledge of the system to be analyzed) or very complex quantitative studies (requiring specialized training, computer programs and experience). The expense of performing such studies increases proportionally to the complexity and scope of the effort. Therefore, selective judgment is needed in planning the analytical effort to the initiated, to assure that its cost is justified by the hazard risk being evaluated.

The deductive approach begins with a defined undesired event, usually a postulated accident condition, and systematically organizes in a graphic form all known events, faults and occurrences (within the context of the system mode established) which could cause or contribute to the occurrence of the undesired event. The data organized within the inductive PHA and DSA matrix formats described above provide both the basis for selecting the undesired event to be evaluated, and the causative factors needed to complete the analysis.

SYSTEM: AZ			DESIGN SAFETY ANALYSIS MATRIX			PREPARED BY: J. Jones Pg. 1 OF 1			
SUBSYSTEM: A						ISSUE DATE: 3 May 71 REV. 0			
1	2	3	4	5	6	7	8	9	10
COMPO-NENT NAME	PART NO.	FAILURE MODE	LIFE CYCLE OR OPERA-TING MODE	EFFECT OF FAILURE ON SYSTEM/SUB-SYSTEM	HAZARD CATEGORY	ITEM NUMBER	ENTRY IN QSS	REMARKS/ NEEDED ACTION	SAFETY CRITICAL-ITY
Battery	(B)	Insufficient Voltage Output	Stand-By	N/A in this mode	I	A1a	No	-	-
			Ready	Tube (T) will not operate	I	A1b	No	-	-
Switch	(S)	Closes Inadvertently	Stand-By	Partially Enables Circuit	II	A2a	Yes	-	I
			Ready	N/A in this mode	-	A2b	No	-	-
		Fails to Close	Stand-By	N/A in this mode	-	A2c	No	-	-
			Ready	Tube (T) will not operate	I	A2d	No	-	-
Tube	(T)	Premature or Delayed Operation	Stand-By	Major System Damage will occur	III	A3a	Yes	FTA should be Developed	III
			Ready	Damage will occur	III	A3a	Yes		
		Fails to Operate	Stand-By	N/A in this mode	-	A3b	No	-	-
			Ready	Sub-A will perform intended function	I	A3c	No	-	-
Relay	(R)	Closes Inadvertently	Stand-By	Partially Enables Circuit	II	A4a	Yes	-	-
			Ready	Tube (T) will operate inadvertently	III	A4b	Yes	FTA Evaluation with Item No. A3a	II
		Fails to Open	Stand-By	N/A in this mode	N/A	A4c	No	-	-
			Ready	Tube (T) will remain operating past intended time	III	A4d	Yes	FTA Evaluation with Item No. A3a	II

Figure 8. Example DSA Matrix Sheet Completed

The most commonly used safety logic model technique within the system safety discipline is the Fault Tree Analysis (FTA). This basic concept and ground rules for the FTA were originated by Bell Telephone laboratory engineers in the early 1960's, and they have been undergoing continuous refinement, particularly in the mathematical evaluation area. For the purpose of this discussion, all illustrations used will be based on the current FTA technique; however, other methods and techniques are being developed and used for specific applications, which are deductive safety analyses but do not conform to the FTA technique methodology as described below.

In essence, the following five basic steps are involved in performing any depth of FTA from qualitative approximation to a rigorous mathematical treatment:

SYSTEM: AZ	DESIGN SAFETY ANALYSIS MATRIX	PREPARED BY: J. Jones	Pg. 1 OF 1
SUBSYSTEM: A	QUANTITATIVE SUPPLEMENT SHEET	ISSUE DATE: 3 May 71	REV. 0

8a	8b	8c	8d	8e	8f	8g	8h	8i
ITEM NO.	HAZARD FAILURE MECHANISM	PRIMARY FAILURE RATE	DATA SOURCE	HAZARD DURATION PERIOD	SECONDARY FAILURE CAUSES	COMMAND INPUTS	LIKELIHOOD OF OCCURRENCE ESTIMATE	REMARKS
A2a	Switch(s) closes during "stand-by"	1×10^{-9}/hr	FARADA	23 hrs per day	Vibration or Shock > 5g	Operator closes switch manually	Primary: 2.3×10^{-8}/day Secondary: 1×10^{-3}/day Command: 1×10^{-2}/day	Secondary likelihood - based on consideration of transport accident. Command-Human error estimate.
A3a	Tube(T) operates inadvertently	No known means of primary failure in this manner		24 hrs per day	Non-known	Early input signal, or delayed removal	Primary: Negligible	Command input likelihood needs evaluation
A4a	Relay(R) closes during "stand-by"	1×10^{-11}/hr	Vendor's Report	23 hrs per day	Vibration or Shock > 5g	Current (> 200 mA) on relay coil	Primary: 2.3×10^{-10}/day Secondary: 1×10^{-3}/day	Command input needs evaluation.
A4b	Relay(R) closes during "ready"	1×10^{-11}/hr	Vendor's Report	1 hr/day	Vibration or Shock > 5g	Current (> 200 mA) on relay coil	Primary: 1×10^{-11}/day Secondary: 4×10^{-5}/day	Command needs evaluation.
A4d	Relay(R) fails to open following intended closure during ready	1×10^{-6}/hr	Engineering estimate based on contacts sticking	1 hr/day	High Temp (> 1000°F) Shock (> 10g)	Coil remains energized	Primary: 1×10^{-6}/day Secondary: 1×10^{-5}/day	Command mode needs evaluation.

Figure 9. Example DSA Matrix Quantitative Supplement Sheet (QSS) Completed.

1. Select the undesired event to be evaluated, and define the system configuration life-cycle mode and environment for the purpose of the study.

2. Obtain data, drawings and functional information available to obtain a thorough understanding of the system to be analyzed.

3. Construct the fault tree logic diagram (see the description of technique below).

4. Evaluate the logic diagram (using either the subjective or objective approaches as discussed below).

5. Prepare a summary conclusion of the FTA findings for management review and appropriate action.

FAULT TREE LOGIC DIAGRAM CONSTRUCTION

Although sufficient for many applications, the following discussion of fault tree logic diagram construction must be recognized as being very elementary in nature. Many refinements and tricks-of-the-trade needed for moderate to highly complex systems may be obtained from careful study of applicable references and from trial-and-error experience.

The diagram is started by drawing a rectangle at the top-center of a blank sheet, into which is written the concise statement of the undesired event ("top event") to be evaluated. At the next lower level, note in separate rectangles the various independent life cycle and/or time phases which are to be evaluated. Below each of these events are placed the input factors, which (either unassisted or in combination) are necessary and sufficient to cause that event to occur. This is done using specific symbols and ground rules.

For the purpose of this example, assume that a DSA is being performed on System AZ. The safety analyst has prepared the DSA matrix forms covering Subsystem (SubA) as shown in Figures 8 and 9. The analyst has determined that Sub A has two operating modes of interest, i.e., "stand-by" for 23 hours a day, and "ready". for 1 hours a day.

In view of the critical nature of the hazard associated with inadvertent operation of tube (T), a quantitative FTA was determined appropriate. As seen in the example DSA matrix (Figure 8), three safety-critical (Hazard Category III) failure modes were identified in Sub A, all relating to the same basic fault as described in Item No. A3a. Therefore, this fault condition was designated as the undesired event to be evaluated by FTA.

The DSA matrix quantitative supplement sheet (Figure 9) was completed for all item numbers assigned a Hazard Category of II or greater. Upon completion of this sheet, the analyst would next begin construction of the FTA logic diagram, as illustrated in Figure 10.

The diagram shown in Figure 10 consists of various shaped symbols, in or around which are placed words and/or numbers. To the experienced FTA analyst, this diagram provides a graphic description of all the ways the top event can occur, its probability of occurring, the critical path by which it most likely can occur,

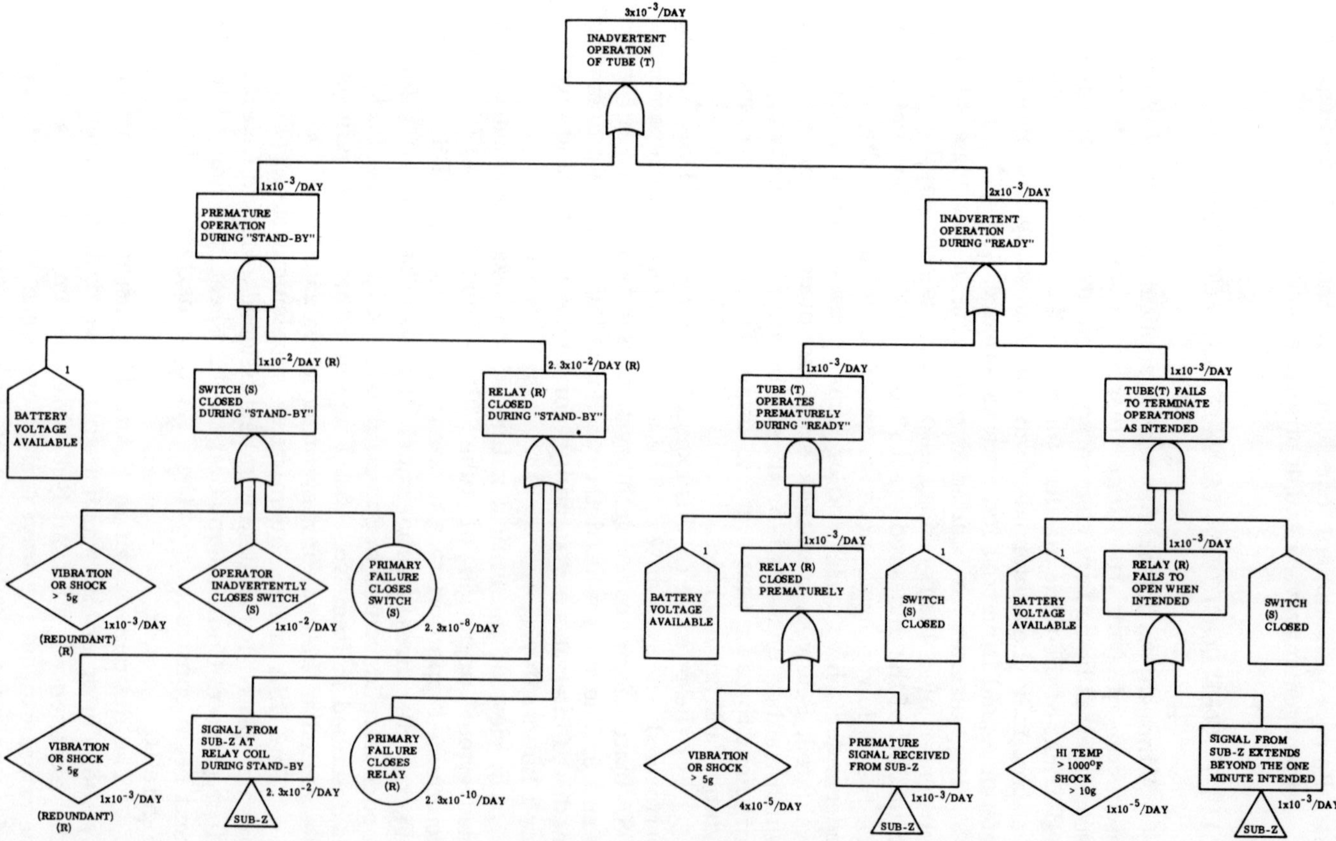

Figure 10. Example Fault Tree Diagram

how the probability of hazard can best be reduced and the various possible solutions which would *not* effectively reduce the probability of hazard.

The diagram illustrated is the most elementary form which can be used, and is shown here for concept description only. For almost any system requiring a quantitative FTA or other inductive technique, many other specialized symbols and mathematical refinements must be used to obtain full benefit of this analytical approach.

The key to preparing and evaluating an FTA logic diagram is an understanding of the logic symbols used. The basic symbols used in this illustrative example are defined in Figure 11.

In the most simple form, this probability relationship can be treated as follows: Probabilities are added when passing up through an "OR" gate, and multiplied when passing through an "AND" gate. This provides a reasonable approximation mathematically only if the probabilities are small (10^{-2} or so) and redundant events are noted and accounted for in the mathematical treatment. The use of Boolean Algebra reduction, or simulation techniques, is needed for more complex and involved FTA's, and where more than numeric approximation is desired.

With this basic background, the method by which this simple logic diagram (Figure 10) was constructed can now be explained. Note that the undesired (Top) event selected did not specify the operational mode. Therefore, the initial step was to identify all the signs at system modes which must be considered. In this base there are two: Stand-by and Ready. These were drawn so as to be connected to the top event through an "OR" gate, since the top event could occur in either mode. Next the left-hand placed event was developed (as a general rule, the various modes are placed left to right on the sheet in relation to the sequence of their normal occurrence in the system life span).

Operation of tube (T) during stand-by mode requires three conditions to co-exist: Voltage available from battery (B), switch (S) closed, and relay (R) closed. These three conditions are therefore entered into the appropriate event symbols on the diagram, and connected through an "AND" gate to the previously defined event rectangle.

(Note: For simplicity of this illustration, potential short circuits, externally-induced currents, bent pins and similar fault modes which must be considered in detailed quantitative DSAs of electronic equipment are excluded.)

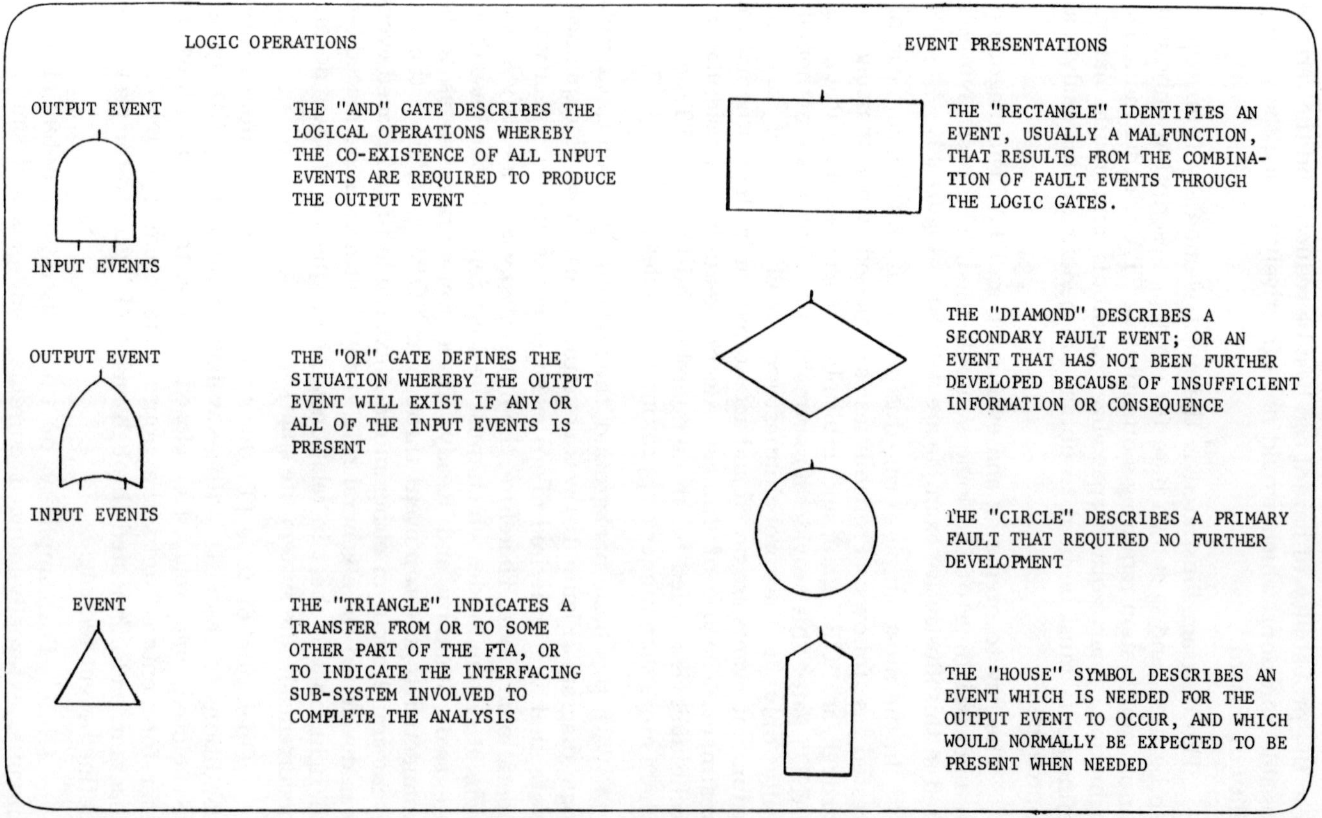

LOGIC OPERATIONS

OUTPUT EVENT ⌂ **INPUT EVENTS**

THE "AND" GATE DESCRIBES THE LOGICAL OPERATIONS WHEREBY THE CO-EXISTENCE OF ALL INPUT EVENTS ARE REQUIRED TO PRODUCE THE OUTPUT EVENT

OUTPUT EVENT ⌂ **INPUT EVENTS**

THE "OR" GATE DEFINES THE SITUATION WHEREBY THE OUTPUT EVENT WILL EXIST IF ANY OR ALL OF THE INPUT EVENTS IS PRESENT

EVENT △

THE "TRIANGLE" INDICATES A TRANSFER FROM OR TO SOME OTHER PART OF THE FTA, OR TO INDICATE THE INTERFACING SUB-SYSTEM INVOLVED TO COMPLETE THE ANALYSIS

EVENT PRESENTATIONS

THE "RECTANGLE" IDENTIFIES AN EVENT, USUALLY A MALFUNCTION, THAT RESULTS FROM THE COMBINATION OF FAULT EVENTS THROUGH THE LOGIC GATES.

THE "DIAMOND" DESCRIBES A SECONDARY FAULT EVENT; OR AN EVENT THAT HAS NOT BEEN FURTHER DEVELOPED BECAUSE OF INSUFFICIENT INFORMATION OR CONSEQUENCE

THE "CIRCLE" DESCRIBES A PRIMARY FAULT THAT REQUIRED NO FURTHER DEVELOPMENT

THE "HOUSE" SYMBOL DESCRIBES AN EVENT WHICH IS NEEDED FOR THE OUTPUT EVENT TO OCCUR, AND WHICH WOULD NORMALLY BE EXPECTED TO BE PRESENT WHEN NEEDED

Figure 11. Basic Fault Tree Symbols

In completing the various branches of the diagram, events placed in rectangles are subsequently developed, revealing the logical interrelationship of system fault events which can cause them. The branch is completed with all input fault events either placed in a circle (indicating primary basic fault), a diamond (indicating a secondary fault or condition which is not feasible to further development) or rectangle, to which has been assigned a transfer or reference triangle. In general, it is helpful to consider each fault as the possible result of primary, secondary and commanded failures. These causative faults are placed in appropriate symbols and connected to the event block through an "OR" gate, since any of the three fault conditions by itself could cause relay (R) to close. Since the faults which would cause the relay coil to become energized must be determined from an analysis of the interfacing subsystem (Sub Z), a transfer symbol was attached to this rectangled event.

In actual practice, this transfer symbol would initiate an effort to conduct an FTA on Sub Z, with the top undesired event being the situation described in the event rectangle having the transfer symbol attached. In complex systems the FTA will comprise many pages of diagrams, interrelated by appropriate use of some transfer coding system established in advance. For the purpose of this example, only the assumed numeric probability which resulted from the FTA of Sub Z is indicated.

The right-hand branch of this diagram or "tree" was developed in similar manner, with the basic differences being that during the "ready" mode, switch (S) is normally closed and the inadvertent operation of tube (T) could occur either because of improper initiation or failure to terminate operation after the intended one minute had elapsed.

Note that the data collected and organized in the QSS supplement to the DSA matrix are specifically oriented to support the preparation of the logic diagram, and for the numeric evaluation which follows the construction of the diagram.

FAULT TREE EVALUATION

In many instances, the mere process of developing and/or subjectively reviewing the logic diagram provides important insights relative to weak points in the system design. Fault conditions not previously considered important may be shown to be critical factors; and, conversely, other aspects often receiving much more attention are revealed not to be as significant, as a result of controlling factors provided in the design. The number of independent faults/conditions needed to cause some undesired event, as revealed in the diagram, is a gross means for evaluating the relative safety of alternate paths or design approaches. However, whenever feasible, the evaluation process should be as objective as practicable, utilizing the best available probability estimates for the various individual fault conditions identified. There are two basic approaches used to quantify fault tree probabilities. These are (1) calculation and (2) simulation. In both methods, the key factor is the assignment of the most realistic estimates of failure probability possible to the final failure mechanisms/conditions identified within each branch of the diagram.

The calculation approach can be used for fault trees which do not involve reparable faults (i.e., fault conditions which may be detected and corrected during the time covered by the analysis) or other time-related dynamic conditions which cannot be adequately accounted for by the logic diagram. For these more sophisticated evaluation requirements, the simulation approach is suggested.

For logic diagrams suitable for calculation or deterministic evaluation, the classical probability approach may be used, where each logic gate indicates the operation to be performed on the probability estimate inputs (i.e., union for "OR" gates and intersection for "AND" gates.) Fig. 12 provides the elementary Boolean Simplification.

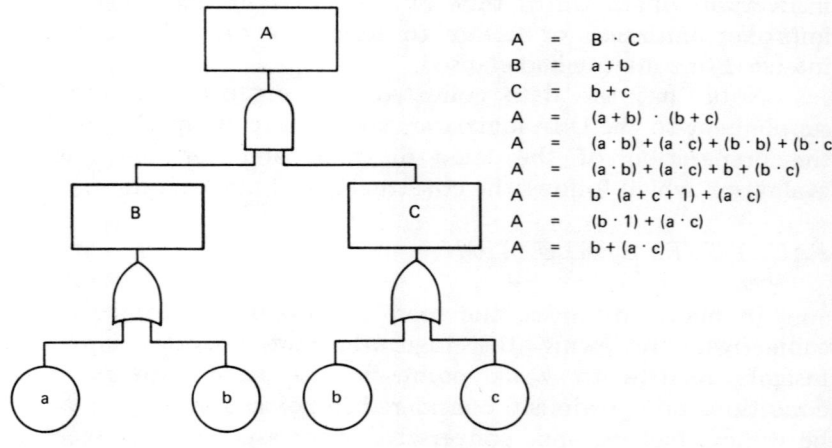

Figure 12. Boolean Simplification

Two major considerations which must be resolved when performing the calculation evaluation method are redundancies and consistent time base.

Redundancies, in FTA terminology, refer to events, faults, conditions, etc., which are similar enough in time and nature that, if they were to occur in real life, they would satisfy or significantly alter the likelihood of satisfying more than one entry within a given fault tree diagram. In situations where two or more redundant events exist within a fault tree, in a logical relationship such that they must pass through an "AND" gate before reaching the top event, a serious error could result in calculation of the top event probability unless the redundancy is removed or treated.

Multiplying the probabilities of the same occurrence will give an unrealistic low estimate of hazard probability. To avoid this error, the analyst must identify and remove redundant probabilities before performing his calculation. This can be done by direct observation of the logic diagram, or by reducing the fault tree to an algebraic expression (by assigning a code system to each noted event and cause) and then operating on this expression by Boolean algebraic theorems to remove redundancies, as shown in Figure 12.

SIMULATION

In the simulation approach, a fault tree is represented on a computer and the failures are simulated over a given mission length. The computer prints out the failure which leads to the undesired event, and the probability is calculated. The simulation approach has all the advantages of the calculation approach except for the greater amount of computer time needed to simulate fault trees with small probabilities. Simulation offers several additional advantages, i.e., the critical paths are listed and the computer can solve larger fault trees (often as much as 10 times larger by "calculation"). Simulation has gone through many stages of development; in its early stages, the amount of computer time required became prohibitive. However, special techniques (importance sampling) have reduced greatly the computer time needed.

The number of trials (a trial represents the predefined mission length of the system) required for an acceptable statistical confidence is reduced. With fewer trials required, computer time is recompensated for by calculating weight factors. Overall, simulation offers more potential and has proven to be more effective in calculating accurate answers than the computation method.

In addition to the design-oriented analytical procedures described above, other applications of system safety analysis techniques indicate methods which can be used to provide before-the-fact hazard evaluation of various types of industrial process operations.

Further refinements of these concepts and techniques should prove invaluable for effective before-the-accident controls in all forms of industrial operations.

RESOURCES

The following information is by no means intended as a complete bibliography on the subject, but rather as a selection of those resources which seem most applicable for the non-aerospace/military-related safety professional wishing to increase his understanding of either general or specific topics in the SSE field beyond what has been covered in this chapter.

Government Publications

Selected references on System Safety are noted below, with information on the availability of these and other publications and services. Write to: Superintendent of Documents, U.S. Government Printing Office, Washington, D.C. 20402.

Air Force Systems Command Design Handbook DH 1-6 "System Safety." Provides the most comprehensive coverage on safety design principles, information, guidance and criteria for aerospace (DOD and NASA) usage currently available. It also contains extensive reference information and a discussion on the application of SSE as a technical discipline. Periodically updated. May be obtained from ASD (ASNPS-40), Wright Patterson AFB, Ohio 45433.

Military Standard 882. "System Safety Program for Systems and Associated Subsystems and Equipment." Provides the requirements for system safety programs performed under military contracts. May be obtained from the Naval Publication and Forms Center, 58-1 Tabor Avenue, Philadelphia, Pa. 19120.

NASA Safety Manual 1700.1 and System Safety Requirement for Manned Space Flight (MSA Safety Program Directive No. 1A.). Contact NASA HQ. Director of Safety, Code DY, Washington, D.C. 20546, for availability of these and related NASA documents.

System Safety Conference Proceedings

Numerous technical papers have been presented on various aspects of System Safety. The following listed publications contain reprints of many of these under a single cover, which will be of assistance in obtaining a representative resource library in this area:

Government-Industry System Safety Conferences, periodically sponsored by the NASA Safety Office. The first was held May 1-3, 1968, and the second May 26-28, 1971. Papers were presented on a variety of topics, including Government Agency Programs, Specific System Safety Programs, Safety Analysis, Safety Information Data Centers, Risk Management, Land Transportation Safety and Product Safety. For sale by Superintendent of Documents, U.S. Government Printing Office, Washington, D.C. 20202 (Price $3.75.)

USAF Industry System Safety Conference, sponsored by Directorate of Aerospace Safety, USAF, February 25-28, 1969. Available from Deputy Inspector General for Inspection and Safety, USAF, Norton AFB, California.

Proceedings, 1969 Professional Conference, published by the American Society of Safety Engineers, 850 Busse Highway, Park Ridge, Illinois 60068.

Associations

Electronic Industries Association (EIA). The G-48 "System Safety" Committee is one of the 16 committees comprising the Government Products (G) Panel of the Engineering Department of EIA. The G-48 Committee currently has about 50 members who are the system safety experts within EIA member companies and participating government agencies. The G-48 provided the primary source of coordinated comments and recommendations from the system safety community available to industry and government. In addition to reviewing proposed government standards, policies and manuals related to system safety, the G-48 Committee prepares EIA Bulletins. To date, the following noted system safety publications have been released by EIA. These are available by contacting:

EIA Headquarters
2001 Eye Street, N.W.
Washington, D.C. 20006
Attn: Engineering Department

1. Safety Engineering Bulletin No. 1,
   "System Safety Education and Training Guide"
   August, 1968, Price $1.10.

2. Safety Engineering Bulletin No. 2,
"System Safety Bibliography"
August, 1970, Price $3.95.

3. Safety Engineering Bulletin, No. 3,
"System Safety Analytical Techniques"
May, 1971, Price $3.15.

System Safety Society — A professional society established to promote the development and dissemination of knowledge and techniques related to the system safety discipline. For further information on membership and publications, write to System Safety Society, 5630 Borwick Avenue, South Gate, California 90280.

SYSTEM SAFETY EDUCATION

Formal courses in system safety technology are increasing in availability and numbers. The following institutions and organizations currently provide such courses, and can provide further information, such as costs, requirements and curriculum:

1. George Washington University
   School of Engineering and Applied Science
   Continuing Engineering Education
   Washington, D.C. 20006
   Telephone (202) 676-6080

2. National Safety Council
   425 North Michigan Avenue
   Chicago, Illinois
   Telephone (312) 527-4800

3. University of Washington
   Office of Short Courses and Conferences
   336 Lewis Hall
   Seattle, Washington 98105
   Telephone (206) 543-5041

4. University of Southern California
   Institute of Aerospace Safety and Management
   University Park
   Los Angeles, California 90007
   Telephone (213) 746-2311

5. Northeastern University
   Center for Continuing Education
   360 Huntington Avenue
   Boston, Massachusetts

6. Texas A & M University
   Department of Industrial Engineering
   College Station, Texas 77840

XVII. A Loss Control Program For Alcoholism

Alcoholism, as well as addiction to other drugs (whether the "hard" drugs, or pills legitimately prescribed by a physician), poses a growing problem for industry. The executive who drinks his lunch, the accountant who can only get to work in the morning with the aid of a pep pill, the secretary who takes diet pills for months on end — all are drug abusers, and their dependence on chemicals means present or potential trouble on the job.

This chapter focuses on what a company can do about employees with the disease of alcoholism. The concepts presented here apply equally well, however, to the early identification of employees suffering from any form of drug addiction, as well as any other personal or medical problem that adversely affects job performance.

THE ALCOHOLIC EMPLOYEE

In 1968 the National Council on Alcoholism made an estimate of the prevalence of alcoholism among employees of American industry and civilian government. Using data available from representative employer organizations, NCA estimated that not less than 5.3 percent of the employee population studied was suffering from the disease of alcoholism in its early, middle or late stages.

Application of this percentage to a labor force of eighty million employees indicates that approximately four million employed men and women are suffering from this illness. The NCA study also found that a conservative estimate of the alcoholism-related losses to the employer ranged between $1,500 and $2,000 per year for each afflicted employee, for a total direct annual cost to employers of six to eight billion dollars a year.

From the standpoint of safety management alone, there are tremendous hidden hazards inherent in this group of at least four million employees. The increased probability of on-and-off-the-job accidents for the thousands of employees who operate plant machinery, are engaged in hazardous occupations or who drive company vehicles (including airplanes) is a major concern to safety engineers and to management.

There has been an increasing recognition of the magnitude and scope of the problem of employee alcoholism. Yet, despite an impressive body of evidence (see Bibliography at end of chapter), readily available and widely disseminated in the business press, most top American managers refuse to believe that at least four to eight percent of *their employees* suffer from alcoholism.

Since management normally devotes considerable time and attention to the objective study of other suspected areas of loss, the widespread reluctance to study employee alcoholism on a factual rather than speculative or emotional basis is puzzling. Some factors that might account for it are:

1. The archaic (but widely-accepted) notion that alcoholism is a self-inflicted moral problem, rather than an illness.

2. The powerful social and moral stigma associated with alcoholism, which makes companies reluctant to believe that any of *their* employees could have the problem.

3. The extraordinary skill of the victims in concealing and denying their illness.

4. The reluctance to accept or recognize the employee's problem until its terminal stages, when his visible symptoms begin to meet the criteria of the skid-row stereotype.

5. The widespread acceptance of alcoholic beverages as a social and business lubricant.

6. The fact that significant costs and losses caused by alcoholism are not visible in cost analysis reports. They are included, but are buried in the loss totals attributed to absenteeism, spoiled materials, accidents and other familiar categories.

7. A lack of information or awareness that tested systems are now available which have been remarkably effective in reducing and controlling the costs of employee alcoholism.

The extent of management's avoidance of this problem was expressed by Lewis F. Presnall in a talk reprinted in the *Congressional Record* (116:80, October 13, 1970):

". . . . . In spite of the fact that we know how to achieve an annual net saving of from five to ten dollars annually for each control dollar spent, probably not more than 250 to 300 corporations (out of more than 1,600,000) in the entire country have done anything toward installing control methods. Furthermore, most of these have what James Davidson, Executive Director of the Alcoholism Council of Greater Los Angeles, has called "pop-up toaster" programs. That is, if an alcoholic reaches the late middle stage of the illness and happens to pop up where he is visible, they grab him and try to get him to accept treatment.

. . . .

"The supreme irony of these facts is that for a decade the knowledge has been readily available, and demonstrated in practice, that enables us to extend the useful lives of employees and reduce these huge losses by a net, above control measure costs, of about 30%. If labor and management across the country were to apply the best technical knowledge now available, this would mean a net gain of about 3 billion dollars annually."

The discussion that follows will attempt neither to convince the reader that the problem exists nor to document any further its cost and magnitude. The primary focus here will be on the components of an effective system of control.

THE DISEASE CONCEPT OF ALCOHOLISM

Perhaps the most important point to be made about the kind of alcoholism prevalent in American industry is that it is a disease, recognized as such by the medical profession. The following is extracted from a book review of *The Disease Concept of Alcoholism*, by E. M. Jellinek (New Haven: Hillhouse Press, 1966). In this review, Ruth Fox, M.D., says:

"After defining alcoholism broadly as any use of alcoholic beverages that causes any damage to the individual or society or both, Dr. Jellinek goes on to describe the various types or *species* of alcoholism and, out of a minute scrutiny of the characteristics of various types and patterns of drinking, attempts to determine which might legitimately be considered as denoting a disease process . . . . .

"*Alpha alcoholism* is described as a continual dependence, psychological in nature, on the effects of alcohol to relieve bodily or emotional pain. This type of drinking usually does not denote a progressive process, nor does it lead to 'loss of control' or 'inability to abstain.' . . . . .

"*Beta alcoholism* is that species of alcoholism in which there is a resulting physical damage such as polyneuropathy, gastritis, or cirrhosis of the liver. These are complications and are probably largely due to the poor nutritional habits of the excessive drinker . . . . .

"*Gamma alcoholism* is one of the two species of alcoholism which can be considered as true diseases. In this type of drinking (1) a true tissue tolerance develops, (2) there is adaptation of the cell metabolism, (3) withdrawal symptoms due to physical dependence on alcohol cause a 'craving,' and (4) there is loss of control. This form of alcoholism shows a definite progression from psychological to physical dependence and there are marked behavior changes. It is gamma alcoholism which represents the typical form of alcoholic drinking on the North American continent, as well as in other Anglo-Saxon countries. This type of alcoholic can abstain from time to time, 'going on the wagon,' but he cannot control his intake once he starts to drink . . . . .

"*Delta alcoholism*, ... characteristic of certain wine drinking countries such as France, ... (is) the constant and daily drinking of small amounts of alcohol (with no actual loss of control) . . . . .

"*Epsilon alcoholism* has been used to designate a periodic drinking . . . . . The 'pseudoperiodic drinking' of the gamma or delta alcoholic is due to 'slips' or relapses in a person who is trying to abstain."

Of the five types of alcoholism described, only the gamma and delta types are true diseases, and only the gamma type is characterized by loss of control. This "typical form of alcoholic drinking on the North American continent" is the concern of this chapter.

The gamma alcoholic is the individual who cannot predict what will happen after he or she takes the first drink. Estimates vary, but it is generally agreed that one out of every twelve to fifteen people who drink any alcoholic beverage in any quantity or for any reason is prone to gamma alcoholism. It is this element of loss of control that causes the gamma alcoholic to become a problem employee. The gamma alcoholic is usually utterly incapable of arresting his disease without help.

There is a major difference between the disease of gamma alcoholism and all other diseases, which is difficult for the nonalcoholic to understand, and it must be dealt with if the disease is to be successfully arrested. *The gamma alcoholic does not want his disease treated.*

Before we assume that this is prima facie evidence of the alcoholic's general worthlessness, let's examine some of the results:

1. The social and moral stigma which has been associated with alcoholism throughout the years.

2. The pervading stereotype notion of the alcoholic as a moral degenerate, an irresponsible weakling, or a skid-row bum.

3. The contempt and scorn which characterize the most common attitude toward the victim of alcoholism.

4. The power of the physical addiction and the intense suffering caused by the withdrawal syndrome, which often results in death in the final stages of alcoholic convulsions or delirium tremens.

5. The alcoholic's sense of his helplessness to stop, often so powerful that it leads to suicide as the only solution.

Typically, the alcoholic is so terrified at the many (real or imagined) consequences of accepting treatment that he will not do so until the *consequences of not accepting treatment become more intolerable than accepting it.*

Such a feeling should be readily understandable. After all, we all know that the logical thing to do when a cavity develops in a tooth is to have it filled as soon as possible. Yet how many put off treatment because of terror of the dentist's drill? Only when the pain of the tooth becomes more intolerable than accepting treatment do most of us go to the dentist.

Thus, the most important aspect of a successful recovery from alcoholism is the *motivation to accept treatment.* This is not to imply that treatment and treatment facilities are not important. But experience has demonstrated that exposure to the finest facilities and the most effective treatment modalities are no help in the absence of motivation. On the other hand, a high degree of motivation often leads to a stable, long-term recovery, even where local treatment facilities are mediocre or inadequate.

INDUSTRY'S ADVANTAGE

In "The Beginning of Wisdom About Alcoholism" (*Fortune*, May, 1968), Herrymon Maurer says: "By putting the body of knowledge about alcoholism to work in company programs, industry is achieving recovery rates as high as 65 to 70 percent — higher than those for other major diseases and far higher than any imagined only a short time ago."

How can we account for this recovery rate? *Industry has the most effective motivational tool known to date: the desire of the employee with alcoholism to hold his job.*

The most effective system of employee alcoholism control developed to date requires no technical or medical knowledge on the part of management; it requires rather the utilization of management skills to motivate employees to accept treatment. Since alcoholism has been formally recognized as an illness by the American Medical Association, and by many other scientific bodies, it should be handled like any other illness; that is, it should be diagnosed and treated by qualified professionals.

Unfortunately, many pioneer programs failed, wholly or in part, because they assumed it was necessary to train supervisors to become "experts" on alcoholism so that they could become diagnosticians or counselors. This idea is still widespread today, and it should be discarded at once by any company that wants to install an alcoholism control program.

COOPERATION BETWEEN LABOR AND MANAGEMENT

The cooperation of both labor and management, in a totally coordinated effort to achieve mutual objectives, is vital to the success of any alcoholism control system. Successful rehabilitation requires high employee motivation, which neither labor nor management can accomplish, working unilaterally.

Labor, because of its relationship with its members, can give understanding and a sympathetic offer of assistance, counseling and treatment. Yet there is overwhelming evidence that alcoholics rarely respond favorably to sympathy. Their most common reaction is denial, accompanied by intense resentment of the implication that they may have alcoholism.

Management has an effective approach through its legitimate

right to initiate corrective action when the employee's job performance falls below reasonable minimum standards. At this point, maximum motivation can be encouraged by giving the employee a clear choice between accepting the standard administrative consequences of his unsatisfactory job performance or accepting confidential counseling and diagnostic services.

This opportunity is lost if the union contests management's action by grievance procedures. No matter who wins, the employee is the loser. A union victory in effect kills the member with kindness, since he is freed from the necessity to choose the professional help that may save his life. A management victory, in the absence of an enlightened program, would probably result in punitive action rather than in urgently-needed treatment.

Only when labor and management work together in good faith toward the common goal of the employee's restoration to good health and productivity can maximum results be achieved.

AN EFFECTIVE SYSTEM OF EMPLOYEE ALCOHOLISM CONTROL

We will outline the essential components and discuss the general principles involved in each step of a proposed alcoholism control program to furnish a practical set of guidelines and a frame of reference, which can be smoothly integrated with the existing operational and personnel policies of the particular organization, and which can be tailored to its unique needs and characteristics.

The essential factors and considerations are: (1) Separation of Management and Treatment Functions; (2) Assignment of Administrative Responsibility; (3) Formulation of Policy; (4) Definition of Management and Labor Functions; (5) Development of Procedures to Implement Policy; (6) Training; (7) Referral for Treatment; (8) Insurance Coverage; (9) Record-keeping and Program Evaluation; (10) Role of Program Administrator.

SEPARATION OF MANAGEMENT AND TREATMENT FUNCTIONS

The total system of alcoholism control falls naturally into two major functional subdivisions: (1) management functions and (2) functions to be accomplished by medical and paramedical professionals. For the sake of clarity, these will be referred to as "pre-treatment" and "treatment," respectively.

Pre-treatment: This phase focuses on functions to be performed by the company, and these should be completely divorced from treatment for the disease. All pre-treatment functions fall within the basic framework of standard managerial and supervisory responsibilities. There is absolutely no necessity to be concerned with identification of alcoholism as such ... or with any aspect of counseling, diagnosis, or treatment success or failure. The success of the pre-treatment phase depends only on basic principles of sound management and good labor relations.

Treatment: This phase is wholly the concern of the qualified professional. Management's only concern with this phase is establishment of an adequate referral system.

ASSIGNMENT OF ADMINISTRATIVE RESPONSIBILITY

Since the pre-treatment phase of the control program is concerned solely with performance of managerial functions, administrative responsibility for this program component should be assigned to an appropriate line executive.

Administrative responsibility for the treatment phase should be assigned to a qualified medical officer or administrator. Many companies do not have medical departments, but this is no bar to the effectiveness of the program. It merely means assignment of treatment responsibility to an appropriate existing qualified resource or facility, in the same way such a company refers employees with other illnesses to appropriate sources of treatment.

No program is complete with only a treatment phase, though many believe that this is true. The only programs that achieve anything close to their maximum effectiveness are those that give careful and meticulous attention to the pre-treatment phase.

FORMULATION OF POLICY

The most effective company control programs are those that have adopted a formal policy on alcoholism. Why, if alcoholism is to be treated like any other disease, should we write a special policy for this illness?

It is necessary because in many companies there is a tacit, unwritten policy on alcoholism, which is well understood by employees. It goes something like this: "The company will award cash and other economic premiums to any employee who can successfully conceal his alcoholism from the attention of management. Such premiums will include sick leave pay, job security, fringe benefits and promotional opportunities. When the employee can no longer conceal his alcoholism, his employment will be terminated."

This unwritten policy is the result of the effects of the social and moral stigma associated with alcoholism on the customs, attitudes and actions of the company. It has proved very difficult for a company to convince employees or unions that the company intends to reverse this policy, unless it is willing to put this intention in writing.

Many companies are fearful that a constructive written policy will adversely affect their public image. The theory is that the public would assume the company was loaded with alcoholics. This underestimates the intelligence of the public. Most companies report extremely favorable reaction to such programs. A full-page multi-color advertisement in most of the country's leading periodicals about the Bethlehem Steel alcoholism program, for example, received overwhelmingly favorable public response.

Another common error in the formulation of a company policy is preparation of a statement of rather pious generalities about alcoholism, avoiding precise statements or clear-cut commitments. Such statements are painfully obvious, deceive no one and accomplish nothing.

An effective statement of company policy on alcoholism should provide a clear and unmistakable outline of the company's purpose, position, attitude and intended actions with respect to employees with alcoholism. If full fringe benefits are to apply, this should be spelled out; if not, this too should be clear. The too-prevalent tendency to philosophize or editorialize should be avoided.

The following sample statement of policy includes the major points that should be covered:

Company Statement of Policy

1. The company recognizes alcoholism as an illness that is treatable.

2. The purpose of this policy is to assure that any employees having this illness will receive the same careful consideration and offer of treatment that is now extended to all our employees having any other illness.

3. The social stigma often associated with this illness has no basis in fact. It is expected that a realistic acceptance of this illness will encourage employees to take advantage of the available treatment when needed.

4. The company's concern with alcoholism is limited strictly to its effects on the employee's performance on the job. It is not concerned with social drinking. Whether an employee without alcoholism chooses to drink or not to drink socially is of concern only to the individual.

5. For the purposes of this policy, alcoholism is defined as an illness in which an employee's consumption of any alcoholic beverage definitely and repeatedly interferes with his job performance and/or his health.

6. It will be the responsibility of all supervisors to implement this policy and to follow the procedures assuring that no employee with alcoholism will have his job security or promotional opportunities jeopardized by a request for treatment.

7. It is recognized that supervisors do not have the professional qualifications to permit them to judge whether any employee has alcoholism. Necessary referral for diagnosis and treatment will be based strictly on unsatisfactory job performance which cannot be corrected through the company's standard corrective procedures or through the employee's own efforts to improve his performance.

8. An employee's refusal to accept diagnosis and treatment, or failure to respond to treatment, will be handled exactly as similar refusals or treatment failures are handled for all other illnesses, when the results of failure or refusal continue to affect job performance.

9. It is expected that through this policy, employees who suspect that they may have an alcoholism problem, even in its early stages, will be encouraged to seek diagnosis and to follow through with prescribed treatment, if necessary.

10. The confidential nature of the medical records of employees with alcoholism will be preserved in the same manner as all other medical records.

11. Implementation of this policy will not require, or result in, any special regulations, privileges or exemptions from the standard administrative practices applicable to job performance requirements.

12. The illness of alcoholism will receive the same financial benefits and insurance coverages provided for other illnesses under established employee benefit plans.

Such a policy is an adequate frame of reference for the guidance of all personnel charged with the responsibility of implementing the policy.

DEFINITION OF MANAGEMENT AND LABOR FUNCTIONS

In general terms, the functions of labor and management respectively in the pre-treatment phase of the program are:

Management Functions

1. To create a company-wide climate by all available means, including employee education, which will gradually eliminate the effects of the social stigma associated with alcoholism, which acts as a barrier to constructive corrective action.

2. To enlist the cooperation and support of labor in implementing the pre-treatment program. Cooperation is essential, for the objectives are neither pro-management nor pro-labor, but rather "pro-people."

3. To bring the full capabilities of the supervisory staff, from top management to front-line foremen, to bear on the early identification and motivation to treatment of possible cases of alcoholism or other problems that affect job performance.

4. To assign clear-cut responsibility to all levels of supervision and labor in implementation of pre-treatment procedures and to develop a positive program of follow-up to assure that such procedures are consistently followed on a continuing basis.

5. To schedule initial and follow-up orientation and training meetings for all supervisory personnel, including top management, labor leaders, front-line foremen and other labor representatives.

6. To maintain adequate, continuing written records covering day-to-day patterns of absence, unsatisfactory work performance, formal and informal disciplinary action taken and any other relevant data that indicate developing employee problems.

Union Functions

The union functions in the pre-treatment phase are entirely consistent with the established basic functions of the shop steward and other union representatives. In implementing the agreed-upon policy and procedures, the union will represent the members' interest through the assurance that:

1. The employee's job security and promotional opportunities are not jeopardized by a request for diagnosis and treatment.

2. The focus of corrective interviews is restricted to the issue of job performance rather than judgments on alcoholism (unless a violation of work rules against drinking on the job is involved.)

3. The confidential nature of medical records is preserved.

4. Any employee having this illness will receive the same careful consideration and offer of treatment extended to employees who have any other illness.

5. All other rights and privileges inherent in the agreed-upon policy and procedures are protected.

Cooperation between shop stewards and supervisors working together to motivate employees to accept needed treatment should significantly reduce alcohol-related grievances. The shop steward is also in an excellent position to encourage the earliest possible utilization of union counseling services for members with a problem and to stimulate self-referrals for treatment.

Development of Procedures to Implement Policy

1. The chief operating executive of the company should designate a top-ranking executive from the *line* organization to assume responsibility for the effective implementation of the management functions of the pre-treatment program. He

will serve as one of the co-chairmen of a committee charged with developing the procedures to be followed by all levels of supervision in implementing policy.

2. The union will be asked to delegate a top-ranking labor representative to be responsible for implementing the union functions of the pre-treatment program. He will serve as the other co-chairman of the committee.

3. The co-chairmen will appoint the other members of the committee, who might include representatives from industrial relations, personnel, medical staff, labor and others involved in implementation of the program.

4. The committee would be charged with developing procedures for all involved personnel, to implement the policy. Procedures should be developed that:

   a) Establish minimum supervisory requirements for documentation of unsatisfactory work performance and recording of corrective action taken.

   b) Outline action to be taken by supervisors in offering professional counseling and diagnostic services to employees unable to satisfy minimum performance requirements or to correct job deficiencies without assistance.

   c) Outline the action to be taken by the supervisor with respect to employees whose performance continues to be unsatisfactory but who refuse professional assistance.

   d) Establish channels of communication with higher authority, union representatives and staff, to assist supervisors in cases where they need assistance or guidance on particular problems.

All the procedures developed must have the full support of both labor and management. Even though some of the procedures (such as keeping management records) must be administratively accomplished by the supervisor, it must be understood that all procedures are cooperatively followed to accomplish the identical goal of labor and management: to motivate the employee to accept the treatment necessary to restore him to full health and productivity.

TRAINING

The word "supervisor" as used in this discussion refers to all supervisory personnel, from the chairman of the board and the president through all levels, including the front line foreman. It also includes all appropriate labor representatives who are involved in implementing the alcoholism program.

It is important for the training program to begin with the top levels of supervision and continue through all levels. The training should include:

1. A thorough review of the company policy, its intent and specific provisions.

2. Explanation of the distinction between the pre-treatment and treatment components. It must be made absolutely clear that all procedures in the pre-treatment phase are restricted to basic management and supervisory functions and exclude all treatment functions, including counseling and diagnosis.

3. A general orientation, to provide better understanding of the nature of the problem of alcoholism. This should cover the following basic points:

    a) The key to the successful motivation of an employee with alcoholism lies in the supervisor's use of his authority in a fair and constructive manner.

    b) The social and moral stigma associated with alcoholism makes the sufferer reluctant to admit his problem.

    c) In most cases, an employee with alcoholism is aware that his drinking is unlike that of most of his friends. As the disease progresses, he becomes increasingly aware that his drinking is becoming more uncontrollable.

    d) Such an employee generally knows that if he continues to drink he will continue to have job problems. If he *could* control his drinking, he *would*. He tries and fails repeatedly. Lectures and threats are as useless as lectures to a tubercular patient about his coughing; logic will arrest neither illness. Neither lectures nor threats motivate the employee toward the one essential element: specialized treatment.

e) Experience with thousands of cases has demonstrated that a mere offer of treatment is also ineffectual, for it does not, by itself, outweigh the intense fear of the effects of social stigma. Both assurance that the acceptance of treatment will not result in job loss and pressure to seek treatment are needed to outweigh this fear.

f) Industry has a legitimate tool ... its right to expect satisfactory job performance ... which is now being used successfully by a number of corporations to give this needed pressure. It must be exerted through realistic and uniformly firm utilization of existing procedures.

Many other illnesses also affect job performance, as do other serious behavioral-medical problems. Within the focus of a realistic company program, these can also be detected. The company will benefit by taking corrective action when any illness adversely affects job performance. There is no need for the supervisor to know the nature of the illness.

The basic job responsibilities of all levels of supervision in the pre-treatment phase of an alcoholism program are:

1. To be alert, by continuing observation, to changes in the work patterns of all personnel under their supervision.

2. To document all instances when an employee's performance fails to meet minimum established standards, or where his individual pattern of performance appears to be deteriorating.

3. To conduct a corrective interview with the employee when such action is warranted by his record of unsatisfactory performance. At the end of this interview, the supervisor will inform the employee that the company offers confidential counseling and diagnostic services in case his poor performance is caused by any personal problem. Upon acceptance, the employee will be referred to a designated source of professional screening and diagnosis.

4. If the employee refuses help and his performance continues to be unsatisfactory, he is given a firm choice between accepting company help through diagnosis and treatment, or accepting whatever existing procedures are followed in all cases of unsatisfactory job performance.

Any deviation from a firm and consistent administration of the procedures, based on misguided feelings of sympathy, can lead only to a serious delay in treatment, which is hazardous to the employee's health and ultimate recovery and well-being. *An employee will rarely accept treatment unless the consequences of not accepting treatment create an alternative that is more intolerable to him than his fear of the results of exposure.*

The above procedures are acceptable to supervisors because they:

1. Make involvement with personal or embarrassing discussions about alcoholism unnecessary.

2. Restrict all disciplinary action to its legitimate function of corrective action for unsatisfactory performance.

3. Give the supervisor specific directions for dealing effectively and comfortably with a type of problem that has hitherto caused him great difficulty.

This orientation should be followed with a clear definition of the specific responsibilities of the supervisors and/or labor representatives in implementing the program and the procedures.

Charts and other visual aids may help by:

1. Illustrating the forms for documenting unsatisfactory performance, records of corrective action, etc.

2. Review of job standards and minimum performance requirements.

3. Review of standard corrective procedures.

4. Charting proper channels of communication and coordination with labor representatives, higher line authority, industrial relations, personnel, medical and other staff personnel.

REFERRAL FOR TREATMENT

In this step, the responsibility for action is transferred from the pre-treatment administrative authority to the treatment administrative authority.

It must be kept in mind that the referral is based solely on two factors: (1) The employee's performance has fallen below minimum acceptable standards; and (2) he has been unable to upgrade his performance by his own efforts, the supervisor's efforts or normal corrective interviews, which focus on an objective view of substandard performance factors.

The employee's deteriorating job performance may be caused by any of a broad range of personal or medical problems other than alcoholism. No member of the supervisory management team is qualified to diagnose alcoholism (or any other illness) as the cause of the work problem.

It follows that the first referral must be made to a facility that is staffed and equipped to provide not only a complete medical examination but also broad diagnostic services, rather than to a facility specializing in alcoholism.

One crucial point must be emphasized here. Alcoholics are extraordinarily successful, particularly in the early or middle stages of the illness, in convincing even physicians, nurses, psychologists, social workers and other professional and paramedical specialists that they do not have alcoholism. This is not true when the professionals to whom they are sent have training and extensive experience in dealing with alcoholism. Such training and experience is unfortunately not common. It is important, therefore, to be sure that there is at least one person on the diagnostic staff of the treatment facility who is thoroughly familiar with alcoholism, impervious to the alcoholic's skill in concealment and able to diagnose and prescribe the best course of treatment for the individual.

Extra Benefits of the Program. The focus on job performance, rather than alcoholism per se, promotes early identification of *any* personal or medical problem that adversely affects job performance, including all types of drug addiction. Clearly, then, it is in the interest of both labor and management to provide the employee with the confidential diagnostic and counseling or medical services he needs to restore him to full health and productivity, regardless of the nature of his problem — medical, psychological or situational. His problem might be neurosis, psychosis, emotional distress, marital trouble, financial difficulty or other medical or personal problems. To focus this program solely on the disease, alcoholism, would be to needlessly exclude the broad benefits to all parties that come with early identification of all problems that adversely affect the employee's productivity, health and well-being.

Treatment. The *only* involvement of the employer with the treatment phase is the responsibility for activating referral systems for the diagnosis and treatment of alcoholism which are as effective as the referral procedures for other illnesses.

The first step is to identify and evaluate the effectiveness of all existing community alcoholism treatment facilities and resources. The next is to establish a good personal liaison and working relationship between the diagnostic and counseling service and the community treatment facilities. Such resources will include a roster of professional people selected on the basis of their experience, skill and competence in dealing with alcoholism. They will include selected physicians, psychiatrists, clergymen and other qualified counselors. Other resources will include hospitals and clinics with physicians and other staff members who have training in alcoholism, institutions specializing in detoxification and alcoholism treatment, local voluntary councils on alcoholism, state alcoholism agencies, state hospitals qualified in the treatment of alcoholism, group-psycho-therapy facilities, family service agencies, union counseling services and Alcoholics Anonymous.

Alcoholics Anonymous can make a major contribution to the success of the treatment program, both in terms of cooperation and skilled services its members are willing to give, and in the accomplishment of the long-term continuing supportive therapy essential to stable recoveries.

The community services divisions of the AFL-CIO, UAW and other unions have developed effective alcoholism counseling and treatment referral services in many communities.

A person should be designated by the company as responsible for identifying all qualified alcoholism treatment facilities and establishing liaison and good working relationships with them.

The National Council on Alcoholism can provide additional information and assistance in this area through its national and regional labor-management services and its national network of affiliates and associate organizations located in metropolitan areas throughout the United States.

INSURANCE COVERAGE

If company policy recognizes alcoholism as a "legitimate" illness, its hospital and medical coverages should provide benefits consistent with those provided for other illnesses. It is common, however, to find clauses in group hospital and medical insurance coverages that exclude "alcoholism, drug addiction or other 'self-inflicted' problems."

In a Position Paper on Alcoholism Treatment Insurance Coverage, a task force of the state of Wisconsin stated:

> "Theoretically, an insurance company will write any kind of coverage requested by the purchaser, including coverage for treatment of alcoholism. In reality, many policies are written either with the exclusion of alcoholism or with reduced benefits. The reasons for this are included among the following issues: (1) lack of willingness on the part of the purchaser to pay the additional costs required for the extended benefits; or (2) because of a reluctance on the part of the insurance company to recommend coverage, fearing that the excessive costs may cause the contract to be less competitive. When coverage is provided, it is often limited by the following: (1) coverage only in a *general hospital*, with a limited number of days allowed which may vary from 15 to 30 days and may limit this to only once a year; (2) coverage limited under the *sanitoria* clause providing for a predetermined per diem rate and a designated number of days; or (3) coverage limiting not only the number of days for hospitalization but also the percent of cost coverage."

Alcoholism exclusions are based on several assumptions, now known to be invalid. The first is that gamma alcoholism can be controlled or arrested without treatment by any "responsible" employee, and that its elimination is a matter of simple choice. Hence the term "self-inflicted."

The second, possibly more important, assumption is that the insurer would save money by excluding payment of benefits for alcoholism. Yet studies of many company claim records establish that they regularly pay hospital and medical claims actually resulting from alcoholism, but disguised under diagnoses such as gastritis, anxiety neurosis, cirrhosis, virus and other covered illnesses — not to mention the many on- and off-the-job accidents and personal injuries resulting from alcoholism, where the true cause does not appear in the records.

A third assumption is that anyone with alcoholism requires extended psychiatric treatment, which is very costly. While it is true that many victims of alcohol (in common with many other people) need psychiatry, experience has shown that the majority of recoveries from alcoholism are achieved without psychiatry, or after extended psychiatric treatment has proved ineffective and the psychiatrist has referred the patient to Alcoholics Anonymous or some other form of long-term treatment.

It is not surprising, therefore, that there is no evidence of increased claim costs after alcoholism exclusions have been removed.

The same fallacies apply to coverages that specify reduced hospitalization periods or reduced benefit payments for alcoholism claims, on the theory that this will reduce such claims. This approach is self-defeating, for it merely transfers the payment of these claims back to acceptable headings, such as "gastritis."

Experience has also demonstrated that hospitalizations for alcoholism under disguised diagnoses result in "revolving door" cases, whereas dealing with alcoholism frankly — as alcoholism — results in a much higher probability of permanently arresting the disease.

We have seen, then, that (1) a policy that recognizes alcoholism as an illness must provide benefits consistent with those provided for other illnesses and therefore must eliminate alcoholism exclusions; and (2) no monetary loss is involved in removing alcoholism exclusions, since such claims are being paid anyway under different terminology.

There is a further opportunity to reduce alcoholism claim costs: by negotiating with the carrier to cover the costs of in-patient treatment in selected institutions that specialize in treatment of alcoholism. There are two major advantages:

1. These institutions are staffed with specialists trained in alcoholism, who focus not only on restoration to physical health but also on the specialized treatment techniques that are essential in permanently arresting the patient's dependence on alcohol. Such institutions have significantly higher percentages of patients who achieve long-term, stable sobriety.

2. The cost of in-patient care in these specialized institutions is significantly below that of general hospitals, because of greatly reduced overhead and equipment requirements.

Some insurance companies are capitalizing on these advantages by evaluating and qualifying such specialized alcoholism facilities for inclusion in client coverages.

RECORDKEEPING AND PROGRAM EVALUATION

Adequate records are important, not only as a source of the essential information needed with respect to the history, treatment and progress of the individuals concerned, but also as a basis for continuing evaluation of the overall program.

Two primary requisites of a good system of records are: (1) adequate recording of essential and relevant information; and (2) the careful preservation of the confidential nature of all records pertaining to the employee's disease.

A good example is provided in the guidelines for Federal Civilian Employee Alcoholism Programs, developed by the Occupational Health Division, Bureau of Retirement, Insurance and Occupational Health, United States Civil Service Commission.

These guidelines are set forth in FPM Letter No. 792, Section VIII:

"VIII. RECORDS AND REPORTS

"1. Maintenance of Records on Individuals

"General supervisory documentaion of employee job performance and actions taken to motivate correction of job deficiencies should be maintained, as with all employee records, in a strictly confidential manner. The responsibility for developing a responsive and useful job performance documentation system rests with agency officials.

"Records on employees who have been referred for counseling, whether by medical, or personnel of other counseling specialists, should be maintained in the strictest confidence and accorded the same security and accessibility restrictions provided for medical records.

"Records containing medical information and reports must be maintained according to requirements prescribed in FPM Chapter 293, Sub-chapter 3-3.

"Official Personnel Folders shall not include information concerning an employee's alcohol problems or efforts to rehabilitate him except as they apply to specific charges leading to disciplinary or separation actions.

"2. Statistical Reports

"Agency Program Administrators should compile sufficient statistical data to provide the basis for evaluating the extent of alcoholism problems and the effectiveness of the counseling program. Reports will be prepared on a regular basis; a report will also be submitted to the Civil Service Commission annually.

"The report to the Commission will include for each fiscal year beginning with Fiscal Year 1972: (1) the number of employees counseled by medical, personnel, or other counseling specialists where the counselor concluded that problem drinking was an issue and (2) the number of employees identified as having been helped through the alcoholism program. (This is followed by details of due dates of reports and to whom they will be submitted.)

Care should be taken that such records are purely statistical and DO NOT IDENTIFY INDIVIDUALS."

The data made available through an adequate system of records can be used as the basis for developing additional data valuable to both labor and management. The largely hidden costs of alcoholism can gradually be identified. The impact of alcoholism on such problems as absenteeism, accidents, quality control, spoiled materials and hospital and medical costs can be identified. This can lead to improved testing and measurement of various motivational and problem-solving approaches.

The development of improved data and more precise information on the scope and effect of alcoholism within the employer organization will lead to the development of more effective solutions.

ROLE OF PROGRAM ADMINISTRATOR

The preceding discussion has focused on the general principles, essential components and design of a system of loss control management of employee alcoholism.

However, the effectiveness of any such system, no matter how well conceived, requires that adequate provision be made to assure its implementation. The most successful existing company programs have assigned responsibility for such implementation to a full-time Program Administrator.

Once a policy has been approved, and procedures have been developed to implement the policy, it is important that there be an assignment of responsibility for achieving all program objectives on a continuing basis.

An example of this function is provided in the guidelines of the United States Civil Service Commission in the previously quoted FPM Letter No. 792:

> "Once a policy and plan has been approved, it is important that there be continuing coordination and assessment of program activities. To accomplish this, a Program Administrator should be designated at the headquarters level to direct the program on an agency-wide basis. Additionally, an individual should be designated at each field installation to coordinate local operations of the program. Individuals selected for such assignments should be allotted sufficient official time to effectively implement the agency policy and program, including bringing education and information to the work force, arranging or conducting supervisory training, developing and maintaining counseling capability (personnel, medical or other counseling resources), establishing liaison with community education, treatment and rehabilitation facilities, and evaluating the program and reporting to management on results and effectiveness.
>
> "..... There is no need to seek out recovered alcoholics to assume key roles, although some recovered alcoholics perform in an excellent manner because they are strongly motivated and knowledgeable in this area. However, if a recovered alcoholic is assigned as a Program Administrator or Program Coordinator, he should be familiar with treatment methods other than the one that was successful for him. It is just as essential that the individual selected be an experienced and effective administrator."

As the Civil Service guidelines have recognized, the person selected must be *an effective administrator*. This will include his development of follow-up systems that will assure maximum implementation of designated procedures at all levels of supervision, maintenance of essential records, development of collection of data which will be useful in identifying hidden alcoholism, related costs and program evaluation, and assuring the continuing cooperative support of the program at all levels of management and labor.

The individual selected should ideally have enough management experience to be familiar with the problems of management in production, labor relations, cost control, personnel, group insurance and other management areas. He should also be capable of dealing tactfully and effectively with people at all levels of labor and management.

Another important consideration is placing this individual in the table of organization so that he will have maximum mobility in communication. Alcoholism occurs with equal frequency at all levels and involves, particularly at the higher levels, organizationally sensitive situations.

His effectiveness will be seriously hampered if his activities must "go through channels" or if he must operate within a "chain of command." This type of immunity from the usual structured channels of communication is also essential to the confidential nature of many of his activities.

Some companies have solved this problem by having the Alcoholism Program Administrator report directly to the chief operating executive. If this is not feasible, it is important to arrange that he is provided with the latitude for "crossing lines," regardless of his formal designation within the table of organization.

SUMMARY

The foregoing discussion is not offered as an exhaustive treatment of Loss Control Management of Employee Alcoholism. Rather, it attempts to outline in broad general terms some of the basic information, general principles, and the considerations which will serve as a basis for development of sound decisions and constructive labor and management action.

The writer's purpose will have been accomplished if the material is useful as a starting point to those executives in labor and management who are aware of the major impact of alcoholism, and who wish to pull it out from under the rug and expose it to constructive remedial action.

Assistance with specific questions or problems is available on request from the Labor-Management Services Division of the National Council on Alcoholism.

BIBLIOGRAPHY

"The Alcoholic Executive," *Fortune*, January, 1960.

Alcoholism and the Federal Employee: Report on a Training Conference, (U.S. Public Health Service Publication No. 2020), Washington, D.C.: U.S. Health Services and Mental Health Administration, Division of Federal Employee Health, 1970.

Alcoholism in Industry: Collection of Selected Articles for the Guidance of Physicians in Industry. Chicago, Illinois: Industrial Medical Association, 1966.

Ball, John, *Alcoholism in Industry: A Self-Instructional Course for Supervisors and Managers.* East Lansing, Michigan: Pergamon Press, 1967.

"Business Copes with Alcoholics," *Business Week*, October 26, 1968.

Coffey, C.W., "Alcoholism in Industry: A $100 Million Hangover for Texas." *Texas Business Review*, 43:6, pp. 1-8, 1969.

A Cooperative Labor-Management Approach to Employee Alcoholism Programs. New York, New York: National Council on Alcoholism, Labor-Management Services Department, 1970.

Dana, Allan H., "Problem Drinking in Industry: Study of Industry's Implications of Alcoholism in Florida." *Research Reports in Social Science*, Institute for Social Research, Florida State University, 6:1, 1963.

Employees with Drinking Problems: A Management Guide. New York, New York: American Telephone and Telegraph Company, Long Lines Department, 1964.

"Folklore and Facts about Employees with Alcoholism." *Journal of Occupational Medicine*, April, 1967, pp. 187-192.

Habbe, Stephen, *Company Controls for Employees with Drinking Problems.* (Studies in Personnel Policy, No. 218) New York, New York: National Industrial Conference Board, 1969.

Hamilton, Andrew, "Business and the Compulsive Drinker." *Reader's Digest*, November, 1969.

"Hidden Alcoholic Employees," *Progress*, 3:3,4, 1961-1962.

"High-Priced Half Man," *Journal of American Insurance*, January, 1971.

Johnson, Harrison R., "Alcoholism — A Man and Profit Destroyer." *Modern Office Procedures*, December, 1962.

Juster, Jacqueline, "Sorry. Mr. Jones Is In Conference." *New Jersey Business*, July, 1971, pp. 38-41.

Kelley, James W., "Case of the Alcoholic Absentee." *Harvard Business Review*, May-June, 1969.

MacIver, John, *Industry and Alcoholism — Where Do We Go from Here?* New York, New York: National Council on Alcoholism, 1963.

Manes, Peter R., "Management of the Alcoholic in Industry," *Journal of Occupational Medicine*, February, 1966.

Margetts, Susan, "The Staggering Cost of the Alcoholic Executive." *Dun's Review*, May, 1968.

Maurer, Herrymon, "The Beginning of Wisdom about Alcoholism." *Fortune*, May, 1968, pp. 176-178 ff.

Maxwell, Milton A., "Early Identification of Problem Drinkers in Industry." *Quarterly Journal of Studies on Alcohol*, 21:4, 1960, pp. 655-678.

Norris, John L., "A Program for Alcoholics in a Company," in Collins, Ralph T. (ed.), *Occupational Psychiatry*. Boston, Massachusetts: Little, Brown & Co., 1969.

Parks, Seigle W., "Alcoholism and Industry." *West Virginia Medical Journal*, January, 1970.

Presnall, Lewis F., *Alcoholism — An Employee Health Problem*. New York, New York: National Council on Alcoholism, 1962.

"Prevalence of Alcoholism Among Employees in Business." *Labor-Management Services Bulletin*, March, 1971. New York, New York: National Council on Alcoholism.

Rouse, Kenneth A., "The Employee with a Drinking Problem." *National Insurance Buyer*, November, 1964.

Seidel, Leon E., "Alcoholism: Industry $4 Billion Albatross." *Textile Industries*, September, 1970, pp. 55-71.

Sheridan, P.J., "Alcoholism — The $2 Billion Theft." *Occupational Hazards*, April, 1964.

Sheriff, Don R. (ed.), *The Problem Drinker in Industry*. (Conference Series No. 6). Iowa City, Iowa: State University of Iowa, 1963.

Silverman, Milton, *Industry's Stake in U.S. Alcoholsim Plans*. Philadelphia, Pennsylvania: 1967.

"A Study of Absenteeism, Accidents and Sickness Payments in Problem Drinkers in One Industry." *Quarterly Journal of Studies on Alcohol*, 20:2, pp. 302-312, 1959.

Substantial Cost Savings from Establishment of Alcoholism Program for Federal Civilian Employees. U.S. Comptroller General, General Accounting Office, September, 1970.

Thomas, Patrick A., "Alcoholism Gets a Treatment from Business." *Business Insurance*, December, 1970, and January, 1971.

Trice, Harrison, *Alcoholism in Industry — Modern Procedures*. New York, New York: Christopher D. Smithers Foundation, 1962.

Von Wiegand, Ross A., "The Problem of Alcoholism." *Environmental and Safety Management*, March, 1971.

Wooley, Doris E., "The American Alcoholic — a Handicapped Worker." *Employment Service Review*, August-September, 1967, pp. 4-7.

Zentner, Arnold S., "Alcoholism and the Job," in Collins, Ralph T. (ed.), *Occupational Psychiatry*. Boston, Massachusetts: Little, Brown & Co., 1969.

Rehabilitation .... XVIII.

Although money awarded to an individual who is disabled as a result of a serious accident or illness is a legal and moral responsibility, it is often regarded by some in our society as an unnecessary expenditure. Most people have little sympathy for employers or insurance companies who pay thousands, even millions of dollars, in medical costs to an injured party. They believe that if money is to be spent, it should be directed toward controlling and possibly eliminating accidents and illnesses which could result in lifelong disability. While it is true that loss prevention and control are vital, rehabilitation is also important. Both disciplines are interdependent. Loss prevention helps to eliminate the causes of the accident "before-the-fact", while rehabilitation helps to reduce the "after-the-fact" economic and emotional consequences. A well-integrated combination of the two can effectively help management control costs that are incurred as a result of serious mishap. A unified program of loss prevention and rehabilitation can also assist in enhancing the corporate image, by demonstrating management's concern and involvement in the welfare of our society. Thus, loss prevention and rehabilitation are partners, helping to eliminate, control or reduce accidents and resultant losses.

WHAT IS REHABILITATION AND WHAT ARE ITS DETERRENTS?

Despite the numerous and varied definitions of rehabilitation, the most widely accepted is offered by the National Rehabilitation Association: "Rehabilitation is the process of restoring the handicapped individual to the fullest physical, mental, social, vocational and economic usefulness of which he is capable".

The end product of rehabilitation, therefore, is productive, dignified work. In order to more clearly understand this, perhaps it would be best to briefly discuss the ideology of the rehabilitative process. Rehabilitation is opposed to malingering rewarded by empathy and handouts of a welfare "do-gooder" nature. Experience has proved that there are few disabled malingerers who have not been created by laws, handouts of various benefit programs, and legal largesse. Since it is basic to the nature of most individuals to work, and subsequently to receive the fruits of their labor, it may be justifiably stated that most humans would rather contribute than act as parasites upon society. Rehabilitation can be the key to realizing these personal goals. It is not, however, an easy undertaking. Disabled people encounter many obstacles in their attempt to achieve true rehabilitation. For example, a fellow employee may object to working with a handicapped person because he "feels sorry" for him, whereas the disabled worker wants no sympathy, just an equal opportunity to compete.

Society has also created architectural barriers that sometimes prove insurmountable. To a wheel chair victim, for example, curbs, steps, toilets and public transportation create obstacles that may hamper even the most determined individual. Difficulties arise when a blind person rides in a self-operated elevator not equipped with braille floor buttons. Research has proved that most barriers of this nature can be aesthetically and economically eliminated at the time of construction. Moreover, such architectural changes can often contribute to the safety of the average "non-handicapped" person ... a secondary benefit rarely considered by the contractor or his client.

Another area which illustrates the need for effective rehabilitation involves a growing number of individuals who react to disability with dollars. They believe that the only way to help the handicapped is through financial assistance. Although this claim may appear exaggerated, this "welfare" culture, if developed and exploited, could result in a world of mislabeled disabled individuals, screened off from the rest of society in villages for the handicapped and supported by the efforts and wages of a diminishing few. Evidence for an increasing trend in this direction can best be seen in awards brought about by workmen's compensation laws, jury and court verdicts in the torts area, and the numerous governmental and industrial benefits programs. All programs should be geared toward the speedy and merciful recovery of the injured person.

THE REHABILITATION PROCESS

Modern, aggressive, enlightened rehabilitation is the blending by a full-time specialist of financial, medical, psychological and vocational techniques into an efficient recovery process. The person responsible for such an effort must be dedicated to his work. He must have a personality that is warm and outgoing, while at the same time he must believe in the dignity of man. This trait does not necessarily mean that such a person is socialistic in nature. To the contrary, he must convey to his patients the lack of punitive reaction in successful rehabilitation. Patience mixed with hard work produces results. No amount of formal education can overcome deficiencies in these areas. Knowledge, however, in the financial, medical, behavioral, and vocational testing areas is an essential tool of the rehabilitation trade. It is the responsibility of the professional, therefore, to keep abreast of the latest information in these various disciplines so that he is fully qualified to apply the newest rehabilitative techniques to the injured party.

The rehabilitation specialist should work with and enlist the full support of any individual or group who can contribute to the patient's complete financial, medical and vocational recovery. Since rehabilitation is a team effort, it requires proper timing and direction of each member. No discipline can be overlooked, nor can any facet of the process be excessive. This delicate blending of pride and procedure requires the utmost skill, tact and diplomatic salesmanship.

Early identification of potential cases and *early application* of the rehabilitative procedures is vital for successful rehabilitation. Words such as "Rehabilitation begins in the ambulance on the way to the hospital" are not merely axioms of after-dinner speakers, but have real meaning in the practical sense. Most professionals are cognizant of the fact that the margin of failure drastically increases each day the rehabilitative process is delayed. This phenomenon could be more easily understood if it were possible to fully comprehend the fears, tensions, misunderstandings and bitterness that the patient encounters following a near-tragic accident or illness. The specialist must begin immediately to eliminate these traumatic effects before they destroy the patient's character. As part of the total identification process, the rehabilitation specialist should also consider the patient's economic resources. Financial counselling of the patient and his family can help by demonstrating the concern the professional has, not only for the victim, but also for his loved ones. No one working on a case should overlook or minimize the importance of the family as an important part of the rehabilitation team. Quite often, funds are readily available that could help to alleviate any mental anguish felt by the patient as a result of financial strain. For example, why give a person thousands of dollars years later when a few hundred at an early date may accomplish the same goal.

In order to achieve successful rehabilitation, it may be best to combine the latest proven medical techniques with the tested methods of the past. The professional practitioner can often help decrease the degree and duration of the disability. Thus, both the professional and his methodology have a vital impact on the rehabilitative process. No case should be handled by an unqualified person, nor should it be administered in an ill-equipped facility. The medical results hinge directly upon these conditions. In its truest sense, then, the rehabilitative process is similar to a procurement function in which the best product available is purchased by and for the customer.

In addition to existing personnel, a consulting medical professional should also be retained to handle certain problems involving medical ethics and decisions. Apart from securing the services of various medical experts, general hospitals, extended care, nursing homes and comprehensive rehabilitation facilities are also necessary. Sheltered workshops, in certain cases, should be utilized in the vocational area. All of these medical and paramedical facilities vary greatly. Whether or not the program is successful is directly related to such factors. It is, therefore, the specialist's duty to be cognizant of the strengths and weaknesses of such centers.

Psychological counselling and vocational rehabilitation, the next important step, begins with the medical recovery program. Although certain goals for full restoration build hope within the individual and his family, it is vital that these vocational goals be realistic and attainable. If a goal is impractical, it should be immediately supplanted by a similar meaningful plan. Placing a disabled person in a vocation beyond his ability will most likely result in failure, and failure often brings about despair and lack of confidence. To plan vocation goals and action, the specialist will need a great amount of practical imagination, coupled with a deep conviction that any disabled person, properly placed, can perform productively. Furthermore, it should always be remembered that realization of effective vocational goals can produce a much happier future life for the patient. No vocational objectives should ever be established without thoroughly investigating past employment, skills and duties. This investigation should also include an analysis of body functions relative to past employment performance, particularly in the area of safety and productivity. The best vocational goal for any patient is his past position, since it should be easier for him to return to an environment in which he is knowledgeable and comfortable. Many disabled employees are not offered former positions because company management is uncertain whether or not they can function as well as they did in the past. Not only is a qualified employee denied his "right to work", but management also suffers because a new employee must be trained at additional expense.

Another factor contributing to error in vocational placement results from a lack of knowledge by the medical advisor regarding the physical and mental prerequisites of the job. It is the responsibility of the rehabilitation specialist to insure that all parties are informed.

THE IDENTIFICATION PROCESS

Before an attempt is made to identify possible rehabilitation cases, it may be best to mention the various disabilities and the sources from which they arise. The three sources include: (1) Congenital; (2) Accidents or (3) Illnesses incurred during one's lifetime.

Congenital Handicaps

Governmental agencies can be a great asset in helping to rectify the problems of the congenitally handicapped. From birth, such a person is often deprived of opportunities experienced by the "normal" child. When he reaches maturity, he may not only have a physical handicap, but his educational and emotional growth may likewise be impaired. Thus it is essential that local, state and federal government agencies subsidize certain programs that will enable the handicapped child to receive the full benefits of a formal education. If this occurs, chances are that he will be able to compete when the time is right.

Accidents or Illnesses Incurred During One's Lifetime

Although a more detailed discussion of congenital disabilities may be in order, this text will concentrate on the other two categories: *accidents* and *illnesses* suffered during one's lifetime. This is primarily because they directly involve the benefits, workmen's compensation, or torts area; therefore, identification and response to these disabilities will directly affect management's control of losses. If potentially disabling injuries and illnesses are not identified promptly by the insured or his insurer, then no effective action can follow. Fortunately, the majority of work-related injuries and non-work-related injuries or illnesses are of short duration and of minimal involvement. If remedied within a reasonable period of time, disabilities of this nature can allow such persons to return to productive work. Herein, however, lies the trap that lulls many employees and insurers into a lack of timely action. Past experience indicates that attitudes such as over-optimism, lack of adequate information and "let sleeping dogs lie" block identification of the real problem and often result in many expensive "creepers". Lack of timely identification, moreover, leads to a non-uniform application of the principles and practice of rehabilitation. Often a phone call to a hospital or to a doctor will bring prompt recognition of a potentially catastrophic case. The following is a list of conditions which lend themselves to rehabilitation procedures:

I. Disease

    A. Cardio-Vascular System

        1. Angina Pectoris
        2. Arterio-Sclerotic Heart Disease
        3. Coronary Insufficiency and Thrombosis
        4. Myocardial Infarction
        5. Phlebitis

    B. Brain

        1. Stroke
        2. Petit Mal and Grand Mal Epilepsy
        3. Psycho-Neurosis and Psychosis

    C. Nervous System

        1. Paralysis
        2. Sclerosis — Spinal Cord Disease

    D. Respiratory System

        1. Emphysema
        2. Silicosis
        3. Sarcoidosis

    E. Ear and Eye

        1. Labyrinthitis
        2. Meuniere's Disease
        3. Glaucoma

    F. Gastro-Intestinal Tract

        1. Cholangitis
        2. Duodenitis
        3. Liver Disorders

    G. Kidney — Renal Disorders

    H. Skeletal and Muscular System

        1. Ankylosis
        2. Degenerative Arthritis or Osteoarthritis

  3. Spondy Lolisthesis
  4. Rheumatism and Fibrositis

II. Injuries

 A. Second and third degree burns on 25% or more of the body and/or the surfaces of extremities.

 B. Head injuries, except slight skull fractures and/or concussion, as defined by the attending physician and requiring less than 48 hours hospitalization. Even slight head injuries should be referred which result in dizziness and other symptoms, or which prevent the man from returning to work within a few days of accident.

 C. All fractures except lineal fractures. Lineal fractures should be referred if they involve a joint area such as wrist, elbow, shoulder, ankle, knee and hip.

 D. Loss of one or both eyes.

 E. Loss of more than 50% of vision.

 F. Severe back conditions wherein there is objective evidence of injury.

 G. All crush injuries, including crush injuries to the foot or hand.

 H. Any other minor or serious injury which is complicated by an underlying ailment, such as diabetes, epilepsy, etc.

 I. Any injury resulting in substantial scarring, particularly to exposed areas of the body where the patient becomes self-conscious about public exposure.

 J. All amputations of extremities except the fingers and toes.

 K. All injuries resulting in paralysis (quadriplegia, paraplegia, and hemiplegia).

III. Other General Guidelines

 A. All cases where psychological, social or psychiatric problems are in evidence, regardless of the insignificance of the initial injury or illness.

 B. Any injury or illness where the period of disability exceeds the original prognosis by the treating psysician. In those cases where the physician states the period of disability to be "indefinite", rehabilitation and medical management should most certainly be considered early, particularly if the disability extends beyond three months.

REHABILITATION AND MEDICAL MANAGEMENT

Rehabilitation and medical management can contribute substantially to the control and reduction of loss. This process cannot be successfully performed by part-time technicians. To be effective, it must be undertaken by a specialist who has the full support and cooperation of senior management. This person must coordinate all contributing forces and disciplines in a timely and efficient manner. It is true that cost savings and benefits will far exceed expenditures, even though many feel rehabilitation is expensive. Perhaps the only valid method of measuring success is to examine several cases which were treated promptly over a reasonable period of time. Anyone who has given modern rehabilitation a fair chance under such circumstances realizes its value as a modern management process.

It is also interesting to note the fringe benefits derived from such a program. Employee morale is favorably affected. Once fellow employees realize the help extended to an individual by management, they feel that if a similar mishap affects them they will also be helped. Public relations, in the form of stories in trade publications and news releases, places the organization in a favorable and enviable light. Handicapped workers who have received the benefits of a rehabilitation program are usually more productive, more loyal, and, in fact, are more safety-conscious. Perhaps this is because they have already witnessed first-hand the results of unsafe working habits and they are more determined it won't happen again.

Today any company or insurer, no matter how large or how small, can have an effective rehabilitation program. The government program of vocational rehabilitation can be utilized, with business supplying funds and trained personnel in certain areas. In addition, the services of International Rehabilitation Associates, Incorporated, a Philadelphia-based company founded in 1970, are available to anyone who wishes to procure them. No matter what path a company chooses to follow, it is urged that they utilize some method to redeem the benefits (to employee and employer) or a rehabilitation program.

BIBLIOGRAPHY

Arthur, Julietta K., *Employment for the Handicapped: a Guide for the Disabled, Their Families and Their Counselors.* Nashville, Tennessee: Abingdon Press, 1967.

"Bibliography of Vocational Rehabilitation with Emphasis on Work Evaluation," by Research and Training Center in Vocational Rehabilitation, University of Pittsburgh, Johnstown, Pennsylvania, 1968.

Chiet, Earl F., *Injury and Recovery in the Course of Employment.* New York: John Wiley & Sons, Inc., 1961.

Chiet, Earl F., and Gordon, Margaret S. (eds.), *Occupational Disability and Public Policy.* New York: John Wiley & Sons, Inc., 1963.

Conley, Ronald W., *The Economics of Vocational Rehabilitation.* Baltimore, Maryland: Johns Hopkins Press, 1965.

Covalt, Donald A. (ed.), *Rehabilitation in Industry.* New York: Grune & Stratton, 1958.

Dawis, Rene V., Lofquist, Lloyd H., and Weiss, David J., "A Theory of Work Adjustment." Minneapolis, Minnesota: University of Minnesota, 1968. Minnesota Studies in Vocational Rehabilitation XXIII, Bul. 47, April, 1968.

"Directory of Organizations Interested in the Handicapped." Washington, D.C.: Committee for the Handicapped, 1218 New Hampshire Ave. N.W.

"Directory of State Agencies for the Blind" and "Directory of State Offices of Vocational Rehabilitation." Washington, D.C.: Department of Health, Education and Welfare, Rehabilitation Services Administration.

European Seminar on Sheltered Employment, August 31 — September 8, 1959. Available from International Society for Rehabilitation of the Disabled, 219 E. 44th St., New York, N.Y. 10017.

Gellman, William, *Adjusting People to Work*, 2nd ed. Menomonie, Wisconsin: Materials Development Center, Dept. of Manpower and Rehabilitation, Stout State University, 1957.

Jaffe, A. J., Day, Lincoln H., and Adams, Walter, *Disabled Workers in the Labor Market.* Totawa, New Jersey: Bedminister Press, 1964.

"Job Redesign for Older Workers; Ten Case Studies," Bul. 1523. Washington, D.C.: U.S. Government Printing Office, U.S. Bureau of Labor Statistics, 1967.

Journal of Rehabilitation (bimonthly). Washington, D.C.: National Rehabilitation Association.

Luchterhand, Elmer, and Sydiaha, Daniel, with the assistance of Lapierre Edmond, and Winn, Alexander, *Choice in Human Affairs; An Application to Aging — Accident — Illness Problems.* New Haven, Connecticut: College and University Press. 1966.

Malikin, David, and Rusalem, Herbert (eds.), *Vocational Rehabilitation of the Disabled: An Overview.* New York: New York University Press, 1969.

McBride, Earl D., *Disability Evaluation and Principles of Treatment of Compensable Injuries.* Philadelphia, Pennsylvania: J. B. Lippincott Co., 1963.

McGowan, John F., and Porter, Thomas L., *An Introduction to the Vocational Rehabilitation Process; A Training Manual.* Washington, D.C.: U.S. Government Printing Office, 1968. (Rehab. Serv. 68 — 32)

Occupational Characteristics of Disabled Workers, by Disabling Condition; Disability Insurance Benefit Awards Made in 1959-1962 to Men Under Age 65. Washington, D.C.: U.S. Government Printing Office, 1967.

Overs, Robert P., *Employment and Other Outcomes After A Vocational Program in a Rehabilitation Center.* Milwaukee, Wisconsin: Curative Workshop of Milwaukee, 1971. (Milwaukee Media for Rehab. Research Reports, No. 11)

Rehabilitation Counseling Bulletin (quarterly). Washington, D.C.: American Rehabilitation Counseling Association.

Rehabilitation in Canada (3 issues per year). Ottawa, Ontario: Dept. of Manpower and Immigration of Canada.

Rehabilitation Literature (monthly). Chicago, Illinois: National Easter Seal Society for Crippled Children and Adults.

Smith, Wilmer Cauthorn, *Principles of Disability Evaluation.* Philadelphia, Pennsylvania: J. B. Lippincott Co., 1959.

TOWER: Testing, Orientation and Work Evaluation in Rehabilitation. New York: Institute for the Crippled and Disabled, 1967.

Viscardi, Henry, Jr., *The Abilities Story.* New York: Paul S. Eriksson, Inc., 1967.

"Vocational Assessment and Work Preparation Centres for the Disabled." Geneva, Switzerland: International Labour Office, 1970.

Vocational Guidance Quarterly. Washington, D.C.: National Vocational Guidance Association.

Wright, George N., and Trotter, Ann Beck, *Rehabilitation Research.* Madison, Wisconsin: University of Wisconsin, 1968.

XIX. Security

Part I General Discussion of Security

INTRODUCTION

    Protection of property is instinctive, and as man has evolved so have his means of protecting his property, from primitive devices to electronic technology.
    Possession of goods and property is the single greatest distinction between the world's "Haves" and "Have-Nots". When unable to better their material situation through the normal means, people may resort to illegal or immoral means to satisfy their real or imagined needs.

Security is the protection of property, tangible or intangible, against deliberate, loss-producing acts.

    The cost of recovering loss from loss of property is much higher than the cost of the property lost.
    Many firms believe that they are already sufficiently secure. This misconception is the major obstacle to providing adequate universal security.
    Some of the concerns of management and the person responsible for security are theft and espionage; vandalism, arson and bombs; looting (following a fire or natural disaster); assault; and sabotage.

Adequate control of people is essential for thorough security.

    Although any breach of security could be committed by an employee or non-employee, during or after the work day or during the day or night, such breaches can be prevented if there is adequate and conscientious application of security techniques.
    There is a market for all kinds of stolen goods and they do not have to be precious metals or consumer goods in order to be readily disposable. However, they usually must be readily moveable and of high unit value as a burglar will attempt to balance his profit against his chances of being caught.

Theft by employees or visitors can be reduced by eliminating temptation.

CASE HISTORIES

Gate Pass

A punch press operator who had been taken off his job for mental retardation treatment returned to the plant to visit his friends. During a rest break, he put his head in the die zone of a punch press and operated the two-hand controls. He was killed instantly as his skull was crushed.

This company has an elaborate security system, but failed in this instance to take back this employee's gate pass.

Theft – An Old Story

A gate guard had been puzzled for some time by a cleaning crew that left at the end of each day with wheelbarrows full of sawdust and debris. He was suspicious, but could not find anything in his careful searchings through the saw dust, etc. One day it finally occurred to the guard that the cleaning crew was stealing wheelbarrows.

Computer Room Security

An actual experiment was conducted with a truck that was equipped with a sensitive receiver. It was parked close to an unshielded computer room where signals were received from the computer. These signals were used to provide the same data in the truck as was being produced in the computer room.

The above illustrates that all computer rooms should be properly shielded to protect against information espionage.

TYPES OF SECURITY PROBLEMS

Theft and Espionage

Both of these involve stealing. Theft is the removal of equipment or material while espionage involves the stealing of plans, designs or ideas. Either may be committed by employees or non-employees. In espionage, the one committing the act could be a normal type of employee, or one who has deliberately used employment to gain access to secrets.

Vandalism, Arson and Bombs

Vandalism is the indiscriminate, undirected damage of property. It is usually inflicted from the outside. Arson is the deliberate setting of fire, and its occurrence will likely depend upon the presence or absence of flammable materials inside or outside the building. Explosive bombs may be planted inside the building by a disgruntled employee or for political or other reasons.

Looting

Additional loss may occur after a fire, flood, earthquake or other disaster through looting, due to the exposure rendered to equipment and material from the disaster.

Assault

Occassionally a dissatisfied customer, employee or former employee or other person, who thinks that he has a grievance, may physically attack an official of the firm in an effort to relieve his frustration.

Sabotage

This is the deliberate mechanical destruction of or damage to equipment or buildings by an employee or outsider, in order to interfere with production.

PROBLEM LOCATIONS

1. *Public Entrances*

   Inadequate lighting, locks or visitor control. Lack of security control for company and employee property.

2. *Employees' Entrance*

   Falsified time cards. Carelessness in checking packages and lunch boxes as employees leave.

3. *Accounting Department*

   Forged or falsified documents. Stolen confidential information.

4. *Personnel Department*

   Hiring undesirable employees because of false employment references, with little or no background investigation.

5. *Supply Room*

   Theft of tools and other office or plant supplies. Lack of adequate material controls. No check of personnel entering and leaving.

6. *Off the Job*

   Employees enter into association with professional receivers of stolen goods or unscrupulous competitors.

7. *Shipping Area*

   Collusion with dishonest truck drivers or those preparing products for shipment. Lack of adequate material controls. Failure to double check deliveries.

8. *Production Department*

   Theft of products and parts. Sabotage by disgruntled employees.

9. *Warehouse Area*

   Inventory losses. Theft of products and parts.

10. *Perimeter Fencing*

    Vandalism and robbery due to improper fencing. Incompetent night watchmen. Inadequate lighting of fenced areas. Inadequate "clear" space.

11. *Gates*

    Careless package checking. Improper inspection of trucks. Lax visitor identification. No log book kept by guards. No parcel or goods passes.

12. *Parking Areas*

    Stolen goods hidden in cars. Inadequate surveillance.

13. *Other Hazards*

Known routine of security officers. Company products sold illegally. Insufficient security training and supervision.

MEASUREMENT OF LOSSES

Establishment of proper loss measuring procedures are of major importance for the following reasons:

1. Such measurements put an actual dollar sign on losses so that expenditures that may be warranted to reduce them can be determined. This is often a "trade-off".

2. They locate the areas where major losses are occurring.

3. They can be used to monitor the effectiveness of organizational security.

Some Useful Methods for Measuring Losses

1. *Theft Reports*

    Employees reporting that tools, supplies, etc. are missing. These should be recorded and investigated.

2. *Inventory Records*

    If the inventories do not correspond with stock records, investigate for possible theft.

3. *Maintenance Records*

    If there are excessive maintenance costs in any particular areas, they may be due to vandalism or sabotage and should be investigated.

4. *Purchasing Records*

    Purchases of an abnormal amount of any particular item should require an explanation.

5. *Quality Control Records*

    If rejects suddenly increase, check for sabotage.

6. *Production Records*

   If production quotas fall, check for vandalism and sabotage. Also, if the number of items produced is less than the number started or the amount of materials and supplies going into production, check for theft.

7. *Scrap Records*

   Abnormal amounts of scrap can indicate vandalism and sabotage. Also, some items may be improperly marked "scrap," and then disappear.

INSURANCE

Many types of insurance are available, such as those against:

Burglary
Vehicle Liability
Fire
Vandalism
Sabotage
Plate Glass Breakage
Flood, Earthquake, Tornado, etc.
Loss of Profit
Product Liability
Public Liability
Lightning
Airplane Crash

However, these many not cover such losses as:

Schedule upsets
Unfilled orders
Interrupted deliveries
Frustrated suppliers
Damaged relations with customers, employees and community
Loss of time as employees discuss event
Physical and mental strain on supervisors
Loss of key personnel and resultant training expenses

Therefore, the cost of losses due to breaches of security are likely to be much higher than the amount covered by insurance.

Part II Security Plans

"Trade-offs" are an important consideration in any security plan. Thus, potential losses and the probability of their occurrence have to be determined. This will dictate the amount of money spent and the extent of security plans.

Security requirements for different companies vary widely, due to the following factors:

1. *Size of Plant*

    Generally, larger plants require greater security measures, because it is more difficult to recognize people and to control their movement around the plant.

2. *Type of Product Being Produced*

    (a) Hazardous products - Plants producing such products as dynamite, flammable liquids and toxic chemicals require stringent security measures which may not apply to other plants.

    (b) High value products - If they are small items such as watches, they could be easily stolen. If they are large items such as refrigerators, trucks or railway cars would be required. Another important consideration is that such items could be readily sold.

    (c) Products requiring strict secrecy in their formulation, such as cosmetics and food flavouring.

    (d) Security risk products - This could include items for military requirements such as rockets and airplanes, where the possibility of sabotage would be a major consideration.

3. *Location of Plant*

    The plant may be located in an area where tornados, hurricanes or floods are more prevalent. Also, the plant may be located in a congested area where adjacent plants could be a hazard, or it could be located in a remote area.

4. *Type of Raw Materials and Supplies Used*

   May be of a type that is useful for off-the-job activities or readily sold.

   Therefore, because of the diversity of plants, raw materials and products, it is necessary to design a security program to fit the individual plant.

Some Methods for Controlling Security

1. *In Hiring*

   Check references thoroughly and also check with previous employers. Check for any potential or existing health problem. If the new employee will be working with confidential formulae, consider an employment contract to safeguard this information. Obtain governmental security clearance where necessary.

2. *Fencing*

   Includes the fencing of the property and also fencing off restricted areas.

3. *Security Officers*

   Should be competent people with good qualifications. They should be properly trained for individual plant requirements.

4. *Passes*

   Depending on the number of employees, employee-picture passes may be used. These should be shown to the security officer when entering or leaving the plant. It is important that employees return their passes if they are leaving the employ of the company or going on extended leave of absence. Visitors should be signed in and out and issued a temporary pass.

5. *T.V. Monitors*

   Useful in surveillance of restricted areas.

6. *Alarm Systems*

   Useful in detecting forced or unauthorized entry during off-shifts and in controlling people entering restricted areas.

7. *Cameras*

   To take photos of anyone entering restricted areas.

8. *Signs*

   Prohibiting people from entering restricted areas.

9. *Insurance*

   Theft, fire, bonding, vehicle, product liability, etc.

10. *Lighting*

    Adequate lighting is important both inside and outside the plant, including property fences. All entrances, exits and parking lots should have proper lighting. Restricted and special security areas should be well illuminated. Emergency lighting should be provided.

11. *Punch Clocks or Stations*

    Should be located in strategic areas and covered during rounds. It is advisable to use a check list which should be employed when the Security Officer is making his rounds. The Security Officer should also check the people and vehicles entering and leaving the plant.

12. *Allocation and Control of Keys*

    Keys should be issued to personnel only when it is absolutely necessary. Care should be taken that all keys are returned if an employee leaves the company or has a change in position.

13. *Safes*

    To safeguard important documents, confidential information, master keys, cash, etc.

SECURITY CHECK LIST

1. *Indicators of Loss*
   a. Inventories - raw materials, finished products, supplies, tools, equipment, etc.
   b. Maintenance records - indicator of vandalism and sabotage.
   c. Theft reports
   d. Purchasing records
   e. Personnel records
   f. Quality control records
   g. Production records

2. *Control of People* (Entering, On and Leaving Premises)
   a. Employees
   b. Visitors
   c. Outside Contractors
   d. Service People
   e. Public Utility Personnel
   f. Truckers, Railway and Ship personnel
   g. Intruders

3. *Control of Systems*
   a. Adequate fencing
   b. Adequate lighting
   c. Signs (restricted areas)
   d. Security Officers properly trained
   e. Allotment and control of keys
   f. Remote television and automatic signals

4. *Control of Undetected Forced Entry*
   Adequate alarm systems in the following areas:
   a. Office
   b. Laboratory
   c. Plant

5. *Control of Confidential Information*
   a. Contracts with employees
   b. Bonding of employees
   c. Security clearance
   d. Patents
   e. Computer data
   f. Safes

6. *Control in Hiring People*
   a. Interviews by more than one person
   b. Check references carefully
   c. Contact previous employers
   d. Other character checks
   e. Medical history
   f. Character tests
   g. Check list

7. *Control of Finances*
   a. Petty cash
   b. Forgeries
   c. Embezzlement

8. *Control Plan for Emergenices*
   a. Evacuation of personnel
   b. Good contact with Police and Fire Departments
   c. Procedures to avoid looting

Part III Specimen Industrial Security Manual

INTRODUCTION

The purpose of the manual is to present objectives of the Security Program and list the duties of the people involved in its execution.

Control of company property can only be maintained through the establishment and attainment of these objectives.

All security will be under the control of the company Security Supervisor.

It is intended that the manual will make each reader cognizant of his specific role in the overall security plans and of his obligations to the company in emergency situations. Periodic review of the "Security Check List" in addition to other relevant security information is instrumental in developing a consciousness of security in each employee.

Courtesy and Self Control

These are absolute requirements and must be practiced when coming into contact with customers, employees and the public, whether in the performance of duties or in assisting others.

Telephone Techniques

When answering or using the telephone, the following instructions and procedures are to be followed:

1. The telephone will be used only for business related to performance of security duties.

2. Answer the telephone as promptly as possible. In answering the telephone, always assume that the call is important and speak accordingly.

3. Speak clearly into the mouthpiece in a moderate tone of voice. Indicate an "at your service" attitude.

4. Identify yourself immediately by giving the name of your company, then use your surname to identify yourself as the speaker.

5. Have pad and pencil handy to take notes. Write clearly and legibly as others may have to read your notes.

6. If the caller does not identify himself, ask "May I have your name, please"; do not say "Who is this?" Ask for the caller's telephone number, in case a return call has to be made or you doubt the origin of the call.

Fire and Fire Equipment

You must be familiar with the operation of all fire fighting equipment and its location in the plant.

Security Supervisor's Duties

1. Supervises patrols, security checks and enforcement of plant rules and regulations; initiates random checks of vehicles.

2. Orients and trains Security Officers.

3. Periodic checks of all *welding and cutting operations in the plant.*

4. Daily fence patrols by the Supervisor or his delegate.

5. Maintain the master lock system.

6. Post indicator valves and isolating valves for the sprinkler system are to be inspected and recorded weekly.

7. Fire extinguishers are to be inspected and recorded weekly.

8. Inspect outside lighting weekly.

9. Conduct tours for governmental and private agencies.

10. Member of Plant Loss Control Committee.

11. Arrange for recharging empty fire extinguishers and sign all receipts and invoices for this service.

12. Maintain operation of communications equipment.

13. Enforce flame permit procedures.

14. Responsibility for snow removal.

Security Officer's Duties

1. Visually check and record passage of vehicles; state whether pick up or delivery; driver is to sign in and out.

2. Inspect trucks as requested by Supervisor.

3. Record all visitors entering and leaving in visitors' log book. Issue visitors' passes.

4. Record all outside contractors entering and leaving in log book.

5. Check all employee badges on entering, record people without badges and people reporting late.

6. Patrol plant as directed by Supervisor.

7. Check "No Smoking Areas" such as boxcars, paint areas, propane storage; check all welding (maintenance and construction) for flame permits and procedure.

8. Report *all* property damage.

9. Check employees for use of safety glasses.

10. Check lunch boxes as directed.

11. Switchboard duty_____ P.M. _____ A.M., and weekends.

12. Record sick call-ins and pass to appropriate person (nurse, etc.).

13. Perform duties of Supervisor as requested.

14. Answer first aid room alarm and administer first aid if necessary.

15. Check fire extinguishers at all locations, and water pressure at all risers during patrols.

16. Keep all telephone business brief, especially during plant operating hours.

17. There will be no reading of books or newspapers during shift change or during the day and afternoon shifts when the plant is operating.

18. Visually check all incoming and outgoing employees; demand passes for all outgoing parcels other than lunch boxes.

19. Keep conversation with employees to a minimum, allow no loitering around the gatehouse.

20. Be sure to read the log book and daily record books for current entries prior to taking over a shift.

21. Keep gatehouse clean and orderly; mop floor, clean windows, etc., as necessary. Each man to leave gatehouse in good order for the man following.

22. Record time and duration of power failures. Restart all equipment according to instructions.

23. Any health problems should be referred to appropriate staff members.

Visitors' Passes

The following procedures will be followed for all visitors:

1. Visitors' passes will be issued in the gatehouse and at the reception desk.

2. All visitors *entering the plant* from the gatehouse will be issued passes which they will fill out and sign. When leaving, the security officer will sign out the visitor and collect the pass.

3. Visitors going to the reception desk, who will enter the plant at a later time, must be issued a visitors' pass. They will sign in and out at the reception desk. The receptionist will collect the pass as the visitors leave, keeping the stubs, which in turn will be collected daily by the Supervisor.

4. People found in the plant without a pass will be taken to the gatehouse for proper clearance.

5. Permanent passes may be issued to certain vendors, whose names will be noted at all entrances.

6. Public utility, government and telephone personnel may not require a visitors' pass as they will be issued a pocket identification card. Their visit will be noted in the log book.

EMERGENCY CONDITION SECURITY PLANNING

Types of Emergencies

    Building collapse
    Earthquakes
    Enemy action (radioactivity, etc.)
    Explosions (flammable liquids, gas, dusts, moisture and hot metal)
    Exposure hazards (from adjacent plants)
    Fires
    Gas or toxic fumes
    Lightning
    Water (flood, rain, snow)
    Wind (cyclones, hurricanes, tornados)
    Bomb threats, riots

Preventive Measures

Any planned measures to prevent emergencies should be an extension of existing plans and facilities.

A periodic survey of the following hazards, in addition to the more frequent fire and safety inspections, is most desirable:

    Process hazards
    Equipment (machinery, etc.)
    Storage and handling methods
    Flammable liquids
    Explosive gases or substances
    Sources of spontaneous combustion
    Static and electrical grounding
    Electrical equipment
    Fire equipment and routine
    Housekeeping
    Pressure relief devices
    Instrumentation
    Exits
    Medical arrangements (supplies, oxygen, etc.)
    Structures
    Personal protective equipment (masks, etc.)

Emergency Procedures

The following procedures will be followed in the event of flood, hurricane or earthquake, or where damage, fire, explosion, trouble in the parking lot, threats of violence or trouble in the plant are suspected.

1. Contact _____ immediately for instructions, or in his absence _____ .

2. Call the local Police and/or Fire Department.

3. Keep a record of what has happened, noting times, names of people involved, etc.

Duties and Responsibilities

Co-ordinator:
Responsible for overall activities and direction of the emergency organization;
Will authorize alarm or delegate;
On a week-end or holiday, will set the emergency plan into operation by phone or delegate;
On a working day, will proceed to the scene and direct operations;
Notify the communications officer.

Assistant Co-ordinator:
Will assist the Co-ordinator or take over his duties, if and when required.

Communications:
Alarm -instructions
    -testing
      -sounding (This must be an unmistakable signal known to all employees.)

Field Communications - messenger service
                          - emergency telephones

General Communications - prepare instructions to switchboard regarding incoming and outgoing calls - referring, routing, screening
-prepare lists of persons to be notified

Public Relations - liaison with community officials and news media representatives
- photographs for plant or community use
- providing waiting room, telephone facilities
- contact Industrial Relations, Public Relations officials
- responsible for contacts with ambulances, doctors, fire department, police, families, hospitals, press, radio, government labor authorities, I.A.P.A., Workmen's Compensation Board.

Departmental:
Appraisal of departmental situation and any necessary actions
Shut-down or hold-safe procedures
Exiting
Reporting status of employees
Assisting where needed

Engineering and Maintenance:
Rescue (extricate from debris, first aid, transportation)
Shut-downs
Stand-by services
Fire fighting (with Industrial Relations)
Building survey and damage control
Salvage
Repairs and replacements
Restoration of services and utilities
Insurance

Industrial Relations:
Evacuation - survey of exits
- procedure for evacuation

Fire - With Engineering and Maintenance. In the event of fire, the local Fire Department should be called in at once and routed to the fire.
- Fire organization (cover all shifts) equipment and drills
- Plant fire action
- Liaison with local Fire Department

Security Officers - Directing ambulances, doctors, fire vehicles or others to the scene.
- Passes
- Gates

- Patrol
- Traffic
- Parking
- Notifying personnel (call-in)
- Call-in other security personnel
- Directing press, public, etc., to the proper person
- Protection against looting and vandalism

Medical - Ambulances (advance arrangements and phone numbers)
- Doctors (advance arrangements and phone numbers)
- First aid (training, supplies, administering)
- Hospital (advance arrangements and phone numbers, instructions)
- Nurses (plant and local)
- Stretchers (ordering, inspecting)

Mutual Aid - Advance arrangements, phone numbers

Safety - Emergency equipment
- Assist medical
- Investigate and report

Welfare and Identification - Identify injured
- Notify families
- Records
- Follow-up
- Bulletin board notices
- Listing injured and relevant personal data
- Inform management
- Support Services (visit, transportation, money, babysitters, assist, write)

Offices
Planned procedures consistent with other emergency procedures.
Assistance in needed areas.

Instruction sheets outlining the various emergency procedures should be issued to each individual, department or unit so that the specified actions may be automatic in an emergency and full attention may be directed to unexpected occurrences.

A FORMAL EMERGENCY ORGANIZATION

Organization Chart

Emergency Co-ordinator (or Temporary Delegate)
Assistant Emergency Co-ordinator (or Temporary Delegate)

Communications	Departmental	Engineering & Maintenance	Industrial Relations	Offices
Alarm	Process	Rescue	Evacuation	Planned
Field	Planned	Shut-down		Measures
General	Measures	Standby Services		
Public Relations		Fire	Fire	
		Building Survey	Security Officers	
		Damage Control	Medical	
		Salvage	Mutual Aid	
		Repair & Replace		
		Restoration of	Safety	
		Services &	Welfare &	
		Utilities	Identification	
		Insurance		

Part IV Bombs and Bomb Security

INTRODUCTION

Few types of emergency situations strike as much fear and wreak such great havoc as do bombs and bomb threats.

It is felt that special attention is needed on this topic due to the widespread use of this particular destructive technique in contemporary society.

While man has been able to minimize the consequences of natural disasters through the application of scientific methods and technology, he is almost defenseless against the "bomber". This person could be motivated by socio-political ideologies or could simply be mentally deranged. Regardless, the low cast and simplicity of many bombs make them a grave threat to any operation.

PLANNING FOR ACTION

Prompt and effective action in case of a bomb threat requires the coordinated effort of many people and a well conceived and thought out plan with adequate training in its application.

Such a detailed plan can best be developed locally, utilizing local management familiarity with a variety of factors that must be considered in the plan.

There are certain essential points that should be covered:

1. *Definition and Delegation of Authority:*

   The responsibilities and the authority for making decisions on conducting a search and on the plant's evacuation must be clearly defined. Delegation of responsibility must be worked out to assure that the absence of an individual does not limit the plan's effectiveness.

2. *Distribution of the Plan*

   The plan should be distributed to all supervisors above a specified level and to key personnel at other levels as well as to those who are to participate in the development, organization and implementation of the program. The "key" personnel would include all those who, by nature of their assignment, might be expected to receive such threats, i.e., telephone operators, receptionists, mailroom personnel, secretaries, etc.

3. *Receiving the Threat*

   The detailed procedures for receiving a threat should include frequent review with potential recipients and should provide for:

   a. Contacting the local telephone company to determine whether tracing is possible and to establish procedures for initiating tracing when feasible.

   b. Recording of the time and the exact words of the message with particular emphasis on the description and the location of the device.

   c. Eliciting as much information as possible from the caller. The caller should be asked to repeat the message and should be questioned as to the reason for making the threat. The line should be kept open as long as possible.

   d. Note the sex of the caller, an impression about his age, any peculiarities of voice or speech such as hoarseness, shrillness, speech impediment, accent, dialect, signs of intoxication, irrationality, etc., and any "pet phrases" or other mannerisms.

4. *Evaluation of the Threat*

   The majority of the bomb threats received are of the crank call variety. Based on the experience of many military and police authorities, there is frequently a clue to the validity of the threat in the message itself or in the attitude and manner of the caller. In general, it has been found that a real or potential threat is:

   a. Almost invariably the work of a deranged person.

   b. The bomber, in placing the call, tends to prolong the call and is willing to furnish some detail as to the location of the device, reasons for planting it, etc.

   c. The call is frequently repeated.

Whereas the crank caller:

a. Tends to be abrupt and hurried in giving the message and seldom can or will provide details regarding the type of device, the location, reasons, etc.

b. Repeats the call less frequently because of the fear of tracing, etc.

5. *Decision to Evacuate*

Aside from the cost considerations, the hasty evacuation of a plant will often result in a rash of threats. Accordingly, unless there is reason to suspect that the threat is valid, a large number of employees should not be evacuated until the threat is proven to be real. The plan should, however, include procedures for evacuation when necessary.

6. *Responding to the Threat*

Plans for a response to a threat should include:

a. Notification of the resident manager.

b. Advising the appropriate local law enforcement agency.

c. A search of the premises will often be unnecessary or ill-advised as most threats are hoaxes. The extent of the search should be based on the evaluation of the threat.

Limited searches might include:

a. The more public areas in the plant, i.e., washrooms, lobbies, stairwells, corridors, etc.

b. A thorough search of key and critical areas such as boiler rooms, gashouses, water towers, computer rooms, etc.

c. A more detailed search in any area(s) specified in the threat.

If an appropriate search fails to confirm the existence of a bomb, emergency crews should be kept on standby until it is reasonably certain that the limit of the threat has passed.

Search activity may give rise to exaggerated rumors among employees. The plan should provide for keeping employees in the immediate area of the search informed.

7. *Discovery of a Bomb*

The plan should provide for provision of action in the event that a bomb, or anything resembling a bomb, is discovered. Such actions might include:

a. *EVACUATE THE AREA.*

b. *DO NOT* attempt to disarm or move the device.

c. Call for the local law enforcement bomb squad.

d. Alert the fire crew to stand by.

e. Alert the medical department to stand by.

Preparedness can eliminate a great deal of confusion and indecision when faced with such a threat.

BIBLIOGRAPHY

1. "Basic Principles of Industrial Security". Oak Security Publications Division—Aurora, Ontario; Madison, Wisconsin: Straus Printing and Publishing Company, 1972.
2. Fletcher, John A., Hugh M. Douglas. *Total Environmental Control.* Toronto: National Profile Limited, 1970.
3. Hemphill, Charles F., Jr., John M. Hemphill. *Security Procedures for Computer Systems.* Homewood: Dow-Jones-Irwin, Inc., 1973.
4. Kakalik, James S., Sorrel Wildhorn. *The Law and Private Police.* Washington: The Rand Corporation, 1971.
5. Kakalik, James S., Sorrel Wildhorn. *The Private Police Industry—It's Nature and Extent,* Vol. IV. Washington: The Rand Corporation, 1971.